MERY

# Selected Studies of Archean Gneisses and Lower Proterozoic Rocks, Southern Canadian Shield

*Edited by*

G. B. MOREY

GILBERT N. HANSON

SPECIAL PAPER
182

# Selected Studies of Archean Gneisses and Lower Proterozoic Rocks, Southern Canadian Shield,

*Edited by*

### G. B. MOREY
*Minnesota Geological Survey, 1633 Eustis Street, St. Paul, Minnesota 55108*

### GILBERT N. HANSON
*Department of Earth and Space Sciences, SUNY, Stony Brook, New York 11790*

**SPECIAL PAPER**

**182**

**THE GEOLOGICAL SOCIETY OF AMERICA**

P.O. Box 9140 · 3300 Penrose Place · Boulder, Colorado 80301

Copyright © 1980 by The Geological Society of America, Inc.
Copyright is not claimed on any material prepared by
Government employees within the scope of their employment.
Library of Congress Catalog Card Number 80-67113
ISBN 0-8137-2182-2


All material subject to this copyright and included
in this volume may be photocopied for the
noncommercial purpose of scientific
or educational advancement


Published by
THE GEOLOGICAL SOCIETY OF AMERICA, INC.
3300 Penrose Place
Boulder, Colorado 80301


Printed in the United States of America

# Contents

Dedication to Samuel Stephen Goldich ............................................................ iv

Preface ......................................................................................... v

Multiphase deformation in the Granite Falls–Montevideo area, Minnesota River Valley ....... *Robert L. Bauer*    1

Archean rocks of the Granite Falls area, southwestern Minnesota
.................. *S. S. Goldich, C. E. Hedge, T. W. Stern, J. L. Wooden, J. B. Bodkin, and R. M. North*    19

Origin of the Morton Gneiss, southwestern Minnesota: Part 1. Lithology
...................... *S. S. Goldich, J. L. Wooden, G. A. Ankenbauer, Jr., T. M. Levy, and R. U. Suda*    45

Origin of the Morton Gneiss, southwestern Minnesota: Part 2. Geochemistry
.................................................... *J. L. Wooden, S. S. Goldich, and N. H. Suhr*    57

Origin of the Morton Gneiss, southwestern Minnesota: Part 3. Geochronology
.............................................................. *S. S. Goldich and J. L. Wooden*    77

Mineral and rock compositions of mafic enclaves in the Morton Gneiss
....................................................... *Bruce V. Nielsen and Paul W. Weiblen*    95

Lead-isotope investigations in the Minnesota River Valley—Late-tectonic and posttectonic granites
.............................................................. *B. R. Doe and M. H. Delevaux*    105

Boundary between Archean greenstone and gneiss terranes in northern Wisconsin and Michigan ..... *P. K. Sims*    113

Tonalitic gneiss of early Archean age from northern Michigan
................................................ *Z. E. Peterman, R. E. Zartman, and P. K. Sims*    125

Early Archean Sm-Nd model ages from a tonalitic gneiss, northern Michigan
..................................................... *M. T. McCulloch and G. J. Wasserburg*    135

Geology and Rb-Sr age of lower Proterozoic granitic rocks, northern Wisconsin
.............................................................. *P. K. Sims and Z. E. Peterman*    139

Penokean deformation in central Wisconsin
............................................. *R. S. Maass, L. G. Medaris, Jr., and W. R. Van Schmus*    147

Chronology of igneous rocks associated with the Penokean orogeny in Wisconsin .......... *W. R. Van Schmus*    159

Boulders from the basement, the trace of an ancient crust?
...................... *Z. Nikic, H. Baadsgaard, R. E. Folinsbee, J. Krupicka, A. P. Leech, and A. Sasaki*    169

**SAMUEL STEPHEN GOLDICH**

This volume is dedicated to Samuel Stephen Goldich, better known as Sam, a very special character. He sets and demands a high standard in anything that he and his coworkers attempt. His career has been marked by an ability to initiate and complete large projects involving groups of people having diverse interests. Such projects include an extensive study of the Archean greenstone belts in northern Minnesota and adjacent Ontario, the dating and analysis of drill cores from the basement of the midcontinent region of the United States, and the geochemistry and geochronology of the ancient gneisses of the Minnesota River Valley. These works have resulted in many important advances in geology and particularly in the systematics of geochronology.

# Preface

This collection of papers on the geology, geochemistry, and geochronology of selected Precambrian rocks of the Lake Superior region was conceived in 1977 to honor Samuel Stephen Goldich, whose work has resulted in international recognition both for himself and for the fundamental significance of the geology and geochronology of the region. Although it is not customary for a volume to contain contributions from the person it is intended to honor, no book on the Lake Superior region could be complete without contributions from Goldich—a testimony to the importance of his scientific work.

In 1961, using a combination of classical stratigraphic and radiometric methods, Goldich divided the Precambrian rocks of the Lake Superior region into three subdivisions—lower, middle, and upper Precambrian—which formed the basis for most discussions of Precambrian time. Subsequent geochronologic studies, also mainly by Goldich and his colleagues, demonstrated that the lower Precambrian or Archean rocks, as they are now called, consist of two fundamentally different terranes—on the south an older gneiss terrane of 3,500-m.y.-old high-grade paragneiss and orthogneiss and subordinate 3,000-m.y.-old granitic rocks; and on the north a greenstone-granite terrane that formed during a relatively short-lived volcanic-plutonic event about 2,700 m.y. ago. The two terranes not only formed at different times but have had different tectonic histories. The gneiss terrane was tectonically mobile throughout much of the Precambrian, as indicated by a succession of events that have been dated at 3,500, 3,000, 2,600, 2,200, 1,800, 1,600, 1,500, and 1,100 m.y. B.P. In contrast, recognized tectonic events in the greenstone-granite terrane took place at 2,700 and 1,100 m.y. B.P.

Much has been published about the greenstone-granite terrane, which is typical of the Superior Province of the Canadian Shield. Relatively little, however, has heretofore been written about the older gneisses, and therefore discussion of their nature, distribution, and relation to younger Archean and Proterozoic events is a common thread throughout this volume. Seven of the fourteen papers provide detailed descriptions of the geology, geochemistry, and geochronology of rocks in the Minnesota River Valley, which is considered to be the type locality for the gneiss terrane. Archean rocks of similar antiquity also have been identified in northern Wisconsin and Michigan by Sims and his colleagues. The results of their studies are presented in the paper by Peterman and others and the paper by McCulloch and Wasserburg. Relations between the ancient gneiss terrane and the younger Archean greenstone-granite terrane—matters of considerable importance to understanding the evolution of the Archean crust—are discussed in a paper by Sims that describes the boundary between the two terranes, mainly in northern Wisconsin and adjoining Michigan.

Much of the earlier Archean gneiss terrane of the region was involved in the Penokean orogeny, a major early Proterozoic tectono-thermal event first documented by Goldich in 1961. The effects of this event greatly complicate geologic and geochronologic studies of the older rocks. The extent to which the Penokean orogeny and somewhat younger events disrupted or reworked the gneiss terrane in central Wisconsin is described in two papers in this volume: a geologic analysis by Maass and others and a geochronologic study by Van Schmus. A similar situation involving the reworking of post-Penokean, Proterozoic rocks in northern Wisconsin is documented by Sims and Peterman and alluded to by Doe and Delevaux in their paper on lead isotope investigations in the Minnesota River Valley. All these papers provide additional evidence for the repeated tectonic mobility of the gneiss terrane.

Although the paper on the geochronology of basement rocks of the Yellow Knife district by Nikic and others deals with an area far removed from the Lake Superior region, it is included in this volume because it demonstrates that the ever-less-elusive earlier Archean basement may be widespread in the North American craton.

Although the studies included in this volume advance our understanding of geologic history and crustal evolution in the Lake Superior region, they also suggest many new problems and questions that can be answered only by additional work. Without doubt the best tribute that can be given to Samuel S. Goldich is the recognition that his pioneering work provides springboards for research by future generations.

**G.B. MOREY**
*St. Paul, Minnesota*

**GILBERT N. HANSON**
*Stony Brook, New York*

# Preface

Geological Society of America
Special Paper 182
1980

# Multiphase deformation in the Granite Falls–Montevideo area, Minnesota River Valley

**ROBERT L. BAUER**
*Department of Geology, Macalester College, St. Paul, Minnesota 55105 and*
*Minnesota Geological Survey, St. Paul, Minnesota 55108*

## ABSTRACT

The lower Precambrian granulite-facies gneisses near Granite Falls and Montevideo have undergone four phases of folding, two periods of metamorphism, and both pretectonic and posttectonic intrusive events. Rare $F_1$ folds occur only in thin quartz-ofeldspathic veins that are at a high angle to the pervasive $S_1$ foliation and banding in the gneisses and that formed during the generation of $S_1$. The $F_2$ folding generated the most prominent structural features in the area—a large-scale, gently plunging antiform near Granite Falls and an inferred synform between Granite Falls and Montevideo. The $F_3$ and $F_4$ fold phases are not evident from mapping of the major $F_2$ structures, but they are well represented by minor structures, and their relation to the $F_2$ folding event is deduced from an analysis of the minor folds and mineral lineations in the gneisses. This analysis indicates a systematic variation in the axial orientations of the minor $F_3$ and $F_4$ folds with their position on the limbs of the major $F_2$ structures.

Analysis of the structures in a 3,050-m.y.-old granitic unit intruding the Montevideo Gneiss suggests the $S_1$ foliation and high-grade $M_1$ metamorphism accompanying $S_1$ were initiated prior to 3,050 m.y. ago. The timing of $F_2$ folding is not well constrained, but $F_3$ folding occurred after 3,050 m.y., and both events were accompanied by $M_1$ metamorphism. Mineral ages of 2,650 m.y. have been interpreted as resulting from the $M_1$ metamorphism and suggest an extreme duration from earlier than 3,050 m.y. to 2,650 m.y. for the high-grade metamorphism.

The $F_4$ folding and intrusion of a set of tholeiitic diabase dikes followed $M_1$ metamorphism but preceded the formation of narrow shear zones about 2,400 m.y. ago. Intrusion of hornblende andesite dikes and a small quartz monzonite pluton 1,800 to 1,850 m.y. ago was accompanied by low-grade metamorphism ($M_2$) without deformation.

## INTRODUCTION TO THE PROBLEM

The gneisses near Granite Falls and Montevideo, Minnesota, form part of a deformed Archean gneiss terrane exposed intermittently within the Minnesota River Valley between Ortonville and New Ulm, Minnesota. The best exposures are in the river valley between Montevideo and Franklin as shown in Figure 1. Gneisses varying in composition from quartz monzonite to tonalite form the most abundant metamorphic rock units exposed in the river valley, but mafic gneisses, aluminous paragneisses, amphibolites, and metagabbro form locally significant units. The isotopic ages of the various gneisses are all greater than 3,000 m.y.

The gneisses throughout the Minnesota River Valley have received considerable attention in the past decade because of their extremely old ages. Rb-Sr whole-rock data from the banded granitic gneiss (Montevideo Gneiss) near Granite Falls and Montevideo suggest that this unit may be as old as 3,800 m.y. (Goldich and Hedge, 1974). An earlier study by Goldich and others (1970) indicated a complex metamorphic history for this area. They recognized a high-grade metamorphic event at about 2,650 m.y. ago and a lower-grade event about 1,800 m.y. ago. The latter event was coincident with or followed shortly after the intrusion of mafic dikes and a small granitic pluton in sec. 28, T. 116 N., R. 39 W. (see Fig. 2), north of Granite Falls, about 1,850 m.y. ago. Goldich and others (1970) found that dating of the Montevideo Gneiss near Montevideo was further complicated by the presence of a more massive phase of the gneiss that locally has discordant relationships with the banded gneiss but commonly occurs as thin concordant layers within the gneiss.

The major structural features in the gneisses between Montevideo and Morton are a series of gently eastward-plunging antiforms and synforms (Himmelberg, 1968; Grant, 1972). The Granite Falls antiform (Fig. 1) is separated from the structures between Sacred Heart and Morton by a section of the valley

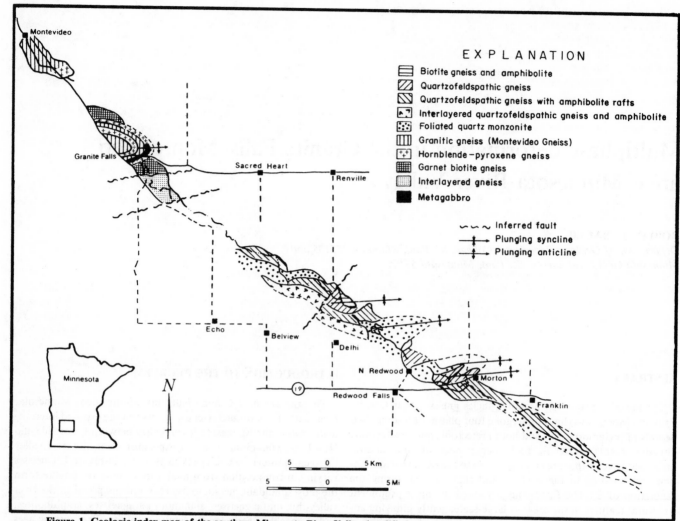

**Figure 1. Geologic index map of the southern Minnesota River Valley (modified after Grant, 1972, and Himmelberg, 1968).**

devoid of outcrop and by inferred faulting. Grant (1972) noted that aeromagnetic and gravity data (Zietz and Kirby, 1970; Craddock and others, 1970) suggested the presence of a major fault zone near the southeastern limit of outcrop south of Granite Falls. Himmelberg (1968) inferred the presence of a second fault, just north of the first fault, between garnet-biotite gneisses on the southern limb of the antiform and a series of interlayered mafic and felsic gneisses farther to the south (Figs. 1, 2). Outcrops closely approaching this second fault comprise a bleached fracture zone with cataclasized but not mylonitized rocks.

Himmelberg (1968) defined the orientation of the antiform in the Granite Falls area from an analysis of the foliation and minor linear structures in the area. In Himmelberg's analysis, most of the minor structures were consistent in orientation with the major structure. However, the minor folds and lineations varied considerably in orientation, and Himmelberg (1973, oral commun.) has suggested this variation may be the result of a second period of deformation in the area. Grant's (1972) analysis of the folding in the gneisses between Sacred Heart and Morton indicated a second generation of minor fold structures younger than the major series, with axes plunging gently to the northeast.

The purpose of the investigation reported here was to examine the possibility of multiple periods of folding in the gneisses exposed between Montevideo and the faulted southern limb of the Granite Falls antiform and to determine the relative ages of deformation, metamorphism, and intrusive events in the area.

## PREVIOUS WORK

Lund's (1956) study of the rocks near Granite Falls and Montevideo is the first extensive work in the area to be published since early exploration. Lund delineated the outcrop exposures and prepared geologic maps and descriptions of the igneous and metamorphic rocks throughout the Minnesota River Valley. His map of the Granite Falls area presents the first evidence of a major fold in the area.

Himmelberg (1968) remapped the Granite Falls–Montevideo area (Figs. 2, 3) and modified Lund's rock terminology, presenting more detailed petrographic analyses of the rock types and distinguishing four major gneissic units in the area: (1) granitic gneiss (Montevideo Gneiss of Lund, 1956; Goldich and others,

1970), (2) hornblende-pyroxene gneiss, (3) garnet-biotite gneiss, and (4) interlayered gneiss.

Himmelberg and Phinney (1967) presented a detailed study of the metamorphic assemblages in the gneisses and made partial analyses of coexisting mafic phases in the hornblende-pyroxene gneiss and the garnet-biotite gneiss. They described the mineral assemblages in the gneisses and assigned them to the granulite facies. The occurrence of pyroxene or hornblende-bearing assemblages at a given locality was attributed to differences in bulk composition of the unit or chemical potential of water, rather than a response to local temperature variations. They found no evidence of mineralogic isograds or metamorphic zoning in the units but did identify retrograde reactions in the gneisses. They attributed partial replacement of orthopyroxene by a "serpentine-like" mineral to reaction during cooling from the high-grade metamorphic event 2,650 m.y. ago, and replacement of both the orthopyroxene and serpentine by cummingtonite to the low-grade metamorphism 1,800 m.y. ago. Alteration of hornblende and clinopyroxene to cummingtonite and actinolite was similarly attributed to the 1,800-m.y.-old metamorphic event.

Recent reviews of field work in the Granite Falls–Montevideo area are contained in Grant and others (1972) and Grant (1972). The history of geochronologic investigations in this area is discussed by Goldich and others (this volume) and will not be considered here.

## LITHOLOGIC UNITS

Except for the unit designation for the Montevideo Gneiss, which has become ingrained in the geologic literature of this area, the names of the rocks units (Figs. 2, 3) suggested by Himmelberg (1968) are used here, rather than those originally proposed by Lund (1956). Himmelberg identified and mapped these units on the basis of their mineral assemblages; therefore the repetition of units in Figure 2 does not necessarily indicate a stratigraphic repetition that must be accounted for structurally, but merely a repetition of rock types.

Himmelberg noted that significant textural and modal variations occur within each of the metamorphic units. The hornblende-pyroxene gneiss as mapped by Himmelberg is particularly variable, and he described three principal variations in the unit: a banded sequence, a massive variety, and amphibolite. These varieties are distinguished as separate and genetically distinct units in this paper and are described as hornblende-pyroxene gneiss, metagabbro, and amphibolite dikes and sills, respectively.

The hornblende-pyroxene gneiss is a distinctly banded unit with alternating leucocratic and melanocratic layers (Fig. 4A). It occurs as two stratiform layers in the Granite Falls area (Fig. 2) and as small, isolated outcrops just south of Montevideo (Fig. 3). The mafic layers of the unit generally contain plagioclase, orthopyroxene, clinopyroxene, and hornblende, with subordinate quartz and biotite. The felsic layers contain abundant quartz and plagioclase with only minor pyroxene and hornblende, but locally with abundant biotite. Layers rich in garnet and magnetite are present in both the mafic and felsic layers of the gneiss. Thin concordant layers of pegmatite and grantic gneiss from less than 1 to several metres thick are particularly common in the inner folded layer of

this unit in sec. 28 and 33, T. 116 N., R. 39 W., near Granite Falls (see Fig. 2). Large layers of granitic gneiss occur at two localities within the hornblende-pyroxene gneiss—a layer about 100 m thick in sec. 19, T. 116 N., R. 39 W., and sec. 24, T. 116 N., R. 40 W., northwest of Granite Falls and a layer about 40 m thick in sec. 4, T. 115 N., R. 39 W., south of Granite Falls (Fig. 2). The contacts of these granitic layers in the hornblende-pyroxene gneiss are sharp and concordant with the gneissic banding. Contacts between the hornblende-pyroxene gneiss and major stratiform layers of the Montevideo Gneiss are not exposed, but the foliation in both units very near the contact is concordant.

The Montevideo Gneiss occurs as two stratiform layers in the Granite Falls area and constitutes most of the outcrop exposed just south of Montevideo. The unit consists of three principal rock types: banded granodioritic gneiss, granite pegmatite, and a relatively massive granitic phase. The banded gneiss contains the tonalite gneiss and the granodiorite gneiss of Goldich and others (this volume), and the massive granitic phase is the equivalent of their adamellite gneiss.

The banded gneiss occurs most commonly as a pink to tan unit with alternating leucocratic and melanocratic layers, but it also occurs as a dominantly leucocratic rock with thin biotite-rich layers defining the foliation in the gneiss. The texture in the gneiss is generally granoblastic inequigranular interlobate. The melanocratic layers of the gneiss range from quartz monzonite to tonalite, whereas the leucocratic layers range from granite to granodiorite. The more granitic leucocratic layers may be small-scale lit-par-lit injections of the granite pegmatite. The most common mineral assemblage in the banded gneiss is quartz-microcline-plagioclase-biotite. The feldspars are commonly microperthitic and microantiperthitic. Garnet occurs as a locally abundant phase in both the leucocratic and melanocratic layers of the gneiss. Blue-green hornblende and hypersthene occur rarely in melanocratic layers of the gneiss in the Granite Falls area but have not been found to coexist with one another. In fresh exposures the hypersthene-bearing layers are olive green in color and contrast sharply with adjacent pink, leucocratic layers.

Locally, abundant pink quartz-feldspar pegmatite veins and lenses of gray quartz occur as both concordant and discordant layers in the banded gneiss. The pegmatite veins contain garnet as an additional phase in the same outcrops that the banded and massive granitic gneiss contain garnet. These pegmatites do not cut the massive phase of the gneiss, although a few orange to pink, quartzofeldspathic dikes, generally less than 10 cm thick, crosscut both of the phases.

The Montevideo Gneiss outcrops just south of Montevideo contain locally abundant veins and layers of what Goldich and others (1970) described as a "red, massive phase" of the gneiss. This massive phase of the gneiss forms concordant and discordant layers within the banded gneiss that range in thickness from a few centimetres to as much as 5 m. The larger units generally form concordant layers within the banded gneiss, and one such layer is traceable about 0.5 km along strike. Some of the thinner units, generally less than 1 m wide, cross the gneissic banding and pegmatite layers at high angles (Fig. 4B).

The mineral assemblage in the massive phase is the same as that in the banded gneiss containing it, with minor biotite and garnet the only mafic silicates. The texture of the massive phase is commonly granoblastic equigranular interlobate but is locally inequi-

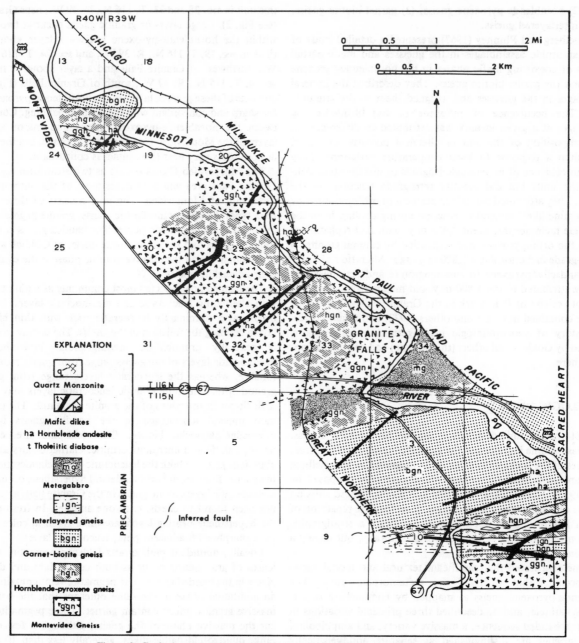

**Figure 2. Geologic map of the Granite Falls area (modified after Himmelberg, 1968).**

granular with elongate quartz grains. The biotite and granulated quartz grains show a preferred orientation that defines a foliation parallel to the banding in the banded gneiss. The banded gneiss and, particularly, the pegmatite and massive phases of the gneiss in the Montevideo area commonly contain feldspars with abundant hematite in both preferred crystallographic orientations and along some grain boundaries and fractures. The hematite-bearing feldspar gives the gneiss a distinctive red color. Textural relationships in the gneisses suggest the hematite formed as a result of oxidation of biotite and magnetite and was distributed along grain boundaries and sites in the feldspars by an oxidizing fluid phase (Bauer, 1974).

Outcrops of garnet-biotite gneiss form a single, outer,

stratiform layer of the Granite Falls antiform. The contact between the hornblende-pyroxene gneiss and the garnet-biotite gneiss on the southern limb of the antiform is concordant and is well exposed in a road cut along Highway 67 in sec. 3, T. 115 N., R. 39 W. The southern contact of this unit on the south limb is a fault contact, as noted previously. The contact between the garnet-biotite gneiss and the hornblende-pyroxene gneiss on the northern limb of the fold is not exposed.

Two types of garnet-biotite gneiss were distinguished by Himmelberg (1968): a dark-gray, medium-grained, well-foliated gneiss, and a light-gray, coarse, granular gneiss. The coarse, granular gneiss occurs as concordant bands as much as 1 m thick in the well-foliated gneiss. The principal minerals in the garnet-

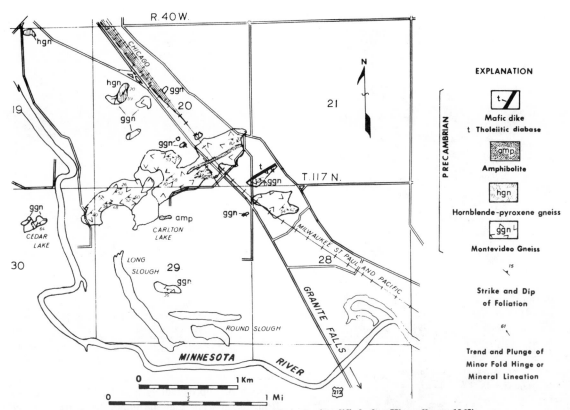

Figure 3. Geologic map of the Montevideo area (modified after Himmelberg, 1968).

biotite gneiss are plagioclase, quartz, biotite, and garnet; orthopyroxene occurs as a major mineral only in the well-foliated gneiss. Potassium feldspar is a common minor phase in both varieties of the gneiss. The plagioclase is commonly microantiperthitic; however, the potassium feldspar is rarely microperthitic. The garnet-biotite gneiss is interpreted as a paragneiss from a graywacke-like protolith.

The massive, metagabbro unit exposed on the southern limb of the Granite Falls antiform is a uniformly gray, medium-grained rock that has a crude gneissic foliation defined by hornblende segregations oriented parallel to the banding in the adjacent gneisses. The texture of the feldspars in the metagabbro is generally granoblastic equigranular, unlike that in the gneisses where the texture is generally granoblastic inequigranular. The most common minerals in the metagabbro are plagioclase, clinopyroxene, orthopyroxene, and hornblende, but biotite and quartz may be present. Exposed contacts of the unit with the banded hornblende-pyroxene gneiss are concordant, but the more massive texture, restricted occurrence, and mineralogy of the unit suggest that it is a metagabbro intrusion.

Amphibolite layers and enclaves are common in the hornblende-pyroxene gneiss and the Montevideo Gneiss, but are less common in the metagabbro, and are not present in the garnet-biotite gneiss. The amphibolites are generally uniformly dark, foliated layers, although locally they have a crude gneissic banding. The principal mineral assemblage invariably contains plagioclase and hornblende, but orthopyroxene and clinopyroxene are abundant, and biotite may occur as a minor phase. Amphibole grains and aggregates define a foliation in the amphibolite layers parallel to the banding in the gneisses. The amphibolite

layers range in thickness from less than 0.5 m to as much as 75 m. The Montevideo Gneiss that forms the inner core of the Granite Falls antiform contains numerous concordant isolated enclaves of amphibolite. In the other gneissic units the amphibolites form longer, continuous layers. The thicker amphibolite layers are completely concordant with the banding in the gneiss, but some of the thinner layers show locally discordant intrusive contacts (Fig. 4C). The amphibolite enclaves in the inner unit of Montevideo Gneiss were probably continuous, concordant layers like those in the outer units, but they have been broken up by the intense deformation in this unit. The amphibolites are interpreted as basaltic dikes and sills that intruded the gneisses prior to high-grade metamorphism. Chemical analyses (Goldich and others, this volume) and significant variations in the modal amounts of plagioclase and mafic phases suggest that more than one generation of amphibolites is represented by this unit.

Younger, igneous intrusive rocks in the area include three varieties of mafic dikes and a small quartz monzonite pluton exposed in sec. 28 (Fig. 2). Himmelberg (1968) classified the dikes as an olivine diabase, a tholeiitic diabase, and a hornblende andesite. Only the tholeiitic diabase and the hornblende andesite occur as mappable dikes (Fig. 2).

## STRUCTURAL GEOLOGY

### General Account

The distribution and orientation of minor structures and their relationships indicate that the rocks in the Granite

Figure 4. (A) Outcrop of hornblende-pyroxene gneiss with well-developed compositional banding. (B) Outcrop of banded Montevideo Gneiss with discordant veins (as shown by arrows) of the massive phase of the gneiss. (C) Hornblende-pyroxene gneiss in sharp contact with a slightly discordant amphibolite dike.

Falls–Montevideo area have undergone four periods of folding. Only the second phase of folding ($F_2$) is evident from mapping the major rock units. The $F_2$ folding formed the most prominent structural features in the area, the Granite Falls antiform (Fig. 2) and an inferred synform between Montevideo and Granite Falls. Although outcrops are lacking in an area between Granite Falls and Montevideo (Fig. 1), the attitude of the foliation in the flanking outcrops suggests the presence of an $F_2$ synformal axis beneath the covered area (Himmelberg, 1968).

An analysis of the foliation and minor linear structures across the antiform by Himmelberg (1968) yielded a fold axis for the foliation pole diagram of N85°E, plunging 15° with an axial surface dipping steeply to the south. Equal-area projections of linear structures from the Montevideo area and northern and southern

Granite Falls areas yielded roughly coincident bearing and plunge maxima of approximately N88°E, 15°. Himmelberg concluded from his analysis that the minor structures and the large-scale folding probably formed during one period of deformation.

The present study demonstrates two periods of folding—$F_3$ and $F_4$—younger than the formation of the major $F_2$ structures, and the presence of rare, minor $F_1$ folds. No large-scale structures were found to be associated with the minor $F_1$, $F_3$, or $F_4$ folds. The orientation of the minor fold axes and mineral lineations associated with the $F_3$ and $F_4$ phases of folding vary systematically with their position on the major $F_2$ fold limbs, and they concentrate along the line formed by the intersection of the average $F_3$ and $F_4$ axial-plane orientations with the local foliation planes.

Although somewhat variable, most $F_3$ axial planes strike northeast and dip moderately to steeply southeastward. Hinges of $F_3$ folds on the northeast-dipping limb of the major antiform plunge gently to moderately east-northeast. Large variations in the axial orientations of $F_3$ folds and $L_3$ lineations occur in the outcrops south of Montevideo and in the southeast-dipping limb of the Granite Falls antiform. These variations are attributed to the low angle of intersection of the $F_3$ axial planes and the foliation in these areas.

The $F_4$ folds are well developed only in the Montevideo area. The axial planes of the $F_4$ folds strike northwest and are inclined moderately to the northeast. The $F_4$ fold axes plunge east-southeast, consistent with the intersection of the dominant foliation orientation in the Montevideo area and the average $F_4$ axial-plane orientation. Sparse north- and northwest-plunging minor folds in the northern Granite Falls area are the only evidence of $F_4$ folding south of Montevideo.

## Methods of Structural Analysis

Field observations on the orientation of minor linear structures and foliation planes form the basis for division of the study area into several relatively homogeneous subareas. Foliation poles and linear data from each of the subareas were plotted as a precontoured pattern on a lower-hemisphere equal-area net using a FORTRAN IV computer program. The program used is a version of that published by Warner (1969), modified to plot equal-area projections. The contouring method is analytically equivalent to the grid method (Turner and Weiss, 1963, p. 61). The uncontoured projections were plotted by hand on a Schmidt net. Average foliation orientations and dominant axial-plane orientations used in the analysis are the planes whose poles lie at the center of the highest concentration contour on the appropriate diagram. Axial planes of minor folds were determined by a visual best fit of the trend and plunge of the hinge line and axial-surface trace of a given minor fold to a single great circle on a stereonet.

## Structural Elements

**Planar Structures.** Alternating layers of variable composition within the gneisses define the principal foliation, $S_1$. The basal planes of biotite and flattened quartz grains parallel $S_1$, and the longest crystal dimensions of prismatic minerals lie in this plane. Some thin, granitic to pegmatitic veins not parallel to the $S_1$ foliation in the gneiss are folded and yield folds ($F_1$) with axial planes parallel to $S_1$ (Fig. 5A).

Calculated axial planes of $F_2$, $F_3$, and $F_4$ folds define $S_2$, $S_3$, and $S_4$, respectively. In areas of intense $F_3$ deformation, elongate quartz lenses and ribbons, biotite, and hornblende grains are oriented parallel to the dominant $F_3$ axial-plane solution and define an $S_3$ foliation. $S_3$ can be readily observed in cut hand specimens (Fig. 5C) and with close observation in the field (Fig. 5B), but a distinct $S_3$ axial-plane foliation is not generally measurable in the field.

Narrow, vertical shear zones, striking in the range west-northwest to north-northeast, are locally well developed in the northwestern Granite Falls area and the Montevideo area. The shear zones range in width from less than 25 cm to 1 m. Both right-lateral and left-lateral displacements were observed across the shears, with left-lateral offset more common in the northwest-trending to north-northeast–trending shears and right-lateral offset dominant in the west-northwest–trending shears. In the Granite Falls area the shears are mylonitic and have a cataclastic microfabric (Fig. 5D). "Shear" zones in the Montevideo area, however, are generally diktyonitic structures (compare Mehnert, 1968), commonly containing a neosome of granitic to pegmatitic material. The microfabric of the neosome is not cataclastic, although it may contain a banding parallel to the direction of shearing as in Figure 5E.

I have not recognized a "shearing" event in the Montevideo Gneiss younger than the high-grade metamorphism ($M_1$) other than that restricted to the narrow zones described above. The pegmatite, massive adamellite, and the banded gneiss contain a locally very pronounced foliation which I attribute to $S_1$ or $S_3$ developed during the $M_1$ metamorphism and not to a younger shearing event (compare Goldich and others, this volume).

**Linear Structures.** Linear structures are present in each of the stratiform units exposed in the Granite Falls and Montevideo areas, but the intensity of their development varies considerably among the units. The Montevideo Gneiss near Montevideo and that forming the core of the Granite Falls antiform contain the

best-developed minor fold structures. The hornblende-pyroxene gneiss, outer layer of Montevideo Gneiss, and garnet-biotite gneiss near Granite Falls contain few minor folds but contain locally well-developed hornblende lineations parallel to either the $F_2$ or $F_3$ fold axes and quartz lineations parallel to $F_3$. $L_3$ quartz lineations occur as fine stripes on $S_1$ foliation planes. These lineations occur most commonly in areas where the quartz forms thin lenses oriented parallel to $S_3$ and are interpreted as $S_1$-$S_3$ intersections. The orientation of these lineations is variable but is generally consistent with the axes of the local $F_3$ folds. Less commonly the quartz lenses yield a more linear fabric and are strongly elongate parallel to the $F_3$ fold axes. $L_2$ quartz lineations are rare and have been observed only in the inner layer of Montevideo Gneiss near Granite Falls. They take the form of elongate quartz rods parallel to $F_2$ fold axes and generally less than 2 cm in length and 3 mm in diameter.

Pinch-and-swell structures are well developed in the Montevideo Gneiss near Montevideo (Fig. 5F). Layers of the massive phase of the gneiss form the pinched competent unit of the structure and the banded gneiss the incompetent unit. The pinched necklines of the structure generally trend south-southeast with a moderate plunge. Rare pinch-and-swell structures are present in the outer layer of Montevideo Gneiss on the Granite Falls antiform. In this area thin, concordant amphibolite layers in the gneiss are pinched with axes plunging gently to the east, consistent with the $F_2$ fold axes.

## $F_1$ Folding

Rare $F_1$ folds are present in the Montevideo Gneiss both near Montevideo and in the core of the Granite Falls antiform. These folds have axial planes parallel to the $S_1$ foliation (Fig. 5A) but do not appear to be a folding of a major, pre-$S_1$ foliaton. The $F_1$ folds occur only as folded, thin, granitic to pegmatitic veins. The folds are believed to have formed in veins intruded at a high angle to the $S_1$ foliation during the latter stages of its formation. They vary from close to isoclinal and are sometimes intrafolial to the $S_1$ banding. Axial orientations of the folds were not generally measurable. $F_1$-$F_2$ and $F_1$-$F_3$ interference patterns were observed in the Granite Falls area but are rare.

The possibility of large-scale $F_1$ folding resulting in the repetition of the Montevideo Gneiss or hornblende-pyroxene gneiss in the Granite Falls area was considered during the field investigation, but no evidence to support this concept was found. In particular, no minor $F_1$ folds were found in outcrops that might contain the axial-surface trace of such a structure and therefore be expected to contain abundant $F_1$ folds. The possibility remains that a major $F_1$ structure, responsible for the formation of the $S_1$ foliation, is concealed by the glacial cover beyond the valley walls.

## Distribution of Post-$F_1$ Folding

Minor $F_2$ folds are best developed in the Granite Falls area; minor $F_3$ folds are well developed in both the Granite Falls and Montevideo areas; and $F_4$ folds are well developed only in the Montevideo area. On the basis of the heterogeneous development of $F_2$ and $F_3$ folds, the Granite Falls area has been divided into the six subareas shown in Figure 6. Contoured equal-area projections of foliation poles and minor linear structures are compiled in

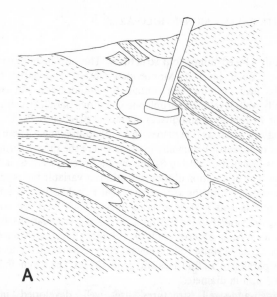

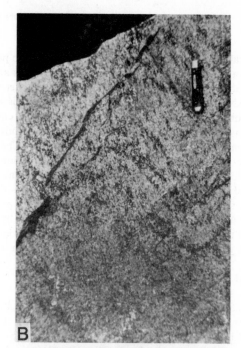

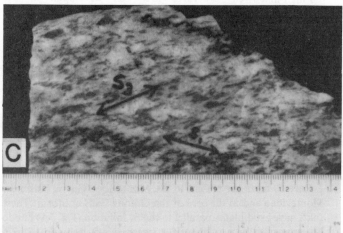

Figure 5. Structural elements in the Montevideo Gneiss. (A) $F_1$ folds in pegmatite veins in the Montevideo Gneiss. (B) $F_3$ fold with $S_3$ axial-plane foliation (parallel to knife) defined by quartz grains and biotite aggregates (Montevideo Gneiss, subarea III). (C) $S_3$ foliation (parallel to arrow) defined by quartz grains (Montevideo Gneiss, subarea III). (D) Shear zone in Montevideo Gneiss (subarea II). (E) Diktyonitic structure with a broad neosomal zone. Note the banding in the neosome parallel to the plane of offset. One distinct band across the zone indicates a displacement of approximately 2 m. (F) Pinch-and-swell structure in the Montevideo Gneiss near Montevideo.

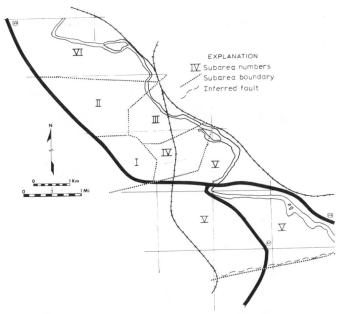

**Figure 6. Index map of Granite Falls structural subareas.**

Figure 7 for these six subareas. The plots of minor linear data include fold hinges and mineral lineations of both $F_2$ and $F_3$ origin. The mineral lineations and fold hinges are plotted together because separate plots of the two indicate they are coaxial. $F_2$ and $F_3$ data are plotted together because individual folds cannot always be assigned to $F_2$ or $F_3$ in the field despite their distinction regionally. Because of the heterogeneous areal development of the fold phases, the characteristics of each fold phase are discussed with reference to the region of their most intense development.

## $F_2$ Folding in the Granite Falls Area

The minor $F_2$ folds are most commonly open to close, upright to steeply inclined folds with subhorizontal to gently eastward-plunging axes. Their geometric fold class is 1C in the competent layers and class 2 or 3 in the incompetent layers by the classification of Ramsay (1967, p. 365). These folds are best developed and preserved along the axial trace of the antiform in the Montevideo Gneiss that forms the inner core of the fold (subarea I in Fig. 6). The $F_2$ folds show significant variations in scale in this area and range from large parasitic folds with wavelengths on the order of 30 m to minor folds a few centimetres in wavelength. The larger folds have symmetry relations consistent with their position on the limbs and hinge area of the major fold, and the symmetry of the smaller folds can generally be related to their position on the larger parasitic folds. Although the $F_2$ folds are generally steeply inclined, their axial-plane orientation varies from upright to recumbent (Fig. 8A, 9A). Nearly recumbent $F_2$ folds with open to close limb angles are common only in subareas II and III in the inner core of Montevideo Gneiss near the contact with the hornblende-pyroxene gneiss. These folds are assumed to be $F_2$ despite their anomolous axial-plane orientation because they are coaxial with other minor $F_2$ folds and their z-symmetry is consistent with their location on the north limb of the major $F_2$ antiform. The relationship of these recumbent folds to the more upright $F_2$ folds is uncertain.

Only subareas I and II in the Granite Falls area are sufficiently unaffected by later deformation and provide enough linear data to demonstrate the relationship of the minor $F_2$ folds to the major antiform. Figures 7B and 7C are equal-area plots of foliation poles and linear features, respectively, from subarea I. The $\pi$ axis of the foliation pole girdle is N80°E at 13°, in general agreement with that suggested by Himmelberg (1968). The center of highest concentration of linear data, at N76°E plunging 13°, is roughly coincident with the $\pi$ axis; however, the concentration contours in Figure 7C are distinctly elongated to the northeast. This elongation is probably an effect of the third period of folding, to be discussed in the following section. A similar distribution of lineations is obtained from the data in subarea II, shown in Figure 7E.

## $F_3$ Folding in the Granite Falls Area

Minor $F_3$ folds are similar to the more upright $F_2$ folds in their geometric form and style, but large $F_3$ folds are much less common than large $F_2$ folds and have been recognized only on the northern limb of the Granite Falls antiform. In the NW¼ sec. 33, T. 116 N., R. 39 W., several broad, open $F_3$ folds with wavelengths on the order of 10 to 15 m plunge at approximately 30°. These folds are the most prominent fold structures in subarea III, which is dominated by the $F_3$ folding. The presence of the open $F_3$ folds is evident in the foliation pole diagram from this area (Fig. 7F), which shows a primary $\pi$ axis, $B_1$, of N87°E at 17° related to the major $F_2$ antiform, and a secondary axis, $B_2$, of N57°E at 30° related to the $F_3$ folding. The plot of all linear elements in this subarea, Figure 7G, yields a maximum concentration at N65°E, 31°, which is in rough agreement with the secondary $B_2$ axis defined by the foliations. Lineations from subarea IV adjacent to the axial trace of the major antiform (Fig. 7I) have an orientation similar to that of subarea III; this orientation indicates a strong $F_3$ influence. However, no large $F_3$ folds or $F_3$ effects on the foliation orientation were observed (Fig. 7H).

Minor folds are sparse in the outer units on the southern limb of the antiform that makes up subarea V, but $L_3$ quartz lineations are fairly common in this subarea. The plot of minor folds and mineral lineations from this subarea (Fig. 7K) shows a large variation in orientation along a great circle striking N54°E, dipping 45° southeast. $F_2$ folds account for the high concentration plunging to the east. $F_3$ folds and lineations most commonly plunge gently to the southwest, but $F_3$ and $L_3$ elements plot all along the great circle.

Distinct interference patterns of $F_3$ folds crossing the upright $F_2$ were not found in any of the subareas. Locally, moderately plunging, open $F_3$ warps fold the limbs of subhorizontal $F_2$ folds (Fig. 8B), but no penetrative refolding of the $F_2$ axial planes was observed. In buckling experiments on intersecting fold patterns, Ghosh and Ramberg (1968) found that if the direction of compression of a second deformation was less than 60° to the first in the plane of the layering, superposed folds were poorly developed and sporadic. At angles of less than approximately 30°, superposed folds did not form in their experiments. They found instead that the most significant effect of the second deformation on the first system was a rotation in trend of the first folds toward that of the second-generation folds. This was accompanied by a considerable increase in amplitude and further tightening of the pre-existing first fold hinges. Many of the minor $F_2$ folds in subareas

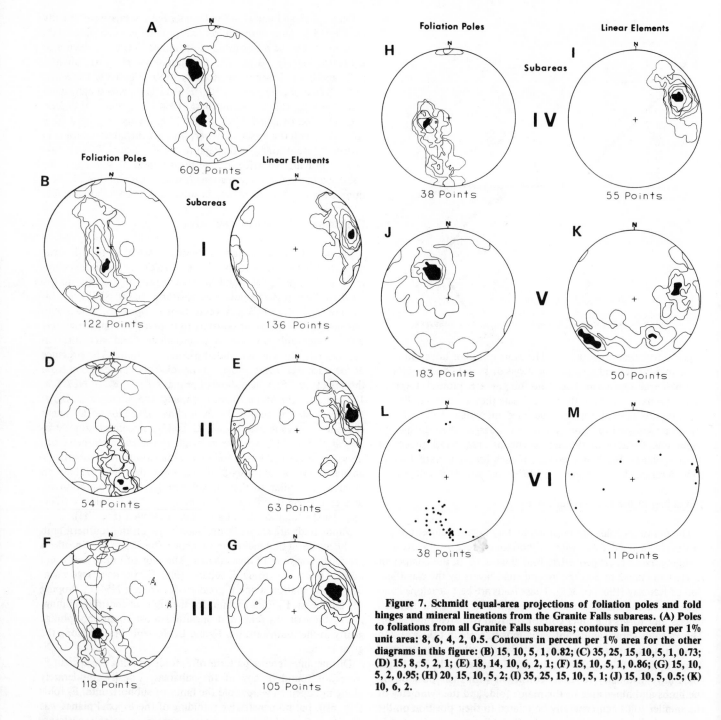

Figure 7. Schmidt equal-area projections of foliation poles and fold hinges and mineral lineations from the Granite Falls subareas. (A) Poles to foliations from all Granite Falls subareas; contours in percent per 1% unit area: 8, 6, 4, 2, 0.5. Contours in percent per 1% area for the other diagrams in this figure: (B) 15, 10, 5, 1, 0.82; (C) 35, 25, 15, 10, 5, 1, 0.73; (D) 15, 8, 5, 2, 1; (E) 18, 14, 10, 6, 2, 1; (F) 15, 10, 5, 1, 0.86; (G) 15, 10, 5, 2, 0.95; (H) 20, 15, 10, 5, 2; (I) 35, 25, 15, 10, 5, 1; (J) 15, 10, 5, 0.5; (K) 10, 6, 2.

III and IV may have undergone a similar response to the deformation that formed the $F_3$ folds and $S_3$ foliation.

## $F_2$ and $F_3$ Axial Planes

Axial-plane solutions for second- and third-period folds are plotted as poles in Figures 9A and 9B. Because of the similarity in orientation and style of the upright $F_2$ and $F_3$ folds, distinguishing between them in the field was not always possible. As a result of this ambiguity, the axial-plane data plotted are separated into two special groups. The first group consists of axial-plane attitudes for folds considered to be definitely $F_2$ folds, that is, folds from the Granite Falls area whose axes are oriented south of N75°E and plunge less than 25°. The data for this group are from subareas I and II and are plotted in Figure 9A. The second group plotted represents axial-plane attitudes for $F_3$ folds and a few folds of uncertain $F_2$-$F_3$ affiliation. These data were collected primarily from subareas III and IV and from the Montevideo area and are plotted in Figure 9B. If, in fact, the response of the minor $F_2$ folds to $F_3$ folding was a rotation of their axial planes toward

Figure 8. (A) Recumbent $F_2$ folds. (B) $F_3$ folds deforming the limb of a minor $F_2$ fold.

parallelism with those of $F_3$, the second group of axial-plane solutions should be representative of $F_3$ axial-plane solutions despite the ambiguity of assignment of some of the minor folds.

The poles to the axial planes of the second group ($F_3$) have a contoured concentration maximum normal to an average axial plane striking N44°E, dipping 57° southeast, but the poles form a diffuse great circle gridle about an $F_3$ $\pi$ axis of N55°E, plunging approximately 20°. In subarea III this fanning of the axial planes can locally be related to the position of the minor folds on larger-scale $F_3$ folds. The axial planes diverge downward in the synforms and upward in the antiforms. Ramberg (1963, 1964) has shown that internal strain, which has been produced by the continued development of the larger wavelength folds, can further deform the small-scale parasitic folds; this results in the reorientation of their axial planes systematically about the axis of the major fold.

Not all the $F_3$ folds can be related to larger-scale folding, however, and a second mechanism may be responsible for some of the variations in axial-plane attitudes. If the $F_3$ phase of deformation caused a reorientation and remobilzation of minor $F_2$ folds, the $F_3$ folds, which formed early in the deformational phase on the limbs of the $F_2$ folds, would exhibit strains associated with this remobilization. Locally, $F_3$ folds on the limbs of larger $F_2$ folds do have slightly curved hinge areas and axial planes; however, such examples are not common. This mechanism was first suggested by Ramsay (1958, p. 292). He was able to demonstrate a change in the orientation of a second-fold axial-plane cleavage where it crossed the axial trace of a major first fold. Although such a systematic relationship cannot be demonstrated in the Granite Falls area, this does not exclude the possibility of varying orientations of the minor $F_3$ folds in response to a remobilization of minor $F_2$ folds.

## Folding in the Montevideo Area

$F_2$ folds in the Montevideo area are difficult to recognize in the field because of the superposition of $F_3$ and $F_4$ folding. The orientation and style of $F_3$ and $F_4$ folds are similar enough to those of the $F_2$ folds so that the latter are recognized only where they are crossed by younger minor folds. Where their recognition is possible, the $F_2$ folds plunge gently to the east.

Minor $F_3$ folds and $L_3$ lineations are present throughout the Montevideo area, but are best developed and least affected by the $F_4$ phase of folding in the area around Lake Carlton (Fig. 3). Most of the $F_3$ folds and mineral lineations plunge gently to the northeast or southwest, but several were found that plunge to the southeast and southwest at moderate angles. Their distribution is similar to that in subarea V (Fig. 7K) on the southern limb of the Granite Falls antiform, where the orientation of the foliation is similar to that near Montevideo.

The axes of pinch-and-swell structures and boudins, which involve more competent layers of the massive phase of the Montevideo Gneiss in the less competent banded gneiss (Fig. 5F),

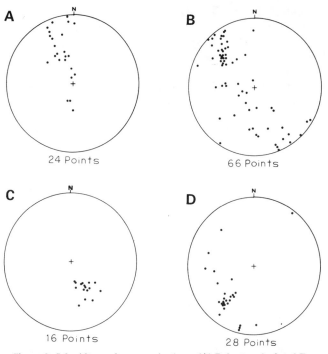

Figure 9. Schmidt equal-area projections. (A) Poles to calculated $F_2$ axial planes. (B) Poles to calculated $F_3$ axial planes. (C) Axes of pinch-and-swell structures in the Montevideo Gneiss near Montevideo. (D) Poles to calculated $F_4$ axial planes.

plunge moderately to the south-southeast (Fig. 9C). These pinched necklines and long axes are interpreted as $L_3$ lineations that
developed normal to the $F_3$ fold axes and $L_3$ mineral lineations.
Not only the orientation of the necklines and long axes, but also
the fabric associated with the $F_3$ folding are consistent with this
interpretation. The locally developed linear quartz fabric in the
gneiss indicates a component of relative elongation parallel to the
$F_3$ fold axes, consistent with the formation of boudin axes normal
to the fold axes.

No large-scale structural features were found to be associated
with the fourth phase of folding. The $F_4$ folds occur most commonly as symmetric open warps of the gneissic banding and
plunge moderately to the east-southeast. The strongest development of $F_4$ folds is in the outcrops east of U.S. Route 212,
southeast of Montevideo. In this area $F_4$ folding is occasionally so
pronounced that the foliation planes of the banded gneiss show a
regular corrugation of crests and troughs of open, minor $F_4$ folds
(Fig. 10A). Several interference patterns of $F_4$ folds crossing gently plunging $F_3$ folds are present in this area (Fig. 10B). Although
the axial-plane solutions of minor $F_4$ folds show some variation in
orientation, they show a distinct concentration, when contoured,
of about N50°W dipping 48° northeast (Fig. 9D).

Contoured plots of poles to foliations and lineations from the
Montevideo area are presented in Figure 11. The axes of minor
folds constitute by far the greatest percentage of the linear data
plotted, but $L_3$ mineral lineations are also included in the
diagram. Although the period of deformation during which the
minor structures developed could commonly be recognized, this
distinction was not always possible and is not necessary in the
analysis which follows.

## $F_2$ Control of $F_3$ and $F_4$ Orientations

The effects of early fold structures on the orientation of
younger structural elements have been demonstrated and discussed by numerous authors and are well summarized by Ramsay
(1967) and Whitten (1966). The analysis is based on the principle
that the attitude of any fold axis is given by the intersection of the
axial surface of the fold and the surface being folded. When folds
with subparallel axial surfaces develop on S surfaces of different
orientations, the trend and plunge of the fold axes will vary as a
function of the attitude of the S surface being folded.

The variation in orientation of the $S_1$ foliation in the Granite
Falls and Montevideo areas, caused by the major $F_2$ folding, has
resulted in the variations in orientation of the linear structures
formed during the $F_3$ and $F_4$ phases of folding. The minor linear
structures of the younger phases of folding show differences in
plunge and trend that result from the varying orientation of the
intersection of the limbs of the major $F_2$ fold structure and the axial planes of the third- and fourth-period folds.

Figure 12A is a schematic illustration of the intersecting planes
shown in the accompanying synoptic diagram (Fig. 12B). These
figures show the spatial relationships of the foliation in the
Montevideo area and the mean or dominant $F_3$ and $F_4$ axial-plane
orientations. The great circle representing foliation in the
southeast-dipping outcrops in the Montevideo area is the plane
whose pole is the center of the maximum concentration of poles
to foliations plotted in Figure 11A. The dominant axial plane of

**Figure 10. (A) Minor $F_4$ folds in the Montevideo Gneiss near
Montevideo. (B) Minor $F_3$ fold with superimposed minor $F_4$ fold in
Montevideo Gneiss near Montevideo.**

the $F_4$ folds (taken as the normal to the point maxima in Figure
9D) intersects this limb of the $F_2$ structure at a point corresponding to the average plunge of the $F_4$ folds in the area.

The dominant $F_3$ axial plane (Fig. 9B) intersects the foliation
plane in the southwest at a very low angle. Because of this low
angle of intersection, the directional stability of the $F_3$ fold axes is
low. Any slight variation in the orientation of the original surface
being folded would result in a great variation in the axial direction
in the $F_3$ folds or $L_3$ lineations formed. The result, as illustrated in
Figure 11B, is the distribution of plotted $F_3$ lineations and fold
axes along a great circle of foliation-$F_3$ axial-plane intersections.
The planar intersections in the synoptic diagram for subarea V
(Fig. 12C), where the orientation of the foliation is similar to that
in the Montevideo area, show a relationship to the distribution of
$F_3$ and $L_3$ lineations similar to those plotted in Figure 7K.

A similar synopsis for the northern Granite Falls area is il-

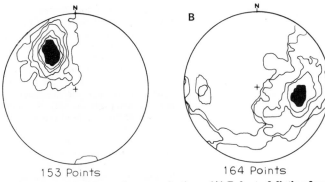

**Figure 11. Schmidt equal-area projections. (A) Poles to foliation from the Montevideo area. Contours in percent per 1% area: 20, 15, 10, 5, 2, 0.65. (B) Minor fold hinges and mineral lineations from the Montevideo area. Contours in percent per 1% area: 12, 7, 4, 1, 0.61.**

lustrated in Figure 12D. In addition to the average $F_3$ and $F_4$ axial planes, two foliation planes are also plotted. The foliation plane $S_{NL}$ (solid line) is the great circle to the pole maximum in Figure 7A for the northeast-dipping limb of the major antiform. This great circle is representative of the average foliation orientation in the outcrops northwest of Granite Falls. The intersection of this foliation orientation and the average $F_3$ axial plane coincides with the general orientation of $F_3$ fold axes on this limb of the antiform. The maximum concentration contours of linear elements from subareas III and IV (Figs. 7G, 7I), plotted about this intersection, illustrate the correlation.

The plane $S_{VI}$ (dashed line) is the average foliation orientation in the outcrops in subarea VI and is plotted to indicate part of the range in foliation orientation on this limb of the antiform. Note that the intersections of the foliation with the average $F_4$ axial plane range along the axial plane between $S_{VI}$ and $S_{NL}$. These intersections are coincident with the few north- and northwest-plunging folds indicated in the linear element plots from subareas III and VI (Figs. 7G, 7M). No crossing relationships were observed between these folds and the $F_2$ and $F_3$ folds, but their coincidence with the foliation-$F_4$ axial plane range of intersections implies they were formed during the $F_4$ folding.

These $F_4$ folds are significantly different from any of the $F_2$ or $F_3$ folds or the $F_4$ folds in the Montevideo area. They are usually large, single folds, roughly concentric in form and contain no higher order parasitic folds. Their amplitudes are generally between 2 and 4 m.

## DISCUSSION: THE SEQUENCE OF GEOLOGIC EVENTS

Abundant radiometric age determinations, field and petrographic observations, and the preceding structural analysis suggest the sequence of geologic events in this area outlined in Table 1. This section discusses the ambiguities remaining in the sequence and the evidence for the geologic history proposed in Table 1. The sequence of geologic events can be divided into three stages of development: stage 1—formation of the gneiss protoliths; stage 2—intrusion, folding, and high-grade metamorphism; and stage 3—postfolding intrusion, shearing, and low-grade metamorphism. The boundaries of these stages are uncertain in some cases, but these divisions mark periods of distinctly

different conditions to which the gneisses were subjected. The principal contributions of the present study concern events that make up stage 2 and the transition to stage 3; therefore, only these two stages will be discussed here. Stage 1 is discussed in detail by Goldich and others (this volume).

### Stage 2: Intrusion, Folding, and High-Grade Metamorphism

Following the formation of the three major gneiss units, the oldest identifiable event is the formation of the gneissic banding and $S_1$ foliation parallel to the banding. During the latter stages of $S_1$ formation, pegmatite veins were emplaced into the Montevideo Gneiss both near Montevideo and in the core of the Granite Falls antiform. The temporal relationship of these pegmatites to formation of the other gneiss units is uncertain. Two pieces of evidence suggest an early emplacement of these pegmatites: (1) they did not intrude the massive phase of the Montevideo Gneiss, and (2) they are locally folded to form the $F_1$ folds with axial planes parallel to the $S_1$ foliation and are therefore pre- or syn-$S_1$.

There is evidence that suggests the compositional banding in the gneisses formed prior to the development of the $S_1$ foliation to which it is parallel. The amphibolites and the massive phase of the Montevideo Gneiss were emplaced after the development of the banding in the gneisses. This is evident from both the concordant and discordant contacts of these units with the gneisses. The concordance of both of these units over substantial distances indicates the influence of the compositional banding on their emplacement. Discordant contacts between these units and the gneisses indicate that they intrude an existing banding in the gneisses. However, both the amphibolites and the massive phase have a foliation parallel to the banding in the gneisses.

In the case of the amphibolites there is no doubt that this is an $S_1$ foliation. The amphibolites are folded along with $S_1$ in the gneisses across the axial trace of the major $F_2$ antiform. The presence of $S_1$ in the amphibolites, whose emplacement was evidently controlled by the banding in the gneisses, suggests two possible relationships: either the banding in the gneisses was present before the development of $S_1$ and the amphibolites were emplaced after banding development and before $S_1$ or the amphibolites were emplaced during the development of $S_1$ and the compositional banding. The former seems more likely because the forces responsible for the formation of $S_1$ would tend to inhabit the emplacement of the amphibolites parallel to $S_1$.

The foliation in the massive phase is generally weak to absent and may be an $S_3$ foliation rather than $S_1$. Because the banding and $S_1$ foliation in the outcrops south of Montevideo are very similar in orientation to $S_3$, foliations $S_1$ and $S_3$ are readily distinguished only where the $S_1$ banding is at a significant angle to $S_3$ and $S_3$ is well developed. Where a foliation is developed in the massive phase, it is defined by sparse biotite grains and flattened quartz grains. The fact that garnet grains and garnet clusters are commonly aligned parallel to the foliation is not a useful criterion for distinguishing $S_1$ and $S_3$ because garnet grains occur parallel to both $S_1$ and $S_3$ in the banded gneiss. The relatively weak foliation in the massive phase, as compared to the banded gneiss, may be a function of the lower biotite content of the massive phase; but the lack of $F_1$ folds in veins of the massive phase crossing the banded gneiss (like the folds present in the crosscutting pegmatite veins) suggests a post-$S_1$ intrusion of the unit.

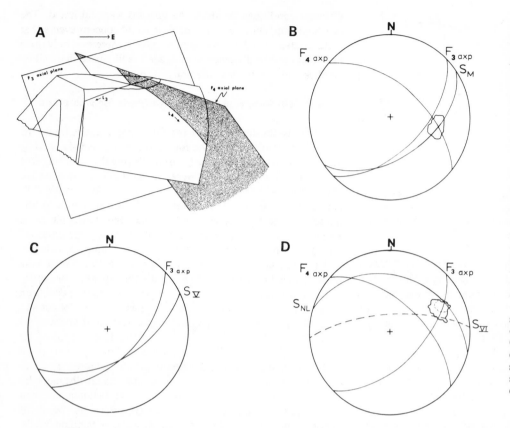

Figure 12. Synoptic illustrations of the relative orientations of the $F_2$, $F_3$, and $F_4$ fold phases. (A) Schematic diagram illustrating the intersection of the southeast-dipping limb of the synform south of Montevideo or the southeast-dipping limb of the Granite Falls antiform and the dominant $F_3$ and $F_4$ axial planes. (B) Synoptic diagram of the structural geometry in the Montevideo area. The contour plotted on the diagram is the maximum concentration contour of linear elements from Figure 11B. (C) Synoptic diagram of the structural geometry on the south limb of the Granite Falls antiform. (D) Synoptic diagram of the structural geometry on the north limb of the Granite Falls antiform. The contours plotted on the diagram are the maximum concentration contours of linear elements from Figures 7G and 7I. $S_M$ = dominant foliation orientation in the Montevideo area; $S_V$ = dominant foliation orientation in subarea IV; $S_{NL}$ = dominant foliation orientation on the north limb of the Granite Falls antiform; $S_{VI}$ = the dominant foliation orientation in subarea VI; $F_{3\ axp}$ = the dominant $F_3$ axial-plane orientation; $F_{4\ axp}$ = the dominant $F_4$ axial-plane orientation.

The assignment of $S_3$ rather than $S_1$ foliation to the massive phase considerably affects estimates of the ages of the deformation periods and the duration of high-grade ($M_1$) metamorphism, because the 3,050-m.y. apparent age of intrusion of the massive phase (Goldich and others, this volume) is the only radiometrically determined age from this area that lies between the gneiss ages and the metamorphic mineral ages attributed to $M_1$, approximately 2,650 m.y. The sequence displayed in Table 1 and the discussion that follows assume the massive phase to be post-$S_1$ in age.

Intrusive contacts between the massive phase and the amphibolites near Montevideo were not observed. However, a post-$S_1$ age for the massive phase requires that the amphibolites be older than the massive phase. The 3,050-m.y. age for the massive phase of the Montevideo Gneiss, therefore, suggests that the emplacement of the amphibolites and the formation of the banding in the gneisses took place prior to 3,050 m.y. ago.

The relative position of the metagabbro unit in this sequence is ambiguous. Wilson and Murthy (1976) reported a Rb-Sr whole-rock age of 2,680 ± 200 m.y. for the unit. Wilson (1976) suggested that this may represent a whole-rock metamorphic age rather than the original age of the rock; he therefore interpreted this as a minimum age for the unit. The presence of an $S_1$ foliation and foliated amphibolites in the metagabbro unit are inconsistent with a 2,680-m.y. primary age if the $S_1$ foliation formed at or before 3,050 m.y. B.P. The concordant position of the metagabbro in the gneisses and its relative lack of banding suggest a syn-$S_1$ or pre-$S_1$ emplacement like that of the amphibolites. The presence of amphibolites within the metagabbro, however, suggests its emplacement prior to that of at least some of the amphibolites.

High-grade regional metamorphism ($M_1$) was syntectonic with the first three fold phases. The preferred orientation of hornblende, biotite, and pyroxenes and the local flattening of garnet in $S_1$ indicate that high-grade metamorphism accompanied the generation of $S_1$. Evidence concerning the relation of $F_2$ and $F_3$ folding to metamorphism suggests that both of these fold phases took place during $M_1$ metamorphism. $L_2$ hornblende lineations in the hornblende-pyroxene gneiss and the local reorientation of quartz lenses, hornblende, biotite, and garnet parallel to $S_3$ are consistent with this interpretation. The variation in orientation of the minor $F_3$ structures on opposing limbs of the major $F_2$ structure indicates the complete development of $F_2$ before $F_3$ folding, but the similarity in style of $F_2$ and $F_3$ minor folds suggests similar competency and probably similar metamorphic conditions in the rocks during both of the fold generations.

The inferred $F_3$ origin for the pinch-and-swell structure and foliation in the massive phase of the Montevideo Gneiss indicates $F_3$ followed the emplacement of this unit There is no direct evidence concerning the relative timing of $F_2$ folding and emplacement of the massive phase because no structures attributed to $F_2$ are identified within or near this unit.

The duration of the $M_1$ metamorphism is another uncertainty in the sequence of events, although some limits can be placed on its extent. The lower (older) age limit is the more difficult to fix. Whereas the massive phase of the Montevideo Gneiss is believed to be post-$S_1$, or at the earliest syn-$S_1$, the high-grade metamorphism accompanying $S_1$ began at least 3,050 m.y. ago. The upper limit on the age of $M_1$ metamorphism is constrained by mineral ages published by Hanson and Himmelberg (1967) and Goldich and others (1970) at approximately 2,650 m.y. The possible range for this event is uncommonly long; however, there is no evidence

TABLE 1. GEOLOGIC EVENTS DURING PRECAMBRIAN TIME IN THE GRANITE FALLS-MONTEVIDEO AREA

| Stage | | Structural events | Metamorphic and rock-forming events |
|---|---|---|---|
| 3 | | | $M_2$ Low-grade regional metamorphism |
| | | | Quartz monzonite pluton |
| | | | Hornblende andesite dikes |
| | | | Olivine tholeiite dikes |
| 2 | $F_4$ | Formation of ESE-plunging folds near Montevideo and N to NW-plunging folds in the northern Granite Falls area Formation of WNW to NE-striking shear zones | Tholeiitic diabase dikes |
| | | | Fixing of mineral ages in the gneisses and metagabbro |
| | | Local generation of $S_3$ foliation | Local reorientation of quartz, biotite, hornblende, and garnet in the $S_3$ foliation plane |
| | $F_3$ | Formation of NE-trending folds on the N-dipping limbs of major $F_2$ structures and SW to SE to NE-trending folds on the S-dipping limbs of the major $F_2$ structures | |
| | $F_2$ | Formation of major eastward-plunging antiforms and synforms | Formation of $L_2$ mineral lineations parallel to $F_2$ fold hinges Intrusion of massive phase of the Montevideo Gneiss |
| | $F_1$ | Development of $F_1$ folds in pegmatite veins | Intrusion of pegmatite veins Recrystallization producing high-grade $M_1$ mineral assemblages oriented by $S_1$ |
| | | Formation of $S_1$ foliation | |
| | | Formation of gneissic banding | Intrusion of amphibolites Intrusion of metagabbro |
| 1 | | | Hornblende-pyroxene gneiss Garnet-biotite gneiss Banded Montevideo Gneiss |

of any metamorphic hiatus in the mineral assemblages or textures in the gneisses prior to the retrograde assemblages interpreted to have formed during the 1,800-m.y. metamorphism.

The $F_4$ folding and the shearing event probably took place during the waning stages of $M_1$ metamorphism. There is no evidence as to the relative ages of these two events, but the style of deformation in both the $F_4$ folds and the shear zones indicates a higher ductility at the time of their generation in the Montevideo area than in the northern Granite Falls area. This parallel variation in the style of deformation of both features suggests they were generated under similar conditions and possibly at nearly the same time. Although the $F_4$ folds and shear zones are common only in areas where both are present, preliminary analysis of the orientation and sense of displacement of the shear zones has yielded no compelling evidence for generation of both events by the same stress system. Unpublished Rb-Sr data on samples from the shear zones led Hedge and Goldich (1976) to conclude that the shearing took place approximately 2,400 m.y. ago.

## Stage 3: Postfolding Intrusion, Shearing, and Low-Grade Metamorphism

The younger intrusive rocks of this stage include the three sets of mafic dikes described by Himmelberg (1968) and the small quartz monzonite pluton in section 28 north of Granite Falls. Himmelberg (1968) recognized the relative ages of shearing and dike emplacement from observation of tholeiitic diabase dikes cut by the shear zones and hornblende andesite dikes cutting the shears. Hanson and Himmelberg (1967) obtained K-Ar ages of 2,080 and 1,800 m.y. on hornblende from two samples of the tholeiitic diabase. Himmelberg (1968) suggested that the 2,080-m.y. age possibly resulted from partial loss of argon from the hornblende during the 1,800-m.y. low-grade metamorphism ($M_2$) recognized from biotite ages throughout the area. This interpretation is consistent with the 2,400-m.y. age for shearing suggested by Hedge and Goldich (1976) and indicates an age greater than 2,400 m.y. for the tholeiitic diabase. There is no definitive evidence for the relative ages of $F_4$ folding and emplacement of the tholeiitic diabase dikes. The dike that intruded the outcrops of Montevideo Gneiss just south of Montevideo (Fig. 3) shows no effects of the $F_4$ folding, but it would have been more resistant to development of the small, open $F_4$ warps in the area than the surrounding gneiss.

Hornblende and biotite K-Ar ages for the hornblende andesite dikes (Hanson and Himmelberg, 1967) and radiometric ages for the quartz monzonite pluton obtained by numerous investigators (compare Doe and Delevaux, this volume) indicate ages of approximately 1,800 to 1,850 m.y. for these units. This is essentially coincident with the $M_2$ metamorphism in the area. Retrograde assemblages in the mafic dikes are consistent with their exposure to the low-grade metamorphism. The tholeiitic diabase contains bladed to fibrous green hornblende aggregates replacing clinopyroxene, biotite mantles on opaque oxides, and secondary epidote. The hornblende andesite contains secondary epidote, calcite, sericite, and irregular aggregates of chlorite that have partially replaced primary brown hornblende.

## SUMMARY AND CONCLUSIONS

The rocks exposed in the Minnesota River Valley near Granite Falls and Montevideo have a complex history of intrusive events, deformation and metamorphism. The sequence of events can be separated into three stages of development. During the earliest stage, the three major rock groups were formed: the Montevideo Gneiss, the hornblende-pyroxene gneiss, and the garnet-biotite gneiss.

The second stage of development includes a complex history of intrusion, deformation, and high-grade metamorphism. Early in this stage a distinct compositional banding developed in the Montevideo Gneiss, garnet-biotite gneiss, and the hornblende-pyroxene gneiss. The emplacement of pegmatite veins, amphibolite dikes and sills, the metagabbro unit, and the massive phase of the Montevideo Gneiss was in part controlled by this compositional banding. Development of an $S_1$ foliation parallel to the banding in the gneiss was accompanied by the $F_1$ folding of pegmatite veins that cross the foliation at a high angle. The amphibolite dikes and sills and the metagabbro also took on a pervasive $S_1$ foliation. High-grade metamorphism ($M_1$) accompanied the development of $S_1$ as indicated by biotite, amphiboles, flattened garnet and quartz grains, and thin irregular pyroxene aggregates that lie in this plane.

Large-scale, upright $F_2$ synforms and antiforms with gently eastward-plunging axes developed in the gneisses exposed between Montevideo and Morton. The Granite Falls antiform, an inferred synform between Montevideo and Granite Falls, and associated minor folds are part of the $F_2$ fold series. This event was followed by the generation of minor $F_3$ folds, $L_3$ lineation, and a locally well-developed $S_3$ axial-plane foliation. The orientations of the minor $F_3$ structures vary with their positions on the limbs of the major $F_2$ structure. The absolute timing of these folding events is not well constrained. The massive phase of the Montevideo Gneiss, intruded approximately 3,050 m.y. ago, contains a weak foliation interpreted as $S_3$, and necklines of pinch-and-swell structures interpreted as $L_3$ lineations generated normal to the $F_3$ fold axes and $L_3$ mineral lineations. The assignment of

the foliation in this phase to $S_3$ suggests that the $F_1$ folding, $S_1$ foliation, and all the rock units affected by $S_1$ are older than 3,050 m.y. The relative timing of $F_2$ folding and emplacement of the massive phase is uncertain, but $F_3$ folding is younger than 3,050 m.y. Continued $M_1$ metamorphism during $F_2$ and $F_3$ is evident from $F_2$ and $F_3$ mineral lineations. If the 2,650-m.y. mineral ages obtained from various rock types in the area (see Goldich and others, 1970) represent resetting by the $M_1$ metamorphism, an extreme duration from greater than 3,050 to about 2,650 m.y. for $M_1$ is suggested. There is no textural evidence of a hiatus in metamorphism between $S_1$ and $S_3$.

$F_4$ folding and northwest- and northeast-trending shear zones probably developed in the waning stages of $M_1$ metamorphism. $F_4$ folds in the Montevideo area generally occur as small open warps of the banding and trend east-southeast with no apparent relation to any larger scale structure. The orientations of $F_4$ minor folds also vary as a function of location on the major $F_2$ structure. $F_4$ folds in the northern Granite Falls subareas have axes trending north to northeast. Unlike the $F_4$ folds near Montevideo, these are larger concentric folds. The difference in the style of $F_4$ folds in the two areas suggest a difference in the ductility of the rock at the time of folding. The shear zones in the two areas reflect a similar geographical difference in the ductility of the rock. Those in the Montevideo area are more commonly diktyonitic structures containing a neosome without a cataclastic microfabric. Shear zones in the northern Granite Falls area are invariably cataclastic.

The third stage of development is characterized by the intrusion of basaltic dikes and the small quartz monzonite pluton in section 28. The fact that shear zones, dated by Hedge and Goldich (1976) as being 2,400 m.y. old, cut the tholeiitic diabase dikes indicates that emplacement of these dikes was significantly older than the hornblende andesite dikes and the quartz monzonite pluton, which have ages of about 1,800 to 1,850 m.y. Low-grade metamorphism ($M_2$) about 1,800 m.y. ago developed retrograde assemblages in the gneisses and mafic dikes but had no associated deformation.

## ACKNOWLEDGMENTS

This report is based primarily on a thesis completed at the University of Missouri under the direction of G. R. Himmelberg and G. W. Viele, both of whom provided valuable discussions and criticism and read early versions of the manuscript. I thank S. S. Goldich and P. J. Hudleston for useful discussions and for their comments and criticism of the manuscript. Partial support of the research was provided by the Department of Geology, University of Missouri and the Minnesota Geological Survey, which I gratefully acknowledge.

## REFERENCES CITED

Bauer, R. L., 1974, Structural geology and petrology of the granitic gneiss near Granite Falls and Montevideo, Minnesota [M.S. thesis]: Columbia, University of Missouri, 108 p.

Craddock, C., Mooney, H. M., and Kohlemainen, V., 1970, Simple Bouguer gravity map of Minnesota and northwestern Wisconsin: Minnesota Geological Survey Miscellaneous Map M-10.

Doe, B. R., and Delevaux, M. H., 1980, Lead isotope investigations in the Minnesota River Valley: Geological Society of America Special Paper 182 (this volume).

Ghosh, S. K. and Ramberg, H., 1968, Buckling experiments on intersecting fold patterns: Tectonophysics, v. 5, p. 89–105.

Goldich, S. S. and Hedge, C. E., 1974, 3,800-m.y. granitic gneiss in southwestern Minnesota: Nature, v. 252, p. 467–468.

Goldich, S. S., Hedge, C. E., and Stern, T. W., 1970, Age of the Morton and Montevideo gneisses and related rocks, southwestern Minnesota: Geological Society of America Bulletin, v. 81, p. 3671–3696.

Goldich, S. S., and others, 1980, Archean rocks of the Granite Falls area, southwestern Minnesota: Geological Society of America Special paper 182 (this volume).

Grant, J. A., 1972, Minnesota River Valley, southwestern Minnesota, in Sims, P. K. and Morey, G. B., eds., Geology of Minnesota—A centennial volume: Minnesota Geological Survey, p. 177–196.

Grant, J. A., Himmelberg, G. R., and Goldich, S. S., 1972, Field trip guidebook for Precambrian migmatite terrane of the Minnesota River Valley: Minnesota Geological Survey Guidebook Series no. 5, 52 p.

Hanson, G. N. and Himmelberg, G. R., 1967, Ages of mafic dikes near Granite Falls, Minnesota: Geological Society of America Bulletin, v. 78, p. 1429–1432.

Hedge, C. E. and Goldich, S. S., 1976, Rb-Sr geochronology of the Montevideo Gneiss, Minnesota River Valley [abs.]: Twenty-Second Annual Institute on Lake Superior Geology, p. 31.

Himmelberg, G. R., 1968, Geology of Precambrian rocks, Granite Falls–Montevideo area, southwestern Minnesota: Minnesota Geological Survey Special Publications Series, SP-5, 33p.

Himmelberg, G. R. and Phinney, W. C., 1967, Granulite-facies metamorphism, Granite Falls–Montevideo area, Minnesota: Journal of Petrology, v. 8, p. 325–348.

Lund, E. H., 1956, Igneous and metamorphic rocks of the Minnesota River Valley: Geological Society of America Bulletin, v. 67, p. 1475–1490.

Mehnert, K. R., 1968, Migmatites and the origin of granitic rocks: Amsterdam, Elsevier Publishing Company, 393 p.

Ramberg, H., 1963, Strain distribution and geometry of folds: University of Uppsala Geological Institute Bulletin, v. 42, p. 1–20.

——1964, Selective buckling of composite layers with contrasted rheological properties: A theory for simultaneous formation of several orders of folds: Tectonophysics, v. 1, p. 307–341.

Ramsay, J. G., 1958, Superimposed folding at Loch Monar, Inverness-Shire and Ross-Shire: Geological Society of London Quarterly Journal, v. 113, p. 271–307.

——1967, Folding and fracturing of rocks: New York, McGraw-Hill Book Company, 568 p.

Turner, F. J., and Weiss, L. E., 1963, Structural analysis of metamorphic tectonites: New York, McGraw-Hill Book Company, 545 p.

Warner, J., 1969, FORTRAN IV program for construction of Pi diagrams with the Univac 1108 computer: Kansas State Geological Survey, Computer Contributions, v. 33, 37p.

Whitten, E. H. T., 1966, Structural geology of folded rocks: Chicago, Rand McNally and Company, 678 p.

Wilson, W. E., 1976, Trace element geochemistry and geochronology of early Precambrian granulite facies metamorphic rocks near Granite Falls in the Minnesota River Valley [Ph.D. thesis]: Minneapolis, University of Minnesota, 140 p.

Wilson, W. E., and Murthy, V. R., 1976, Rb-Sr geochronology and trace element geochemistry of granulite facies rocks near Granite Falls, in the Minnesota River Valley [abs.]: Twenty Second Annual Institute on Lake Superior Geology, p. 69.

Zietz, I., and Kirby, J. R., 1970, Aeromagnetic map of Minnesota: U. S. Geological Survey Geophysical Investigations Map GP-725, scale 1:1,000,000.

RECEIVED BY THE SOCIETY APRIL 25, 1979
REVISED MANUSCRIPT RECEIVED NOVEMBER 26, 1979
MANUSCRIPT ACCEPTED JANUARY 11, 1980

Geological Society of America
Special Paper 182
1980

# Archean rocks of the Granite Falls area, southwestern Minnesota

**S. S. GOLDICH***
*Northern Illinois University, De Kalb, Illinois 60115*

**C. E. HEDGE**
*U.S. Geological Survey, MS 963, Federal Center, Denver, Colorado 80225*

**T. W. STERN**
*U.S. Geological Survey, National Center, Reston, Virginia 22092*

**J. L. WOODEN**
*Lockheed Engineering and Management Services Company, Inc., NASA-Johnson Space Center, Houston, Texas 77058*

**J. B. BODKIN**
*Pennsylvania State University, University Park, Pennsylvania 16802*

**R. M. NORTH***
*Northern Illinois University, DeKalb, Illinois 60115*

## ABSTRACT

Geochemical and geochronologic data are presented for the Archean rocks in the Granite Falls area of the Minnesota River Valley, southwestern Minnesota. The rocks form two major groups: mafic and felsic gneisses. The mafic rocks include layered hornblende-pyroxene-plagioclase and biotite-pyroxene-plagioclase types with variants of each, amphibolite, and metagabbro. The biotite-pyroxene gneiss and some of the hornblende-pyroxene gneiss are metasedimentary with graywacke precursors. The amphibolites are of tholeiitic and basaltic komatiite composition, both of igneous derivation. Younger (Proterozoic) diabase dikes approach the tholeiitic amphibolite in composition.

The felsic gneisses have a wide range of composition and include tonalite, granodiorite, adamellite, and pegmatitic granite, all of igneous derivation and of different ages. Tonalite, granodiorite, and lesser amounts of adamellite gneiss are interlayered on a large scale with the layered mafic gneiss, and the apparent concordant relationships may be, in part, the result of deformation, but may also be interpreted as (1) an original volcanogenic-sedimentary pile or (2) sill-like intrusions of felsic rocks in an older mafic basement complex. The older felsic and mafic gneisses were folded before the intrusion of younger granitic pegmatite and adamellite, and subsequently the region was folded in late Archean time.

U-Pb analyses of zircon give minimum ages of 3,230 m.y. for the granodiorite and related old gneiss, 3,050 m.y. for the adamellite gneiss, 2,600 m.y. for the late Archean high-grade metamorphic event, and 1,800 m.y. for the Proterozoic igneous activity. Rb-Sr whole-rock analyses indicate an age of 3,600 m.y. for the older metamorphic complex and 3,000 to 3,100 m.y. for the pegmatitic granite and adamellite gneisses.

Chemical and petrographic data reveal some of the interactions that occurred during the emplacement of the pegmatitic granite and adamellite and suggest that the magmas were emplaced approximately 3,050 m.y. ago. Mixed-rock and contaminated samples are not easily recognized in the field. These not only reflect interactions between country rock and the invading magmas, but also the high-grade metamorphism 2,600 m.y. ago and subsequent events that involved shearing and hydrothermal alteration.

Although there are marked differences in rock types, metamor-

*Present addresses: Goldich; Department of Geology, Colorado School of Mines, Golden, Colorado 80401, and U.S. Geological Survey, Box 25046, MS 963 Federal Center, Denver, Colorado 80225; North, New Mexico Bureau of Mines and Mineral Resources, Socorro, New Mexico 87801.

phic grade, and structure in the Granite Falls and Morton areas, the geologic history of the migmatitic terranes is similar. In the Granite Falls area the paleosome is largely granodiorite gneiss with minor amounts of adamellite and amphibolite, whereas in the Morton area it is tonalite gneiss and amphibolite with minor amounts of granodiorite. In both areas the neosome is pegmatitic granite and adamellite gneiss. The thick sequences of layered mafic gneiss at Granite Falls are lacking in the Morton area. The younger adamellite-2 and the aplite dikes, approximately 2,600 m.y. old, of the Morton area have not been found in the Granite Falls area.

## INTRODUCTION

### Problem

This paper presents the results of petrographic, chemical, and geochronologic studies of the Archean rocks in the Granite Falls area in the Minnesota River Valley of southwestern Minnesota (Fig. 1). The principal objective is the interpretation of the geologic history in an area of ancient gneisses that have been altered by younger igneous and metamorphic events. The petrographic and chemical studies were directed to identifying and characterizing some of the major rock units and providing insight into the geologic processes that disturbed the isotopic decay systems. Some of the difficulties encountered in earlier investigations (Goldich and others, 1961, 1970) have been resolved. This has resulted in considerable progress in deciphering the geologic history.

### Previous Work

The first detailed petrographic work on the Precambrian rocks in the Granite Falls area was by Lund (1950, 1956), and his maps and rock units were the basis for the sampling in the earlier

radiometric dating (Goldich and others, 1961). The area was again mapped by Himmelberg (1968), who made revisions in Lund's rock units and nomenclature. Papers by Himmelberg and Phinney (1967) and Himmelberg (1968) contain petrographic data and some partial chemical analyses. A few rock analyses were reported by Goldich and others (1970).

Lund's structural interpretations were expanded considerably by Himmelberg and have been further refined by Bauer (this volume). Notwithstanding the geologic studies of Lund, Himmelberg, Bauer, and our own efforts, some fundamental problems remain unsolved. One such problem is the relationship between the mafic and felsic gneisses. Lund (1956) considered the layered mafic gneisses to be an older basement complex into which the precursors of the present-day felsic gneisses were intruded. Himmelberg (1968) suggested that the rocks at Granite Falls represent a metamorphosed "stratigraphic sequence" rather than igneous rocks emplaced in an older basement complex. Wilson and Murthy (1976), however, have returned to Lund's concept and suggest that a very old layered sequence of basaltic rocks and graywacke was intruded by the granitic rocks.

Himmelberg (1968) did not use a formal nomenclature but divided the rocks into four units: (1) granitic gneiss, (2) hornblende-pyroxene gneiss, (3) garnet-biotite gneiss, and (4) interlayered gneiss. Himmelberg and Phinney (1967) recognized a number of mineral assemblages in the mafic gneisses and three in the granitic gneiss. Two samples, collected by Himmelberg from the road cut along Highway 212 just south of Montevideo, the type locality of the Montevideo Gneiss of Lund (1956), were dated by Goldich and others (1970). The two rocks, one a foliated biotite-quartz-microcline-plagioclase gneiss and the other a "massive" quartz-microcline-plagioclase gneiss were shown to be of very different ages, and re-examination of the outcrops revealed discordant relationships. The massive leucocratic quartz-feldspar gneiss has a crude foliation imparted by elongated quartz grains. This younger unit cuts the thin-layered gneiss, which has a well-defined foliation imparted by aligned biotite. In the present paper we recognize three principal rock types within the granitic gneiss of Himmelberg (1968) and the Montevideo Gneiss of Lund (1956): granodiorite gneiss with good foliation, adamellite gneiss (the massive leucocratic phase), and granite pegmatite.

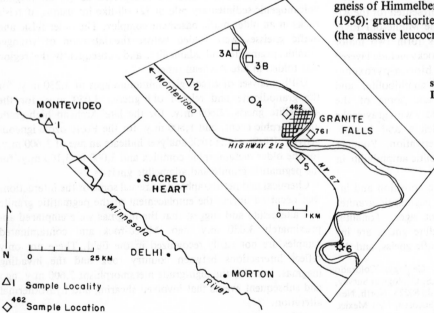

**Figure 1. Index map of the Granite Falls area showing sample localities and sample locations. Locality symbols are used in figures that follow.**

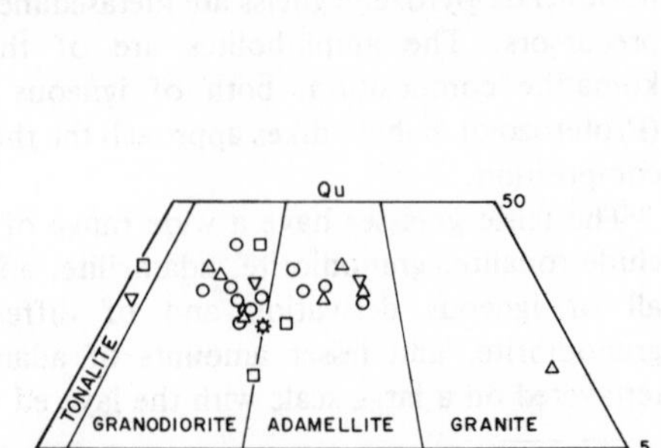

**Figure 2. Modal classification of the felsic gneisses in Johannsen's (1939) system (volume percent). Locality symbols as in Figure 1.**

In the earlier radiometric work (Goldich and others, 1961, 1970), no distinction was made between the phases of the granitic gneiss. Later, Goldich and Hedge (1974) reported a preliminary age of 3,720 m.y. (revised constant, see Table 1 in Goldich and Wooden, this volume) for samples of the foliated biotite granodiorite gneiss. Farhat and Wetherill (1975) published Rb-Sr data for suites of samples obtained from "individual blocks about 2 m in size." The samples were differentiated on the basis of color, that is, gray (D), red (R), and transitional (T or D + R). Their first block (MV-102) samples gave an isochron age of 2,420 ± 22 m.y. [$R_i$ = 0.7121 ± 0.0004 (1$\delta$)], and the second block (MV-100), an age of 2,670 ± 199 m.y. ($R_i$ = 0.711 ± 0.004). They suggested that the linearity of the data points and the high initial ratios indicate a time of metamorphic re-equilibration. They also presented an "errorchron" of approximately 3,045 ± 202 m.y. ($R_i$ = 0.706 ± 0.001) for individual samples from locality 6 (Fig. 1). They concluded that the rocks are approximately 3,250 m.y. old and, locally, have been partially or nearly completely reset during a metamorphic event approximately 2,500 m.y. ago. The data and interpretations by Farhat and Wetherill are discussed in a later section.

## Sampling

Six sample localities and the approximate locations of two individual samples are shown in Figure 1. Most of the samples are from the large rock-aggregate quarry operated by the Green Company, Granite Falls (loc. 4). Old quarry sites, highway and railroad cuts, and small excavations supplied additional samples, and a few were obtained with a sledge from natural outcrops (see App. 1). An effort was made to collect single rock types, away from contacts. The large samples of 80 to 100 kg collected in earlier work generally had more than a single rock type. Samples of selected pieces or larger single pieces of uniform composition, ranging from 10 to 15 kg, were cut normal to the layering with a diamond saw in water. These slabs were then cut into smaller slices, which provided the material for thin sections and rock powders.

## ROCK DESCRIPTIONS

Modal analyses (volume percentages) for 41 samples are arranged in Tables 1 to 7 by rock classification (Johannsen, 1939) without regard to apparent age. Analyses were made by counting 500 to 1,000 points on one to four thin sections. Large samples of coarse-grained gneiss present problems in obtaining representative modes. A number of modal analyses have been reported by Himmelberg (1968), and the reader is referred to his descriptions and geologic map of the Granite Falls area. For our purposes the Archean rocks are divided into two major groups: felsic and mafic gneisses.

### Felsic Gneisses

Although modes are not available for all samples, the data plotted in Figure 2 are representative of the major rock units. Trondhjemitic tonalite and pegmatitic granite gneiss are represented by only two samples and one sample, respectively. Both are impor-

tant rock units, but the tonalite is subordinate to the granodioritic and adamellitic gneisses, and the pegmatite is sheared and hard to sample.

**Tonalite Gneiss.** Tonalite gneiss is not abundant and is not easily recognized in the field. The modes (Table 1) show principally oligoclase, quartz, and biotite, with little or no K-feldspar and minor hornblende, and the gneiss may be classed as trondhjemite. At localities 3A and 6, the dark gray, medium-grained, thin-layered gneiss is similar to the granodiorite gneiss in appearance. At locality 2, the rock is gray, medium grained, and massive to weakly foliated. The gneiss has been sheared and recrystallized to a granoblastic texture. The plagioclase commonly is antiperthitic, and the hornblende is idioblastic. The accessory minerals are magnetite-ilmenite, apatite, zircon, and sphene; alteration minerals include sericite, epidote, and chlorite.

TABLE 1. MODAL ANALYSES OF
TONALITE GNEISS

| Sample No. | 607 | 791 |
|---|---|---|
| Locality No. | 3A | 2 |
| Quartz | 35.1 | 30.4 |
| Plagioclase | 54.9 | 60.4 |
| Microcline | X | 0.8 |
| Biotite | 8.9 | 6.0 |
| Hornblende | X | 1.8 |
| Opaque | X | X |
| Apatite | X | X |
| Sphene | | X |
| Zircon | X | X |
| Chlorite | X | |
| Epidote | X | |
| Sericite | X | X |
| An content | 23 | 30 |

X = Less than 0.5 percent.

**Granodiorite Gneiss.** In general, the granodiorite gneiss is a dark gray to reddish-brown, fine- to medium-grained, thin-layered rock with well-aligned biotite. Modal analyses of samples from localities 1, 2, 3, and 6 are given in Table 2 and from locality 4 in Table 3. Although there are some variations in mineral composition, the samples form a coherent group with an overall average of 31% quartz, 46% oligoclase, 16% microcline, and 6% biotite. The accessory minerals are magnetite-ilmenite, zircon, apatite, and sphene, similar to those in the tonalitic gneiss. The alteration minerals are mainly sericite, chlorite, epidote, and calcite.

TABLE 2. MODAL ANALYSES OF GRANODIORITE GNEISS FROM LOCALITIES 1, 2, 3, AND 6

| Sample No. | 613 | 614 | 608 | 605 | 609 | 620 | 209 |
|---|---|---|---|---|---|---|---|
| Locality No. | 1A | 1A | 1B | 2 | 3A | 3A | 6 |
| Quartz | 33.4 | 35.8 | 28.4 | 32.5 | 40.0 | 15.4 | 31.4 |
| Plagioclase | 45.7 | 48.0 | 48.6 | 41.7 | 39.4 | 44.2 | 46.0 |
| Microcline | 12.6 | 10.4 | 20.4 | 18.1 | 16.6 | 23.0 | 13.4 |
| Biotite | 7.4 | 4.8 | 2.4 | 6.3 | 2.6 | 15.8 | 8.8 |
| Hornblende |  |  |  |  | X |  | X |
| Opaque | X | X | X | X | X | 1.2 | X |
| Allanite |  |  |  |  | X |  |  |
| Apatite |  | X |  | X |  | X | X |
| Sphene | X | X |  | X |  | X | X |
| Zircon | X | X | X | X |  | X | X |
| Calcite | X |  |  | X |  |  |  |
| Chlorite |  |  |  | X | X |  | X |
| Epidote |  |  | X |  |  |  |  |
| Garnet |  |  |  | X |  |  |  |
| Hematite | X | X | X | X | X |  | X |
| Sericite | X | X | X | X | X | X | X |
| An content | 23 | 19 | 21 | 23 | 23 | 23 | 23 |

X = Less than 0.6 percent.

TABLE 3. MODAL ANALYSES OF GRANODIORITE GNEISS FROM LOCALITY 4

| Sample no. | 54GN | 606 | 610 | 611 | 691 | 693 | 695 | 700 |
|---|---|---|---|---|---|---|---|---|
| Quartz | 27.8 | 31.5 | 38.3 | 29.3 | 30.6 | 31.1 | 33.2 | 26.4 |
| Plagioclase | 47.0 | 45.9 | 40.4 | 47.7 | 47.2 | 43.8 | 52.0 | 47.2 |
| Microcline | 15.9 | 13.3 | 12.4 | 20.4 | 17.8 | 21.8 | 11.7 | 19.6 |
| Biotite | 6.6 | 8.2 | 7.5 | 2.5 | 3.4 | 2.1 | 1.6 | 5.4 |
| Hornblende | 0.6 | X | 0.5 |  |  |  |  | X |
| Opaque | X | X | X | X | X | X | X | X |
| Apatite | X | X | X | X | X | X | X | X |
| Sphene |  | X |  |  |  | X |  | X |
| Zircon | X | X | X | X | X | X | X | X |
| Calcite | X |  | X | X | X |  | X | X |
| Chlorite | 0.8 | X | X | X |  |  | 0.6 | X |
| Epidote | X |  | X | X | X |  |  | X |
| Hematite | X | X | X | X |  | 0.7 | X | X |
| Sericite | X | X | X | X | X |  | X | X |
| An content | 27 | 24 | 32 | 19 | 19 | 19 | 20 | 23 |

X = Less than 5 percent.

The granodiorite gneiss is granoblastic, locally, with porphyroblasts of K-feldspar and hornblende. A fine-grained granoblastic variant with a distinctly greenish cast is fairly abundant at locality 4; sample 691 (Table 3) represents this rock. The relationship between the granodiorite and tonalite gneisses is obscure. Veins of granoblastic rock formed by shearing and recrystallization of tonalitic gneiss, similar to those at Morton, have not been seen in the Granite Falls area. The relatively high contents of microcline and biotite and the small amounts of hornblende in the granodiorite gneiss in the Granite Falls area argue against a probable origin by recrystallization of the tonalite gneisses.

**Adamellite Gneiss.** Two types of adamellite gneiss are readily differentiated in the field on the basis of color and texture. At locality 1, southeast of Montevideo (Fig. 1), the adamellite gneiss is a leucocratic quartz-feldspar gneiss of bright brick-red color. Elongated or stretched quartz grains impart a crude structure to the rock, which in the field appears massive in contrast with the interlayered dark gray biotite-granodiorite gneiss. Locally, thin seams of granulated pegmatite are abundant in the interlayered granodiorite and adamellite gneisses. Like the adamellite, the pegmatite is bright red. The two rocks are closely associated, and gradations from adamellite to pegmatite are common, although it is not easy to differentiate between granulated pegmatite and adamellite. Some crosscutting relationships suggest that the pegmatite may be somewhat older than the adamellite; however, both units are definitely younger than the granodiorite gneiss.

At locality 1 the adamellite and granodiorite are interlayered on a scale ranging from a few centimetres to metres. The structure in the granodiorite gneiss was developed before the emplacement of the pegmatite and the adamellite. The magmas invaded the gneiss in lit-par-lit fashion, but locally truncated the structure. Later deformation affected both phases and developed boudins of granodiorite gneiss with adamellite separations.

Modes of the leucoadamellite gneiss from locality 1 (Table 4) show nearly equal amounts of quartz, oligoclase, and microcline. The plagioclase is xenoblastic with alteration to sericite and hematite. The microcline is perthitic with abundant strings, rods, and eye-shaped blebs of plagioclase approaching 50% (mesoperthite). Accessory minerals are magnetite, zircon, apatite, sphene, and poikioblastic garnet, as much as 3 mm across. The brick-red color is caused by hematite formed by pervasive hydrothermal alteration during and possibly subsequent to shearing and granulation, which affected both the pegmatite and the adamellite much more severely than the finer-grained granodiorite gneiss. Where the granodiorite gneiss layers are a few centimetres wide, alteration is apparent as red speckling and small red fractures.

A pervasive oxidation is also apparent in rock at locality 2. A slab of sheared rock (Fig. 3) from this locality consists of dark gray granodiorite gneiss and bright red, granulated pegmatite. The bright red color in this specimen is the result of hydrothermal alteration. Oxidation proceeded centripetally in the larger microcline fragments (Fig. 3), which have many red fractures. Hematite-stained fractures in the granodiorite layers in the lower part of the photograph are not visible. In addition to the oxidation and alteration of the sheared and granulated pegmatite and adamellite and, to a lesser extent, granodiorite gneisses, numerous veinlets of well-crystallized epidote are found in these rocks. These were formed during a younger period of fracturing and hydrothermal alteration, possibly related to the thermal event 1,800 m.y. ago.

At locality 4, the adamellite gneiss is a medium- to coarse-grained pink to red rock consisting of quartz, oligoclase,

TABLE 4. MODAL ANALYSES OF PEGMATITE GNEISS AND ADAMELLITE GNEISS FROM LOCALITIES 1, 2, 3, and 6

| Sample No. | 813 | 413 | 431 | 384 | 793 | 385 | 819 |
|---|---|---|---|---|---|---|---|
| Locality No. | 1A | 1B | 1B | 1B | 2 | 3B | 6 |
| Quartz | 18 | 30.6 | 36.2 | 33.4 | 33.6 | 26.9 | 27.1 |
| Plagioclase | 5 | 33.6 | 27.6 | 28.4 | 26.5 | 42.5 | 46.5 |
| Microcline | 67 | 30.2 | 30.8 | 37.6 | 34.7 | 27.3 | 24.6 |
| Biotite | 8 | 2.0 | X | X | 1.1 | 0.9 | 1.2 |
| Hornblende | | | | X | | X | |
| Opaque | | X | 1.0 | X | X | X | X |
| Apatite | | X | X | | | X | X |
| Sphene | | X | | | | | X |
| Zircon | X | X | X | | X | X | X |
| Calcite | | X | X | | | | |
| Chlorite | | | | X | X | X | X |
| Epidote | | | | | X | X | |
| Garnet | 2 | | 1.4 | X | | X | |
| Hematite | X | 1.3 | X | X | X | X | 0.6 |
| Sericite | | X | X | X | X | X | X |
| An content | 9 | 19 | 19 | 19 | 23 | 25 | 24 |

X = Less than 0.5 percent.

TABLE 5. MODAL ANALYSES OF ADAMELLITE GNEISS FROM LOCALITY 4

| Sample No. | 54GR | 612 | 696 | 738 |
|---|---|---|---|---|
| Quartz | 30.8 | 35.2 | 33.2 | 33.4 |
| Plagioclase | 28.4 | 35.2 | 36.6 | 33.2 |
| Microcline | 37.8 | 25.6 | 26.6 | 29.6 |
| Biotite | 4.0 | 3.0 | 2.2 | 3.0 |
| Hornblende | X | X | | X |
| Opaque | X | X | X | X |
| Apatite | X | | X | X |
| Zircon | X | X | X | X |
| Calcite | | | X | X |
| Chlorite | X | X | X | X |
| Epidote | X | | | X |
| Hematite | X | X | X | X |
| Sericite | X | X | X | X |
| An content | 19 | 19 | 19 | 23 |

X = Less than 0.5 percent.

microcline, biotite, and accessory hornblende, magnetite, apatite, and zircon (Table 5). Alteration minerals are sericite, hematite, epidote, chlorite, and calcite. The grain size ranges up to 5 mm. Myrmekite and antiperthite are common, and the microcline-microperthite is characterized by abundant strings and blebs of plagioclase (mesoperthite). Intergrowths of biotite with feldspar and quartz suggest shearing and recrystallization. Some hornblende crystals, up to 3 mm in size, poikilitically enclose small quartz grains. Garnet has not been found in the adamellite but is abundant in the pegmatite at locality 4.

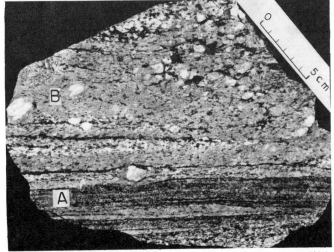

**Figure 3. Photograph of sheared rock, locality 2. The lower part (A) is mostly dark gray granodiorite gneiss; the upper part (B) is brick-red granulated pegmatite. The larger feldspar fragments show lighter colored remnants with hematite-stained fractures.**

Modal analyses (Table 5) of the coarse-grained adamellite are similar to those for the adamellite gneiss from other localities (Table 4). The coarse-grained granoblastic adamellite at locality 4 has not been granulated and altered appreciably in comparison with the finer grained, brick-red adamellite and pegmatite at localities 1 and 2. Pegmatite and adamellite crosscut earlier-formed foliation of the granodiorite gneiss.

**Pegmatitic Granite Gneiss.** Pegmatitic granite, more or less broken up and granulated, occurs at all localities of the felsic gneisses. The pegmatite is closely associated with the adamellite and has been included in the descriptions above. Sampling of the pegmatite is difficult, and thin sections are not satisfactory for arriving at the composition. The mode (Table 4, sample 813) is estimated from hand specimens and optical work on crushed fragments. Garnet is common in the pegmatite at localities 1 and 4 and occurs in small amounts in the adamellite and in traces in the granodiorite gneiss.

## Mafic Gneisses

The mafic gneisses were studied in some detail by Himmelberg (1968), who reported a number of modal analyses similar to those of Table 6. The age of these rocks has been investigated by Wilson (1976), and some of the results summarized by Wilson and Murthy (1976) are reviewed in the section on geochronology. We have examined only a few samples in search of some with a low Rb/Sr ratio to anchor the lower end of the isochron, a problem that complicated the isochron for the granodiorite-gneiss samples (Goldich and Hedge, 1974). For this purpose, samples of layered gneiss (54CGN and 624) were collected from the Green quarry, and sample 462 was obtained from an excavation for a basement in the northwestern part of Granite Falls (Fig. 1). Because neither site would be available for later inspection, a few samples of the mafic gneisses from Memorial Park (loc. 5, Fig. 1) have been included. The mafic gneisses are divided into three groups: (1)

TABLE 6. MODAL ANALYSES OF LAYERED MAFIC GNEISSES

| Sample No. | 464M | 464F | 54CGN | 624 | 815 | 816 | 817 | 462 |
|---|---|---|---|---|---|---|---|---|
| Locality No. | 3 | 3 | 4 | 4 | 5 | 5 | 5 | 5 |
| Quartz | 28.6 | 35.4 | 6.6 | 22.0 | 7.6 | 22.0 | 25.8 | 27.4 |
| Plagioclase | 50.0 | 58.2 | 50.4 | 55.8 | 77.4 | 55.2 | 49.4 | 59.0 |
| Microcline | 0.4 | 0.6 | X | X | X | | X | 1.8 |
| Biotite | 18.2 | 5.2 | X | 10.1 | X | 13.2 | 13.8 | 5 6 |
| Opaque | X | X | 7.0 | X | 2.8 | X | 1.0 | X |
| Hornblende | 2.4 | 0.4 | 11.4 | 3.8 | 0.8 | | | |
| Orthopyroxene | | | 4.8 | X | 1.4 | 3.2 | 3.0 | X |
| Clinopyroxene | | | 18.8 | 1.3 | 8.8 | 6.2 | 3.6 | X |
| Apatite | X | X | X | X | X | X | X | X |
| Sphene | | | | | X | | X | X |
| Zircon | X | X | | X | X | X | X | X |
| Amphibole* | | | | 5.8 | | | | 4.6 |
| Garnet | | | | | | | 3.4 | |
| Calcite | | | | | | | | X |
| Chlorite | | | | | X | | | |
| Sericite | X | X | X | X | X | X | X | X |
| An content | 32 | 32 | 44 | 30 | 30 | 41 | 38 | 36 |

*X = Less than 0.5 percent.*
*\* Secondary, i.e., cummingtonite, actinolite, etc.*

layered gneiss, (2) amphibolite, and (3) metagabbro. This system is a compromise between those of Lund (1956) and Himmelberg (1968).

The term "layered gneisses," as used here, refers to layering on a small scale and should not be confused with the regional layering of the felsic and mafic gneisses shown on the geologic map of Himmelberg (1968). Himmelberg distinguished two types of mafic gneiss which he called hornblende-pyroxene gneiss and garnet-biotite gneiss. Both types show compositional banding or layering. They were further subdivided on the basis of mineral assemblages (Himmelberg and Phinney, 1967).

With the exception of 464M and 464F, dark and light layers, respectively, the samples classed as layered gneiss (Table 6) are characterized by orthopyroxene and clinopyroxene, although the range, from less than 0.5% to 23% of total pyroxene, is considerable. Himmelberg (1968) identified the opaque minerals as magnetite and ilmenite and the fibrous secondary amphibole as cummingtonite. He also noted an unidentified blue-green amphibole. Most of the samples are quite different in appearance and in modal and chemical composition from the felsic gneisses and the amphibolites, and, as will be discussed in a later section, these rocks probably are metasedimentary. Samples 464M and 464F have an average composition which is close to that of the tonalite gneiss (Table 1), and possibly these samples are improperly classed with the mafic gneisses.

Amphibolite of tholeiitic composition and a low-alumina variety of the composition of basaltic komatiite both occur in the Granite Falls area (Table 7). These are similar to the amphibolites in the Morton area (Goldich and others, this volume). The mode of a sample of metagabbro (Table 7, sample 761) is dominated by

the large percentage of labradorite, and the rock, before metamorphism, was anorthositic gabbro. Lund (1956, p. 1478) gave a mode of gabbro gneiss from this locality (Fig. 1) of different composition: 75% plagioclase ($An_{71}$), 19% pyroxene, and 6% hornblende. Lund's sample is less altered than sample 761.

The modes of two Proterozoic diabase dikes that were intruded into the Archean gneissic complex are included in Table 7 for comparison with the older amphibolites.

## GEOCHEMISTRY

Thirty-seven bulk and four partial chemical analyses are included. Thirty-one of the bulk analyses were made by a combination of instrumental and conventional methods; three of these were previously published (Goldich and others, 1970). Six conventional analyses can be recognized in the tables by $SiO_2$ and $Al_2O_3$ values reported to two decimal places. The concentrations of Rb and Sr and the ratio $^{87}Sr/^{86}Sr$ were determined by isotope-dilution procedures in the laboratories of the U.S. Geological Survey, Denver. Ba was determined by X-ray fluorescence; Ni, Cr, V, Cu, Y, and Zr were determined by emission spectrography on amphibolite, metagabbro, and diabase samples.

The chemical analyses are tabulated (Tables 8 to 15) without regard to apparent age. The groupings correspond to the petrographic classification with a few exceptions that, in part, result from inadequate thin-section coverage. The tonalite gneiss

TABLE 7. MODAL ANALYSES OF AMPHIBOLITE, METAGABBRO, AND DIABASE

| | Amphibolite | | Metagabbro | Diabase | |
|---|---|---|---|---|---|
| Sample No. | 763 | 777 | 761 | 775 | 588 |
| Locality No. | 4 | 4 | 5B | 4 | 3B |
| Quartz | | | 8.8 | | X |
| Plagioclase | 3.7 | 52.0 | 63.4 | 56.0 | 51.5 |
| Hornblende | 89.2 | 14.8 | 19.8 | | 15.8 |
| Pyroxene | 3.6 | 30.8 | 7.8 | 25.2 | 26.5 |
| Olivine | | | | 2.8 | |
| Biotite | 2.5 | 0.6 | X | 3.8 | 5.5 |
| Opaque | X | 0.8 | X | 7.0 | X |
| Apatite | X | X | X | X | X |
| Rutile | X | | | X | |
| Sphene | X | | | | |
| Zircon | | X | | | |
| Amphibole* | | X | X | X | X |
| Garnet | X | | | | |
| Calcite | X | X | X | X | |
| Chlorite | X | X | | 2.2 | X |
| Sericite | X | X | X | X | X |
| Serpentine | | | | X | |
| An content | 34 | 54 | 67 | 55 | 60 |

*X = Less than 0.5 percent.*
*\* Secondary.*

TABLE 8.  CHEMICAL ANALYSES OF TONALITE GNEISS

| Sample No. | 792 | 607 | 821 |
|---|---|---|---|
| Locality No. | 2 | 3A | 6 |
| wt % | | | |
| SiO2 | 71.9 | 70.05 | |
| Al2O3 | 15.4 | 15.91 | |
| TiO2 | 0.24 | 0.32 | |
| Fe2O3 | 0.71 | 0.74 | |
| FeO | 1.53 | 1.86 | |
| MnO | 0.04 | 0.03 | |
| MgO | 0.63 | 0.83 | 0.94 |
| CaO | 3.18 | 3.06 | 2.69 |
| Na2O | 5.20 | 5.15 | 4.66 |
| K2O | 1.03 | 1.27 | 1.60 |
| P2O5 | 0.09 | 0.09 | |
| H2O+ | 0.17 | 0.38 | |
| H2O- | 0.10 | 0.12 | |
| CO2 | 0.00 | 0.11 | |
| F | 0.03 | 0.02 | |
| Total* | 100.24 | 99.93 | |
| g/cm3 | | | |
| D | 2.661 | 2.688 | |
| ppm | | | |
| Rb | 34.9 | 41.5 | 59.7 |
| Sr | 295 | 379 | 260 |
| Ba | 295 | 585 | 380 |
| ratios | | | |
| K/Rb | 245 | 253 | 223 |
| Rb/Sr | 0.118 | 0.110 | 0.230 |
| Sr/Ba | 1.0 | 0.65 | 0.68 |
| Ba/Rb | 8.5 | 14.1 | 6.4 |

* Totals corrected for F.

TABLE 9.  CHEMICAL ANALYSES OF HIGH-CALCIUM GRANODIORITE GNEISS

| Sample No. | 54GN | 790 | 738 | 701 | 694 | 605 | 606 | 695 | 614 | 613 |
|---|---|---|---|---|---|---|---|---|---|---|
| Locality No. | 4 | 2 | 4 | 4 | 4 | 2 | 4 | 4 | 1A | 1A |
| wt % | | | | | | | | | | |
| SiO2 | 71.8 | 71.5 | 71.6 | 73.0 | 72.0 | 72.26 | 72.87 | 74.1 | 72.4 | 71.7 |
| Al2O3 | 14.7 | 15.0 | 14.5 | 14.4 | 14.7 | 14.50 | 14.62 | 14.3 | 15.3 | 15.1 |
| TiO2 | 0.28 | 0.33 | 0.23 | 0.23 | 0.24 | 0.25 | 0.22 | 0.15 | 0.19 | 0.28 |
| Fe2O3 | 0.84 | 0.46 | 0.52 | 0.50 | 0.81 | 0.56 | 0.59 | 0.49 | 0.72 | 0.72 |
| FeO | 1.57 | 1.80 | 1.63 | 1.07 | 0.97 | 1.58 | 1.13 | 0.56 | 0.66 | 1.50 |
| MnO | 0.04 | 0.03 | 0.04 | 0.03 | 0.03 | 0.03 | 0.02 | 0.01 | 0.02 | 0.03 |
| MgO | 0.82 | 0.75 | 0.82 | 0.51 | 0.51 | 0.74 | 0.59 | 0.29 | 0.30 | 0.71 |
| CaO | 2.77 | 2.69 | 2.37 | 2.32 | 2.27 | 2.27 | 2.28 | 2.16 | 2.18 | 2.14 |
| Na2O | 4.39 | 4.58 | 3.86 | 4.20 | 4.24 | 4.29 | 4.23 | 4.14 | 5.12 | 4.82 |
| K2O | 2.53 | 2.29 | 4.00 | 3.02 | 3.69 | 2.72 | 3.21 | 3.27 | 2.47 | 2.72 |
| P2O5 | 0.09 | 0.09 | 0.07 | 0.08 | 0.07 | 0.08 | 0.10 | 0.02 | 0.03 | 0.06 |
| H2O+ | 0.20 | 0.22 | 0.24 | 0.22 | 0.14 | 0.19 | 0.09 | 0.20 | 0.19 | 0.29 |
| H2O- | 0.05 | 0.09 | 0.11 | 0.19 | 0.16 | 0.12 | 0.10 | 0.12 | 0.06 | 0.07 |
| CO2 | 0.04 | 0.02 | 0.12 | 0.19 | 0.25 | 0.07 | 0.00 | 0.14 | 0.03 | 0.07 |
| F | 0.03 | 0.04 | 0.03 | 0.01 | 0.02 | 0.03 | 0.03 | 0.01 | 0.03 | 0.01 |
| Total* | 100.14 | 99.87 | 100.13 | 99.97 | 100.09 | 99.68 | 100.07 | 99.96 | 99.67 | 100.15 |
| g/cm3 | | | | | | | | | | |
| D | 2.666 | 2.674 | 2.624 | 2.648 | 2.637 | 2.628 | 2.671 | 2.632 | 2.644 | |
| ppm | | | | | | | | | | |
| Rb | 75.0 | 114 | 89.5 | 74.7 | 92.1 | 80.0 | 89.1 | 89.9 | 88.4 | 109 |
| Sr | 270 | 243 | 257 | 252 | 245 | 288 | 257 | 245 | 514 | 377 |
| Ba | 590 | 405 | 835 | 815 | 760 | 795 | 720 | 720 | 1440 | 640 |
| ratios | | | | | | | | | | |
| K/Rb | 280 | 167 | 371 | 336 | 332 | 283 | 299 | 301 | 233 | 207 |
| Rb/Sr | 0.28 | 0.47 | 0.35 | 0.30 | 0.38 | 0.28 | 0.35 | 0.37 | 0.17 | 0.29 |
| Sr/Ba | 0.46 | 0.60 | 0.31 | 0.31 | 0.32 | 0.36 | 0.36 | 0.34 | 0.36 | 0.59 |
| Ba/Rb | 7.8 | 3.6 | 9.3 | 11 | 8.3 | 9.9 | 8.1 | 8.0 | 16 | 5.9 |

* Totals corrected for F.

TABLE 11. CHEMICAL ANALYSES OF ADAMELLITE GNEISS

TABLE 11. CHEMICAL ANALYSES OF ADAMELLITE GNEISS

| | | 819 | 413 | 431 | 54GR |
|---|---|---|---|---|---|
| Sample No. | | 819 | 413 | 431 | 54GR |
| Locality No. | | 6 | 1B | 1B | 4 |
| **wt %** | | | | | |
| $SiO_2$ | | | 73.32 | 73.73 | 74.38 |
| $Al_2O_3$ | | | 14.40 | 14.45 | 14.15 |
| $TiO_2$ | | | 0.20 | 0.14 | 0.09 |
| $Fe_2O_3$ | | | 0.78 | 0.64 | 0.24 |
| $FeO$ | | | 0.60 | 0.53 | 0.47 |
| $MnO$ | | | 0.02 | 0.01 | 0.01 |
| $MgO$ | | | 0.25 | 0.18 | 0.24 |
| $CaO$ | | | 1.31 | 1.27 | 1.70 |
| $Na_2O$ | | 4.54 | 3.83 | 4.12 | 3.75 |
| $K_2O$ | | 4.15 | 4.84 | 4.45 | 4.33 |
| $P_2O_5$ | | | 0.04 | 0.03 | 0.01 |
| $H_2O+$ | | | 0.16 | 0.12 | 0.16 |
| $H_2O-$ | | | 0.10 | 0.07 | 0.06 |
| $CO_2$ | | | 0.06 | 0.08 | 0.14 |
| $F$ | | | 0.04 | 0.03 | 0.02 |
| Total* | | | 99.93 | 99.84 | 99.73 |
| **$g/cm^3$** | | | | | |
| D | | | 2.635 | 2.629 | 2.628 |
| **ppm** | | | | | |
| Rb | | 109 | 173 | 139 | 123 |
| Sr | | 295 | 213 | 170 | 259 |
| Ba | | 850 | 835 | 620 | 940 |
| **ratios** | | | | | |
| K/Rb | | 317 | 232 | 265 | 292 |
| Rb/Sr | | 0.37 | 0.81 | 0.82 | 0.48 |
| Sr/Ba | | 0.35 | 0.26 | 0.27 | 0.28 |
| Ba/Rb | | 7.8 | 4.8 | 4.5 | 7.6 |

* Totals corrected for F.

---

TABLE 10. CHEMICAL ANALYSES OF LOW-CALCIUM GRANODIORITE GNEISS

| | 700 | 209 | 384 | 608 | 609 |
|---|---|---|---|---|---|
| Sample No. | 700 | 209 | 384 | 608 | 609 |
| Locality No. | 4 | 6 | 1B | 1B | 3A |
| **wt %** | | | | | |
| $SiO_2$ | 72.4 | 73.6 | 71.6 | 73.7 | 74.3 |
| $Al_2O_3$ | 14.6 | 14.1 | 14.1 | 14.6 | 14.3 |
| $TiO_2$ | 0.25 | 0.21 | 0.26 | 0.12 | 0.15 |
| $Fe_2O_3$ | 0.66 | 0.82 | 0.85 | 0.65 | 0.57 |
| $FeO$ | 1.17 | 1.63 | 1.65 | 0.48 | 0.43 |
| $MnO$ | 0.03 | 0.03 | 0.03 | 0.02 | 0.02 |
| $MgO$ | 0.57 | 0.43 | 0.50 | 0.19 | 0.20 |
| $CaO$ | 1.87 | 1.82 | 1.73 | 1.57 | 1.53 |
| $Na_2O$ | 4.10 | 4.38 | 4.60 | 4.93 | 4.61 |
| $K_2O$ | 3.66 | 2.50 | 3.27 | 3.35 | 3.39 |
| $P_2O_5$ | 0.07 | 0.07 | 0.08 | 0.05 | 0.05 |
| $H_2O+$ | 0.24 | 0.33 | 0.28 | 0.09 | 0.04 |
| $H_2O-$ | 0.11 | 0.08 | 0.05 | 0.09 | 0.07 |
| $CO_2$ | 0.05 | | | 0.00 | 0.00 |
| $F$ | 0.02 | | | 0.02 | 0.01 |
| Total* | 99.79 | 100.0 | 99.0 | 99.85 | 99.63 |
| **$g/cm^3$** | | | | | |
| D | 2.626 | 2.628 | | 2.634 | 2.602 |
| **ppm** | | | | | |
| Rb | 98.4 | 79.4 | 134 | 92.7 | 71.7 |
| Sr | 245 | 204 | 251 | 308 | 339 |
| Ba | 800 | 530 | 300 | 390 | 715 |
| **ratios** | | | | | |
| K/Rb | 309 | 261 | 203 | 300 | 392 |
| Rb/Sr | 0.40 | 0.39 | 0.53 | 0.30 | 0.21 |
| Sr/Ba | 0.31 | 0.39 | 0.31 | 0.79 | 0.47 |
| Ba/Rb | 8.1 | 6.7 | 6.0 | 4.2 | 10 |

* Totals corrected for F.

| Sample No. | 612 | 611 | 793 | 385 | 813* |
|---|---|---|---|---|---|
| Locality No. | 4 | 4 | 2 | 3B | 1A |
| **wt %** | | | | | |
| $SiO_2$ | | | 75.8 | 74.1 | 69.0 |
| $Al_2O_3$ | | | 13.6 | 14.5 | 16.6 |
| $TiO_2$ | | | 0.11 | 0.04 | 0.02 |
| $Fe_2O_3$ | | | 0.21 | 0.38 | 0.64 |
| $FeO$ | | | 0.62 | 0.30 | 0.81 |
| $MnO$ | | | 0.01 | 0.00 | 0.03 |
| $MgO$ | 0.46 | 0.34 | 0.24 | 0.09 | 0.25 |
| $CaO$ | 1.85 | 1.79 | 1.53 | 1.18 | 0.88 |
| $Na_2O$ | 4.30 | 3.80 | 3.72 | 3.86 | 4.03 |
| $K_2O$ | 4.89 | 4.40 | 3.74 | 5.03 | 7.43 |
| $P_2O_5$ | | | 0.01 | 0.00 | 0.05 |
| $H_2O+$ | | | 0.14 | 0.11 | 0.11 |
| $H_2O-$ | | | 0.04 | 0.04 | 0.07 |
| $CO_2$ | | | 0.05 | | 0.09 |
| $F$ | | | 0.02 | | 0.04 |
| Total[+] | | | $99.8_3$ | 99.6 | $100.0_3$ |
| **g/cm³** | | | | | |
| D | 2.624 | 2.628 | 2.604 | | 2.596 |
| **ppm** | | | | | |
| Rb | 125 | 93.9 | 84.8 | 83.8 | 278 |
| Sr | 248 | 260 | 272 | 383 | 258 |
| Ba | 1160 | 1020 | 1010 | 2460 | 1150 |
| **ratios** | | | | | |
| K/Rb | 324 | 389 | 365 | 499 | 222 |
| Rb/Sr | 0.50 | 0.36 | 0.31 | 0.22 | 1.08 |
| Sr/Ba | 0.21 | 0.25 | 0.27 | 0.16 | 0.22 |
| Ba/Rb | 9.3 | 11 | 12 | 29 | 4.1 |

\* Sample 813 is pegmatitic granite gneiss; all others are adamellite gneiss.
+ Totals corrected for F.

TABLE 13. CHEMICAL ANALYSES OF LAYERED MAFIC GNEISSES

| Sample No. | 464M | 464F | 464* | 54CGN | 624 | 815 | 816 | 817 | 462 |
|---|---|---|---|---|---|---|---|---|---|
| Locality No. | 3B | 3B | 3B | 4 | 4 | 5A | 5A | 5A | 5B |
| **wt %** | | | | | | | | | |
| $SiO_2$ | 69.5 | 74.5 | 72.0 | 59.2 | 65.6 | 53.7 | 63.8 | 66.3 | 69.2 |
| $Al_2O_3$ | 15.1 | 14.1 | 14.6 | 14.5 | 15.4 | 17.4 | 15.7 | 15.2 | 15.8 |
| $TiO_2$ | 0.38 | 0.16 | 0.27 | 1.19 | 0.54 | 1.62 | 0.64 | 0.65 | 0.17 |
| $Fe_2O_3$ | 3.03 | 0.62 | 1.83 | 3.31 | 1.01 | 4.02 | 0.64 | 0.47 | 0.47 |
| $FeO$ | 1.15 | 1.10 | 1.13 | 6.38 | 3.68 | 9.02 | 5.27 | 4.96 | 2.55 |
| $MnO$ | 0.06 | 0.02 | 0.04 | 0.19 | 0.08 | 0.19 | 0.08 | 0.06 | 0.06 |
| $MgO$ | 1.34 | 0.50 | 0.92 | 3.28 | 2.47 | 2.44 | 3.35 | 2.80 | 1.75 |
| $CaO$ | 3.31 | 2.78 | 3.05 | 7.84 | 4.65 | 5.96 | 4.61 | 3.43 | 4.46 |
| $Na_2O$ | 4.70 | 4.43 | 4.57 | 2.74 | 2.53 | 4.10 | 3.32 | 3.53 | 4.33 |
| $K_2O$ | 1.47 | 1.67 | 1.57 | 0.73 | 1.54 | 1.08 | 1.58 | 1.96 | 0.71 |
| $P_2O_5$ | 0.09 | 0.04 | 0.07 | 0.12 | 0.20 | 0.32 | 0.19 | 0.13 | 0.03 |
| $H_2O+$ | 0.49 | 0.22 | 0.36 | 0.40 | 0.44 | 0.14 | 0.40 | 0.41 | 0.75 |
| $H_2O-$ | 0.06 | 0.05 | 0.06 | 0.09 | 0.16 | 0.09 | 0.16 | 0.12 | 0.09 |
| $CO_2$ | 0.01 | 0.03 | 0.02 | 0.02 | 0.03 | 0.21 | 0.09 | 0.04 | 0.15 |
| $F$ | | | | | | 0.05 | 0.06 | 0.06 | |
| Total[+] | $100.6_9$ | $100.2_2$ | $100.4_9$ | $99.9_9$ | $100.3_3$ | $100.3_2$ | $100.0_3$ | $100.0_9$ | $100.5_2$ |
| **g/cm³** | | | | | | | | | |
| D | 2.699 | 2.671 | 2.686 | | 2.727 | | 2.769 | 2.775 | 2.702 |
| **ppm** | | | | | | | | | |
| Rb | 39.8 | 25.8 | 32.8 | 8.0 | 54.5 | 17.3 | 59.2 | 55.2 | 7.5 |
| Sr | 254 | 292 | 273 | 298 | 366 | 414 | 447 | 324 | 319 |
| Ba | 475 | 670 | 575 | 350 | 300 | 610 | 730 | 660 | 290 |
| **ratios** | | | | | | | | | |
| K/Rb | 307 | 537 | 406 | 759 | 235 | 518 | 221 | 295 | 785 |
| Rb/Sr | 0.16 | 0.088 | 0.12 | 0.027 | 0.15 | 0.042 | 0.13 | 0.17 | 0.024 |
| Sr/Ba | 0.53 | 0.44 | 0.47 | 0.85 | 1.22 | 0.68 | 0.61 | 0.49 | 1.1 |
| Ba/Rb | 12 | 26 | 18 | 44 | 5.5 | 35 | 12 | 12 | 39 |

\* Average of 464M and 464F.
+ Totals corrected for F.

**TABLE 14. CHEMICAL ANALYSES OF AMPHIBOLITE, METAGABBRO, AND DIABASE**

| Rock Type | Amphibolites | | | Metagabbro | Diabase | |
|---|---|---|---|---|---|---|
| Sample No | 763 | M-14-16 | 777 | 761 | 775 | 588 |
| wt % | | | | | | |
| $SiO_2$ | 46.5 | 47.2 | 51.5 | 53.9 | 47.2 | 47.5 |
| $Al_2O_3$ | 9.68 | 8.76 | 14.4 | 20.3 | 14.9 | 14.1 |
| $TiO_2$ | 1.18 | 1.08 | 0.78 | 0.39 | 1.71 | 2.50 |
| $Fe_2O_3$ | 2.84 | 2.37 | 1.54 | 1.16 | 2.61 | 4.39 |
| FeO | 10.04 | 10.46 | 9.05 | 4.85 | 10.88 | 10.28 |
| MnO | 0.21 | 0.23 | 0.18 | 0.10 | 0.22 | 0.36 |
| MgO | 13.88 | 14.5 | 7.75 | 4.52 | 7.03 | 5.98 |
| CaO | 10.38 | 10.43 | 10.91 | 11.1 | 11.0 | 8.84 |
| $Na_2$ | 1.97 | 1.66 | 2.57 | 2.62 | 2.45 | 2.76 |
| $K_2O$ | 0.75 | 0.76 | 0.40 | 0.59 | 0.53 | 1.47 |
| $P_2O_5$ | 0.09 | 0.09 | 0.03 | 0.04 | 0.26 | 0.32 |
| $H_2O+$ | 1.54 | 1.22 | 0.37 | 0.52 | 0.60 | 1.05 |
| $H_2O-$ | 0.12 | 0.12 | 0.12 | 0.08 | 0.09 | 0.05 |
| $CO_2$ | 0.17 | 0.35 | 0.10 | 0.16 | 0.61 | 0.00 |
| F | 0.03 | 0.08 | 0.05 | 0.04 | 0.05 | 0.19 |
| Total* | $99.3_7$ | $99.2_8$ | $99.7_3$ | $100.3_5$ | $100.1_2$ | $99.7_1$ |
| $g/cm^3$ | | | | | | |
| D | 3.154 | 3.168 | 3.011 | 2.825 | 3.046 | 2.911 |

*Totals corrected for F.*

**TABLE 15. TRACE ELEMENT DETERMINATIONS FOR AMPHIBOLITE, METAGABBRO, AND DIABASE**

| Rock Type | Amphibolite | | | Metagabbro | Diabase | |
|---|---|---|---|---|---|---|
| Sample No. | 763 | M-14-16 | 777 | 761 | 775 | 588 |
| ppm | | | | | | |
| Rb * | 12.8 | 21.5 | 4.4 | 1.92 | 11.8 | 35.1 |
| Sr * | 130 | 72.0 | 126 | 279 | 494 | 269 |
| Ba + | 45 | 60 | 50 | 200 | 360 | 245 |
| Cr | 1300 | 1300 | 500 | 40 | 70 | 120 |
| Ni | 660 | 530 | 130 | 60 | 40 | 60 |
| V | 250 | 220 | 250 | 120 | 320 | 380 |
| Cu | 205 | 165 | 120 | 82 | 110 | <10 |
| Y | <10 | tr | 20 | 18 | 25 | 45 |
| Zr | 100 | 75 | 86 | 100 | 95 | 200 |
| ratios | | | | | | |
| K/Rb | 486 | 293 | 755 | 2550 | 373 | 348 |
| Rb/Sr | 0.098 | 0.30 | 0.035 | 0.007 | 0.024 | 0.13 |
| Sr/Ba | 2.9 | 1.2 | 2.5 | 1.4 | 1.4 | 1.1 |
| Ba/Rb | 3.5 | 2.8 | 11 | 104 | 31 | 7.0 |

*Isotope dilution; +X-ray fluorescence; all others by emission spectrography.*

TABLE 16.   AVERAGE COMPOSITION OF PRINCIPAL ROCK TYPES COMPARED WITH MORTON AREA

| Rock Type | Tonalite gneiss | | Granodiorite gneiss | | | Adamellite gneiss | | Pegmatite gneiss | |
|---|---|---|---|---|---|---|---|---|---|
| Area | Granite Falls | Morton | Granite Falls | Morton | Granite Falls | Morton | Granite Falls | Morton | |
| No. Samples | (2-3)* | (1)+ | (10)** | (5)*** | (4)++ | (3)**** | (5)+++ | (1)***** | (2)++++ |
| **wt %** | | | | | | | | | |
| $SiO_2$ | 71.0 | 68.7 | 72.3 | 73.1 | 70.0 | 73.81 | 74.8 | 69.0 | 72.7 |
| $Al_2O_3$ | 15.7 | 16.8 | 14.7 | 14.3 | 15.0 | 14.33 | 13.6 | 16.6 | 13.8 |
| $TiO_2$ | 0.28 | 0.36 | 0.24 | 0.20 | 0.37 | 0.14 | 0.08 | 0.02 | 0.08 |
| $Fe_2O_3$ | 0.73 | 0.66 | 0.62 | 0.71 | 0.80 | 0.55 | 0.34 | 0.54 | 0.59 |
| FeO | 1.70 | 1.75 | 1.25 | 1.07 | 2.15 | 0.53 | 0.45 | 0.81 | 0.92 |
| MnO | 0.04 | 0.04 | 0.03 | 0.03 | 0.05 | 0.01 | 0.01 | 0.03 | 0.02 |
| MgO | 0.80 | 0.86 | 0.60 | 0.38 | 1.08 | 0.22 | 0.14 | 0.25 | 0.18 |
| CaO | 2.98 | 3.72 | 2.35 | 1.70 | 3.04 | 1.43 | 1.42 | 0.88 | 0.43 |
| $Na_2O$ | 5.00 | 5.39 | 4.39 | 4.52 | 4.27 | 3.90 | 3.09 | 4.03 | 1.86 |
| $K_2O$ | 1.30 | 1.52 | 2.99 | 3.23 | 2.73 | 4.54 | 5.36 | 7.43 | 8.90 |
| $P_2O_5$ | 0.09 | 0.09 | 0.07 | 0.06 | 0.13 | 0.03 | 0.03 | 0.05 | 0.03 |
| $H_2O+$ | 0.28 | 0.26 | 0.20 | 0.20 | 0.27 | 0.15 | 0.14 | 0.11 | 0.16 |
| $H_2O-$ | 0.11 | 0.08 | 0.11 | 0.08 | 0.08 | 0.08 | 0.03 | 0.07 | 0.03 |
| $CO_2$ | 0.06 | 0.00 | 0.09 | | 0.07 | 0.09 | 0.09 | 0.09 | 0.06 |
| F | 0.03 | 0.05 | 0.02 | 0.02 | 0.06 | 0.03 | 0.01 | 0.04 | 0.02 |
| **g/cm³** | | | | | | | | | |
| D | 2.675 | 2.693 | 2.646 | 2.622 | 2.668 | 2.631 | 2.632 | 2.596 | 2.597 |
| **ppm** | | | | | | | | | |
| Rb | 45.4 | 57.2 | 90.2 | 95.2 | 81.0 | 145 | 112 | 278 | 203 |
| Sr | 311 | 478 | 295 | 270 | 375 | 214 | 357 | 258 | 356 |
| Ba | 420 | 400 | 770 | 645 | 660 | 800 | 1770 | 1150 | 3455 |
| **ratios** | | | | | | | | | |
| K/Rb | 240 | 220 | 275 | 280 | 280 | 260 | 400 | 220 | 364 |
| Rb/Sr | 0.15 | 0.12 | 0.31 | 0.35 | 0.22 | 0.68 | 0.31 | 1.08 | 0.57 |
| Sr/Ba | 0.74 | 1.2 | 0.38 | 0.42 | 0.57 | 0.27 | 0.20 | 0.22 | 0.1 |
| Ba/Rb | 9.3 | 7.0 | 8.5 | 6.8 | 8.1 | 5.5 | 16 | 4.1 | 17 |

*Data sources:   This paper:   \*Table 8; \*\*Table 9; \*\*\*Table 10; \*\*\*\*Table 11; \*\*\*\*\*Table 12, No. 813.*
*Wooden and others (this volume):  +Table 2;, No. 657;  ++Table 3;  +++Table 9, adamellite-1;*
*++++Table 9.*

(Table 8) is characterized by relatively low contents of $K_2O$, Rb, and Ba. The samples of granodiorite gneiss are divided into high-Ca (Table 9) and low-Ca (Table 10) varieties. Sample 738, included with the high-Ca granodiorite gneiss, falls with the adamellite gneiss on the basis of the mode (Table 5). The samples of adamellite gneiss are divided into adamellite gneiss (Table 11) and a Ba-rich variety (Table 12) in which the ratio Ba/Rb ranges from 9.3 to 29 and the ratio K/Rb from 324 to 499. These ratios for the adamellite gneiss in Table 11 range from 4.5 to 7.8 and from 232 to 317, respectively. Sample 611 is included with the Ba-rich adamellites (Table 12) but modally is classed as granodiorite (Table 3). The pegmatitic granite gneiss (813) is listed in Table 12.

Chemical analyses of the layered mafic gneisses are given in

Table 13. Bulk analyses of amphibolite, metagabbro, and two samples of younger (Proterozoic) diabase dikes are listed in Table 14, and the trace-element determinations on these rocks are given in Table 15. Some average compositions of principal types are given in Table 16, which includes similar data for rocks from the Morton area. Additional determinations on samples not included in Tables 8 to 15 are given in Table 17, which contains the Rb-Sr data, or in Appendix 1.

## Major and Minor Constituents

The low contents of $H_2O$ and $CO_2$ in the great majority of the samples of the felsic gneisses reflect the low degree of alteration

TABLE 17.  Rb-Sr ANALYTICAL DATA FOR SAMPLES FROM THE GRANITE FALLS AREA

| Sample No. | ppm Rb | ppm Sr | $^{87}Rb/^{86}Sr$ | $^{87}Sr/^{86}Sr$ |
|---|---|---|---|---|
| **Locality 1A** | | | | |
| 613 | 109 | 377 | 0.8398 | 0.7456 |
| 614 | 88.4 | 514 | 0.4993 | 0.7252 |
| 813 | 278 | 258 | 3.159 | 0.8359 |
| **Locality 1B** | | | | |
| 368 | 193 | 194 | 2.880 | 0.8079 |
| 369 | 105 | 291 | 1,045 | 0.7521 |
| 384 | 134 | 251 | 1.546 | 0.7651 |
| 413 | 173 | 213 | 2.370 | 0.8027 |
| 431 | 139 | 170 | 2.388 | 0.8033 |
| 608 | 92.7 | 308 | 0.8735 | 0.7474 |
| **Locality 2** | | | | |
| 605 | 80.0 | 288 | 0.8076 | 0.7434 |
| 790 | 114 | 243 | 1.359 | 0.7689 |
| 791 | 52.4 | 274 | 0.5546 | 0.7371 |
| 792 | 34.9 | 295 | 0.3424 | 0.7273 |
| 793 | 84.8 | 272 | 0.9059 | 0.7520 |
| **Locality 3A** | | | | |
| 607 | 41.5 | 388 | 0.3183 | 0.7206 |
| 609 | 71.7 | 339 | 0.6136 | 0.7324 |
| 620 | 135 | 383 | 1.025 | 0.7523 |
| **Locality 3B** | | | | |
| 385 | 83.8 | 387 | 0.6330 | 0.7300 |
| 464F | 25.8 | 292 | 0.2567 | 0.7201 |
| 464M | 39.8 | 254 | 0.4552 | 0.7291 |
| **Locality 4** | | | | |
| 54CGN | 8.0 | 298 | 0.0774 | 0.7073 |
| 54GN | 75.0 | 270 | 0.8066 | 0.7455 |
| 54GR | 123 | 259 | 1.384 | 0.7648 |
| 606 | 89.1 | 257 | 1,009 | 0.7544 |
| 610 | 78.7 | 257 | 0.8871 | 0.7470 |
| 611 | 93.9 | 260 | 1.049 | 0.7522 |
| 612 | 125 | 248 | 1.467 | 0.7695 |
| 624 | 54.4 | 366 | 0.4326 | 0.7273 |
| 691 | 112 | 288 | 1.133 | 0.7559 |
| 692 | 12. | 242 | 1.476 | 0.7685 |
| 693 | 138 | 238 | 1.688 | 0.7769 |
| 694 | 9..1 | 245 | 1.093 | 0.7574 |
| 965 | 89.9 | 245 | 1.067 | 0.7537 |
| 696 | 118 | 242 | 1.420 | 0.7669 |
| 700 | 98.4 | 245 | 1.169 | 0.7599 |
| 701 | 74.7 | 252 | 0.8629 | 0.7481 |
| 738 | 89.5 | 257 | 1.014 | 0 7551 |
| 763 | 12.8 | 130 | 0.286 | 0.7123 |
| 777 | 4.40 | 12€ | 0.101 | 0.7068 |
| M-14-16 | 21.5 | 72.0 | 0.868 | 0.7359 |
| **Locality 5** | | | | |
| 462 | 7.50 | 319 | 0.0685 | 0.7129 |
| 761 | 1.92 | 279 | 0.0581 | 0.7045 |
| 815 | 17.3 | 414 | 0.1209 | 0.7074 |
| 816 | 59.2 | 447 | 0.3838 | 0.7165 |
| 817 | 55.2 | 324 | 0.4949 | 0.7211 |
| **Locality 6** | | | | |
| 290 | 79.4 | 204 | 1.132 | 0.7619 |
| 819 | 109 | 295 | 1.074 | 0.7549 |
| 821 | 59.7 | 260 | 0.6668 | 0.7424 |

much better than the secondary minerals sericite, chlorite, calcite, and so forth, which are difficult to estimate in thin section. A classification of the felsic gneisses on the basis of normative plagioclase, K-feldspar, and quartz (Fig. 4) shows less scatter of the samples than the modal classification (Fig. 2). Because K in biotite and plagioclase is calculated as normative orthoclase, the tonalite gneiss samples fall in the granodiorite field. The perthitic feldspars result in other divergences; the granitic pegmatite, for example, falls in the adamellite field. With a few exceptions, however, the discrepancies are minor.

A plot of $K_2O$ versus $Na_2O$ (Fig. 5) includes a larger number of samples than can be shown in Figure 4. Two trends can be seen with most of the samples from locality 4 and a few from locality 2 along line A, and the samples from locality 1 and some from locality 3 along line B. The trends shown in Figure 5 are also seen in Figure 6 in which the ratios $CaO:Na_2O:K_2O$ are plotted. Again the samples from locality 4 define a trend (line A), and the samples from locality 1 a second trend (line B). Sample 464, the average composition of the layered gneiss from locality 3, falls near the tonalite samples along line A. Other samples of the layered series plot to the right and along a trend to the CaO apex.

The ratios $MgO:CaO:FeO_t$ (Fig. 7) show similar relationships. Samples of the felsic gneisses have trends similar to those of Figures 5 and 6; however, rock types cannot be distinguished. All the samples of amphibolite from the Morton area have been included, and the low-alumina variety of basaltic komatiite composition is differentiated from the tholeiitic variety. The samples of tholeiitic composition define a line showing enrichment in Fe relative to Ca and Mg. The younger diabase samples fall near the line, but the samples of the layered mafic gneisses show considerable scatter, and the variability supports the conclusion that the layered mafic gneisses probably are metasedimentary rocks, with the exception of sample 464 which falls among the felsic gneisses (Figs. 5, 6, and 7). With the exception of sample 464, the layered gneisses (Table 13) are characterized by Fe contents well above that of the tonalitic gneiss, by MgO and CaO contents below that of the amphibolite but well above that of the tonalite, and by relatively high $P_2O_5$.

**Trace Elements**

The tonalite gneiss can be separated from the more granitic gneiss on the basis of K and Rb contents (Fig. 8); however, the granodiorite and adamellite rocks, which can be separated at a K level of approximately 3.5 wt%, have a similar range in Rb. A number of samples with relatively high contents of K contain less than 100 ppm Rb. The K/Rb ratios for these samples are 350 or more (Fig. 8). Most of the samples from locality 4 are aligned with a K/Rb ratio of approximately 300 in contrast with the samples from locality 1 which have a ratio of 200 to 250. These differences reflect locality and rock type variations which are further considered in the discussion of geochronology.

The variations of Rb with Ba (Fig. 9) show relationships similar to those of Figure 8. Most of the samples from locality 1 have a low Ba/Rb ratio of approximately 5, but a notable exception is the granodiorite sample 614 which plots in the adamellite field (Fig. 9). Samples from locality 4 show an alignment of samples, with a Ba/Rb ratio of approximately 7.5. The alignment of samples across the granodiorite-adamellite boundary, both in

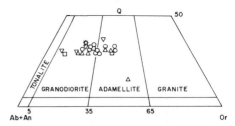

**Figure 4. Chemical classification of the felsic gneisses based on the ratios of normative orthoclase:plagioclase:quartz (weight percent). Locality symbols as in Figure 1.**

Figures 8 and 9, is interpreted to be largely the result of mixed-rock samples formed by reaction between granodioritic (possibly tonalitic) rocks and the younger adamellitic and pegmatitic phases. This mixing may be the result of the high-grade metamorphism 2,600 m.y. ago (Goldich and others, 1970), but more likely it was effected at an earlier time during the intrusion of adamellite and pegmatitic granite magmas into the older metamorphic complex of tonalite, granodiorite, and amphibolite.

## GEOCHRONOLOGY

Radiometric dating of the Precambrian rocks in the Granite Falls area has evolved along two principal lines. The first is analytical and involves the radiometric techniques; the second is geologic and involves sampling methods and the intepretation of the age measurements. Limitations of specific dating methods became apparent in the geologic interpretation of the data (Goldich and others, 1961, 1970), but at the same time the radiometric age determinations indicated that the geologic history as earlier interpreted from mapping and petrologic studies had been greatly oversimplified.

The first effort (Goldich and Hedge, 1974) in resolving the problem introduced by the different ages of the rock components was based on a small number of samples of the foliated granodiorite phase of the Montevideo Gneiss. This work led to an age of 3,720 m.y. for the granodiorite gneiss; however, Farhat and Wetherill (1975) concluded that the gneiss could not be older than 3,250 m.y. and was metamorphosed approximately 2,500 m.y. ago with redistribution and complete homogenization of $^{87}Sr$.

This paper presents new Rb-Sr data for a number of principal rock types and U-Pb data for the adamellite gneiss and summarizes the progress that has been made in determining the geologic history and in developing a time framework for the major events. The disturbance of the different isotopic systems in the different rock types that resulted from the complex series of geologic events also is of considerable interest. All the isotopic ages in this paper are based on the decay constants recommended by the Subcommission on Geochronology of the IUGS (Steiger and Jager, 1977).

### Rb-Sr Age Determinations

The Rb-Sr analytical data for both new and old samples are given in Table 17. Most of the samples are from locality 4, but localities 1 and 6 are considered first because Farhat and Wetherill (1975) have published data for samples from these localities.

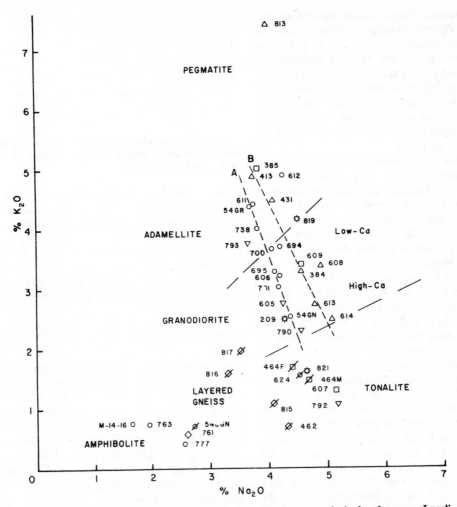

**Figure 5.** Plot of K₂O versus Na₂O showing relationships between principal rock groups. Locality symbols as in Figure 1; samples of layered mafic gneiss from localities 4 and 5 indicated by slash symbol.

**Locality 1.** Our data and the results of Farhat and Wetherill (1975) are included in the Rb-Sr diagram of Figure 10. The isochrons (lines A and B) are mixing lines. Samples 369 (granodiorite gneiss), 384 (mixed rock), and 368 (adamellite gneiss) show roughly the same relationship. Like the gneiss in the Morton area (Goldich and others, this volume), the gneiss at locality 1 is a hybrid rock with a paleosome of granodiorite and a neosome of adamellite and pegmatitic granite; hence, there is a bimodal distribution of samples (Fig. 10) with the mixed-rock samples 384 and 4(T) plotting between the two groups. Radiogenic $^{87}$Sr was redistributed among the minerals not only during the high-grade metamorphic event 2,600 m.y. ago but also during the event 1,800 m.y. ago (Goldich and others, 1970). Diffusion across rock contacts (that is, granodiorite-adamellite) undoubtedly occurred, but we emphasize the effects that were produced by diffusion and infiltration at the rock-magma interfaces at the time of intrusion of the pegmatitic granite and adamellite magmas. These effects were further complicated by the metamorphism 2,600 m.y. ago and by subsequent shearing and alteration.

Samples 413, 431, and 813 (Fig. 10) are aligned and give an approximate age of 2,910 m.y. ($R_i$ = 0.7025). The lower intercept is imprecise because of the long extrapolation, but the age probably

is a minimum value because of the alteration of the rocks as will be discussed in later sections.

**Locality 6.** Locality 6 is a small ledge exposed in the bank of the Minnesota River south of Granite Falls (Fig. 1). Nine samples collected by Farhat are significant and useful. These are plotted in Figure 11 in which they are numbered in order of increasing $^{87}$Rb/$^{86}$Sr. The designations R (red) and D (dark) are from Farhat and Wetherill (1975), and sample 3D + R is a mixed-rock sample. All the available data (Fig. 11) are here treated as two distinct populations. Samples 1, 3, and KA-209 give an age of 3,725 m.y. ($R_i$ = 0.7007), and samples 2, 5, 7, 8, and 9 give an apparent age of 2,970 m.y. ($R_i$ = 0.7059). Samples 819 and 6 fall between the lines, and samples 821 and 4 fall above line A; these samples were not included in the regressions. Line B is a mixing line, and sample 9 is adamellite gneiss. If an initial ratio of 0.7025 (from Fig. 10, loc. 1) is used, the age of sample 9 is 3,055 m.y. If a lower initial ratio of 0.701 is assumed, the age is 3,115 m.y. The data from locality 6 are similar to those from locality 1, and the combined results are reviewed in the following section.

**Localities 1 and 6.** All the data from localities 1 and 6 are combined in Figure 12. A regression for five samples gives an age of 3,675 ± 115 m.y. [$R_i$ = 0.7010 ± 0.0012; 95% confidence

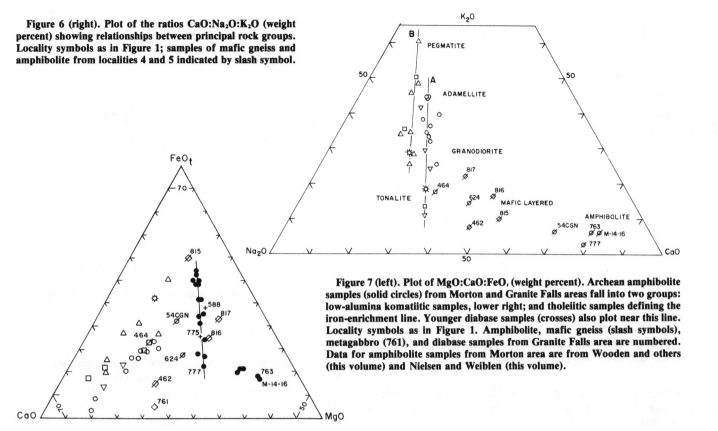

Figure 6 (right). Plot of the ratios CaO:Na₂O:K₂O (weight percent) showing relationships between principal rock groups. Locality symbols as in Figure 1; samples of mafic gneiss and amphibolite from localities 4 and 5 indicated by slash symbol.

Figure 7 (left). Plot of MgO:CaO:FeO, (weight percent). Archean amphibolite samples (solid circles) from Morton and Granite Falls areas fall into two groups: low-alumina komatiitic samples, lower right; and tholeiitic samples defining the iron-enrichment line. Younger diabase samples (crosses) also plot near this line. Locality symbols as in Figure 1. Amphibolite, mafic gneiss (slash symbols), metagabbro (761), and diabase samples from Granite Falls area are numbered. Data for amphibolite samples from Morton area are from Wooden and others (this volume) and Nielsen and Weiblen (this volume).

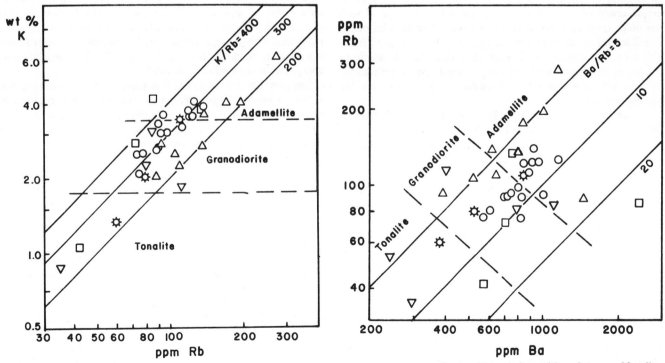

Figure 8. K versus Rb showing variations with rock type and locality. Symbols as in Figure 1.

Figure 9. Rb versus Ba showing variations with rock type and locality. Symbols as in Figure 1.

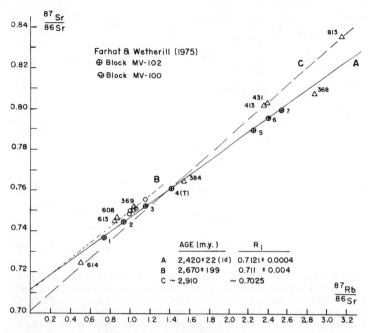

Figure 10. Rb-Sr diagram for all published and new data from locality
1. See Table 17 for data plotted as triangles.

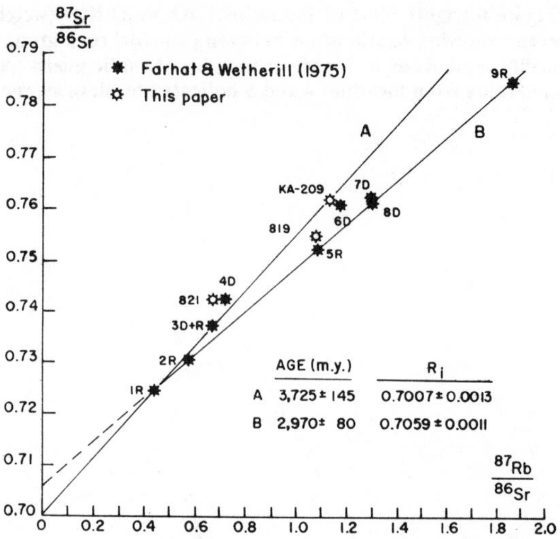

**Figure 11. Rb-Sr diagram for all data from locality 6.**

level(C. L.)]. The dashed line (Fig. 12, line B) further divides the
samples into groups for discussion. Samples 821 and 4D, which
fall above the isochron (Fig. 12, line A), are tonalite gneiss that
has either gained radiogenic $^{87}$Sr or lost Rb. The samples between
lines A and B are contaminated samples that have gained Rb by
infiltration of granitic material or by diffusion at the time of
emplacement of the pegmatite and adamellite interlayered with
the older granodiorite and related gneisses. The samples below
line B are the result of loss of radiogenic Sr during hydrothermal
alteration that accompanied or followed shearing subsequent to
the major deformation of 2,600 m.y. ago. These interpretations
of the age measurements are based in large part on the
geochemical data.

A plot of Rb versus K (Fig. 13) shows some basic differences
between the rocks of localities 1 and 6 and also within each locality. As a group the samples from locality 1 are enriched in Rb and
plot above the samples from locality 6. Samples from locality 6
fall along two lines (Fig. 13, lines A and B); this suggests two rock
series. Sample 385 from locality 3 falls with samples 1 and 2 along
line A. Its Rb-Sr model age (Goldich and others, 1970) and the
$^{207}$Pb-$^{206}$Pb zircon age (Catanzaro, 1963) are both approximately
3,250 m.y. These granitic rocks belong to an older series than the
adamellite samples along line B. The older adamellites have
remarkably high K/Rb ratios of 500 compared to 380 for the
adamellites on line B and 225 for those on line C.

The variations of Sr with K for samples for which K determinations are available are shown in Figure 14. Block sampling produces the horizontal fields of MV-100 and MV-102 in which the
Sr concentrations fall within restricted ranges, but K (and similarly Rb) ranges widely. A number of our samples also fall in the two
fields. A third horizontal field may be added for the samples from
locality 6, but there are exceptions, notably samples 1, 2, and 3.
Samples 1 and 2 are characterized by high contents of Sr and plot
in an area that includes samples 385, 613, and 614 (Fig. 14).

Possibly these high-Sr rocks are related. A significant conclusion,
however, is that a number of specific rock types can be recognized
and are represented in the samples, but many are more or less
contaminated samples that cannot be recognized as such in the
field. These behave similarly to mixed-rock samples (that is, 384
and 3D + R). The younger adamellite and pegmatitic granite
gneisses (Fig. 14) all have K/Sr ratios greater than 150, but the
older gneisses and the contaminated or mixed-rock samples all
have ratios of less than 150. Our interpretation, then, is that
samples 2R, 819, and 6D (Fig. 11) have gained Rb or are contaminated with younger adamellite, but sample 3D + R probably
is a mixed-rock sample with components of older adamellite (1R)
and granodiorite similar to KA-209.

Samples that fall below line B in Figure 12 are all from locality
1. We considered the possibility that these are younger rocks from
magmas derived by melting of the older gneisses. Such an origin
would account for a high initial $^{87}$Sr/$^{86}$Sr ratio of the order of
0.710. The evidence contrary to this interpretation, however, is
compelling. For example, the adamellite and pegmatitic granite
gneisses are strongly deformed and are similar to adamellite-1 and
pegmatitic granite gneisses of the Morton area. These considerations, together with the Rb-Sr model ages, indicate that the
adamellite and closely associated pegmatite are older rocks that
were involved in the high-grade metamorphism 2,600 m.y. ago.
After the event 2,600 m.y. ago, the rocks in locality 1 were
sheared, hydrothermally altered, and oxidized to produce the
bright brick-red color of the adamellite and pegmatite. During
this alteration, some radiogenic $^{87}$Sr was removed.

**Locality 4.** Twenty samples from the large quarry northwest of
Granite Falls (Fig. 1, loc. 4) are plotted in Figure 15. The
restricted range in Rb/Sr in the older gneiss samples (54GN, 701,
606, and 738) presents the problem of anchoring the lower end of
the isochron. If sample 54CGN, a sample of the layered mafic
gneiss from the quarry, is used, the regression age is 3,530 ± 55
m.y. ($R_i$ = 0.7034 ± 0.0003; 95% C.L.) The high initial ratio
makes the age unacceptable, and it is noted that sample 624, a

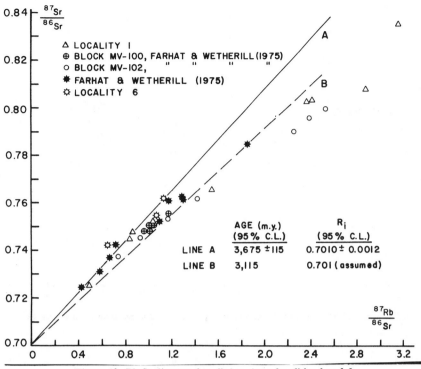

Figure 12. Rb-Sr diagram for all data from localities 1 and 6.

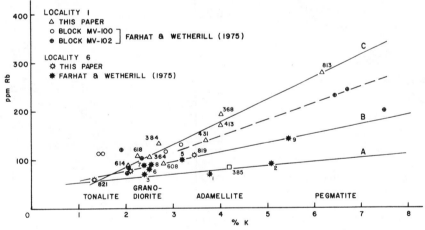

Figure 13. Variations of Rb with K in samples from localities 1 and 6.

foliated metasedimentary rock, also plots above the 3,530–m.y. B.P. line. The Rb-poor rocks have been appreciably disturbed either through loss of Rb or gain of radiogenic $^{87}$Sr. Points to the right of the line represent younger granitic rocks and contaminated or mixed-rock samples that formed by interactions with the older granodiorite and related gneiss. These samples, as can be seen in the plots of Figures 5, 6, 8, and 9, form continuous series across the granodiorite-adamellite boundary.

The model age of adamellite sample 693 (Fig. 15) with an assumed $R_i = 0.701$ is 3,100 m.y., similar to the model age for sample 9R, from locality 6. This age is further supported by a $^{207}$Pb-$^{206}$Pb age of 3,050 m.y. on zircon, which will be discussed in

a later section. Line B (Fig. 15) gives an age of 2,575 ± 80 m.y. ($R_i = 0.7137 \pm 0.0014$; 95% C.L.), and line C gives an apparent age of 2,265 ± 60 m.y. ($R_i = 0.7217 \pm 0.0010$). These are mixing lines. The rocks were affected by an event approximately 3,000 m.y. ago, and a regression of the granodiorite samples (54GN, 701, 606, and 738) gives an age of 3,010 ± 345 m.y. ($R_i = 0.7091 \pm 0.0047$; 95% C.L.), a secondary isochron similar to the fine structure in the isochron for the tonalite gneiss in the Morton area (Goldich and Wooden, this volume). Farhat (1975) obtained similar results of 2,961 ± 101 m.y. ($R_i = 0.7096 \pm 0.0010$; 1$\sigma$) for composites of ten samples each from four blocks (2 to 3 m × 1 to 2 m) from the quarry.

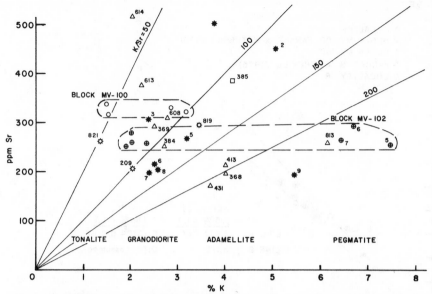

Figure 14. Variations of Sr with K in samples from localities 1 and 6. Symbols as in Figure 13.

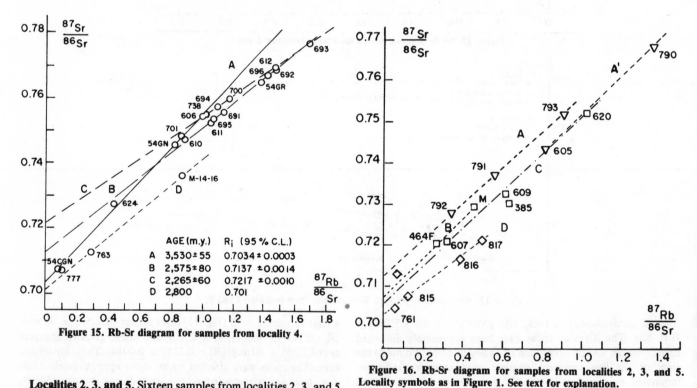

Figure 15. Rb-Sr diagram for samples from locality 4.

Figure 16. Rb-Sr diagram for samples from localities 2, 3, and 5. Locality symbols as in Figure 1. See text for explanation.

**Localities 2, 3, and 5.** Sixteen samples from localities 2, 3, and 5 are included in Figure 16. Sample 605 (loc. 2) and samples 607 and 609 (loc. 3A) were reported by Goldich and Hedge (1974), and sample 385 (loc. 3B) by Goldich and others (1970). The remaining samples were collected for the present study and include four of layered mafic gneiss and one of the metagabbro (761, Fig. 1). The relatively small number of samples should be kept in mind, and it is fair to say that they raise more questions than we can answer. For purposes of discussion, some tentative lines have been drawn for samples of each locality (Fig. 16). Lines A, A′, B, and C give apparent ages and initial ratios of 3,035 m.y. (0.712), 3,190 m.y.

(0.706), 3,250 m.y. (0.708), and 3,045 m.y. (0.706), respectively. None is an original or primary age, and each reflects metamorphic processes with the principal effects produced approximately 3,100 m.y. ago. Line C (samples 607, 609, and 620) has the appearance of a rotated isochron. Samples 464F and 464M (line B) are light and dark layers from a single large specimen from locality 3B.

Segregation of minerals into felsic and mafic layers by shearing and plastic deformation may upset the Rb-Sr isotopic system. If redistribution of radiogenic $^{87}Sr$ had occurred before this segregation, the result would be felsic layers with unsupported [87]Sr and mafic layers with excess Rb. The extent of the movement or displacement between adjacent layers cannot be determined, and sampling across the layers, as was done for samples 464F and 464M, does not represent the original rock.

Three samples from the layered mafic gneisses (line D, Fig. 16) give an apparent age of approximately 2,500 m.y. with $R_i$ approximately 0.703. The metagabbro sample (761) falls near this line, but the field relationships suggest that it is a younger rock intruded in the layered mafic gneisses. Wilson and Murthy (1976) reported an age of $2,620 \pm 200$ m.y. $R_i = 0.7037 \pm 0.0001; 2\sigma$) for the metagabbro and an age of $3,460 \pm 156$ m.y. ($R_i = 0.7003 \pm 0.0009; 2\sigma$) for the layered mafic gneisses from locality 5. Mineral and whole-rock isochrons give a secondary age of 1,760 m.y. The latter is a biotite age and would not reveal the event 2,600 m.y. ago that is clearly recorded in Stern's essentially concordant U-Pb ages (Goldich and others, 1970) on zircon concentrates from the biotite-garnet gneiss from locality 5.

## U-Pb Age Determinations

Two fractions of a zircon concentrate from a composite of adamellite gneiss samples 611 and 612 (loc. 4) were analyzed (Tables 18 and 19). The samples are of similar appearance and were collected from a site in the large quarry. The mode of sample 611 (Table 3) is granodiorite, and that of sample 612 (Table 5) is adamellite. The partial chemical analyses (Table 12), which are more representative than the thin sections, show that the samples are similar, and on the basis of the ratio of normative Or/(Ab + An), the samples would plot on the boundary between granodiorite and adamellite (Fig. 4). There are differences in the

TABLE 18.  U-Th-Pb AGES AND ANALYTICAL DATA FOR ZIRCON FRACTIONS, ADAMELLITE GNEISS

| Sample No. | Description | $\dfrac{206\text{Pb}}{238_U}$ | $\dfrac{207\text{Pb}}{235_U}$ | $\dfrac{207\text{Pb}}{206\text{Pb}}$ | $\dfrac{208\text{Pb}}{232_{Th}}$ |
|---|---|---|---|---|---|
| 611-12A | Dark, 60-120 mesh, non-mag., 1.5 amps. | 2,695 | 2,900 | 3,050 | 2,522 |
| 611-12B | Dark, 60-120 mesh, mag., 1.5 amps. | 2,574 | 2,849 | 3,050 | 2,660 |

| | Pb (ppm) | U (ppm) | Th (ppm) | 204Pb | 296Pb | 207Pb | 208Pb |
|---|---|---|---|---|---|---|---|
| 611-12A | 816.1 | 1,428 | 140.9 | 0.0267 | 78.58₅ | 18.31₇ | 3.07₁ |
| 611-12B | 775.9 | 1,437 | 140.4 | 0.0228 | 78.52₈ | 18.30₇ | 3.14₁ |

*Decay constants:*

$$238_U = 1.551 \times 10^{-10} \; yr^{-1}$$
$$235_U = 9.848 \times 10^{-10} \; yr^{-1}$$
$$232_{Th} = 4.947 \times 10^{-11} \; yr^{-1}$$

*Atom ratio:*

$$238_U / 235_U = 137.88$$

*Corrections for Pb blank:*

$$204 : 206 : 207 : 208 = 1.35 : 25.28 : 21.13 : 52.23$$

TABLE 19.  U-Pb AGES ON ZIRCON FROM ADAMELLITE GNEISS FROM GRANITE FALLS AND PEGMATITIC GRANITE GNEISS FROM MORTON, MINNESOTA

| Sample No. | U (ppm) | $\dfrac{206\text{Pb}}{238_U}$ | Age (m.y.) $\dfrac{207\text{Pb}}{235_U}$ | $\dfrac{207\text{Pb}}{206\text{Pb}}$ | Analyst |
|---|---|---|---|---|---|
| 611-12A | 1,428 | 2,695 | 2,900 | 3,050 | T. W. Stern (this paper) |
| 611-12B | 1,437 | 2,574 | 2,849 | 3,050 | |
| 781A | 830 | 2,550 | 2,840 | 3,055 | J.L. Wooden, (Goldich and Wooden this volume) |
| 781B | 809 | 2,565 | 2,835 | 3,035 | |
| 781C | 988 | 2,535 | 2,815 | 3,025 | |
| 390L | 962 | 2,615 | 2,873 | 3,055 | E.J. Catanzaro (1963) |

concentrations of Rb and Sr; sample 612 falls with the adamellite samples near sample 693, and No. 611 plots with the granodiorite gneiss samples 691 and 695 in the Rb-Sr diagram (Fig. 15).

The two fractions of zircon (A and B) are shown on a concordia diagram (Fig. 17), together with samples from earlier work (Catanzaro, 1963; Goldich and others, 1970). Our interpretation of the data follows: (1) 3,230 m.y. is the minimum age for the older gneisses in the Granite Falls area, based on sample 385 (locality 3B) analyzed by Catanzaro (1963); (2) 3,050 m.y. is the approximate time of emplacement of the adamellite and closely associated pegmatitic granite; (3) 2,600 m.y. dates the major metamorphic event during which the final structure, including the Granite Falls antiform, was developed, on the basis of the nearly concordant U-Pb ages obtained by Stern (Goldich and others, 1970) on zircon from the biotite-garnet gneiss of the layered mafic gneisses (loc. 5), and emplacement of the Sacred Heart Granite of Lund (1956) accompanied or closely followed the folding, based on sample 388 (Goldich and others, 1970); (4) 1,800 m.y. ago is the time of Proterozoic igneous activity and emplacement of small plutons, such as the adamellite (loc. 3B) dated by Catanzaro (1963).

The age assignments for these events are considered in more detail in the discussion that follows.

## DISCUSSION

### Summary of Rb-Sr Age Determinations

Samples from the Granite Falls area show varying degrees of disturbance of the Rb-Sr isotopic system as a result of the complex geologic history of multiple intrusive action and polymetamorphism. The data, however, support the field relationships which indicate that an ancient metamorphic complex of granodiorite and related gneiss and amphibolite was involved in a period of folding before the intrusion of adamellite and pegmatitic granite approximately 3,100 m.y. ago. Two samples of adamellite gneiss and one of pegmatitic granite gneiss from locality 1 give an age of 2,910 m.y. ($R_i = 0.7025$). Other adamellite gneiss samples from this locality give much lower ages. Both the adamellite and the pegmatite are characterized by a bright brick-red color, which is attributed to hydrothermal alteration accompanying or following shearing. The low ages of

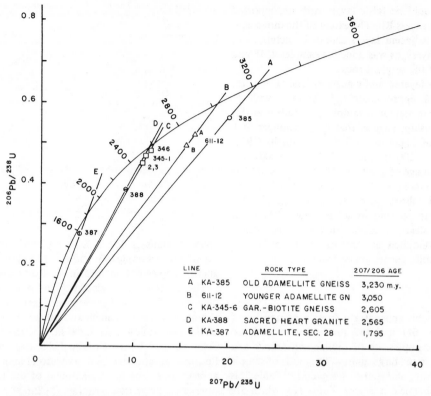

**Figure 17. Concordia diagram of U-Pb analyses of zircon from rocks in the Granite Falls area and from the Sacred Heart Granite. Samples 385 and 387 from Catanzaro (1963); 345, 346, and 388 from Goldich and others (1970); 611-12, new data, this paper.**

the adamellite gneiss samples from locality 1 are explained as the result of loss of radiogenic $^{87}$Sr. Shearing of the rocks at locality 2 is similar.

Veinlets of well-crystallized epidote are common at all localities. These traverse both the bright brick-red adamellite gneiss and the interlayered granodiorite gneiss which has been less affected by oxidation. The epidote in the veinlets, like that of the Morton area, is enriched in $^{87}$Sr, and an analysis of a sample from locality 4 gave 3,600 ppm of Sr, an $^{87}$Sr/$^{86}$Sr ratio of 0.7240, and less than 1 ppm of Rb. Possibly the epidote veinlets are related to the thermal event of 1,800 m.y. ago. They clearly indicate the loss of radiogenic $^{87}$Sr from the gneisses.

Samples of adamellite gneiss from localities 4 and 6 give somewhat older ages than those of locality 1. If the initial ratio of 0.7025 is used, ages of 3,040 and 3,055 m.y. are obtained for adamellite gneiss from localities 4 and 6, respectively. If a lower initial ratio of 0.7010 is assumed, the ages are 3,100 and 3,115, respectively. Samples from localities 2, 3, and 5 (Fig. 16) also show the effects of a metamorphic event approximately 3,100 m.y. ago.

Samples of the older gneisses clearly show the effects of the younger events. Some give very old ages, but many more have low apparent ages that reflect Rb gain, either by infiltration of granitic material or by diffusion. Rb-Sr diagrams of samples taken by block sampling show well-defined lines (Farhat and Wetherill, 1975). Similar alignments are obtained by plotting samples collected individually with regard to rock type from a restricted area. We interpret these as mixing lines.

Earlier Rb-Sr studies (Goldich and others, 1970) demonstrated

that radiogenic $^{87}$Sr was redistributed among mineral phases as a result, not only of the 2,600–m.y. B.P. metamorphism, but also of the 1,800–m.y. B.P. event. Where the granodiorite and adamellite gneiss are intermingled on a small scale, and also at the contacts between the rock types, radiogenic $^{87}$Sr undoubtedly was mobilized during the high-grade 2,600–m.y. B.P. metamorphism, but not on a regional scale. The important mechanism that disturbed the Rb-Sr isotopic system was the introduction of Rb which we ascribe to diffusion of Rb and infiltration of granitic material into the older gneisses at the time of emplacement of the adamellite and granite pegmatite magmas. The effects have been further complicated by subsequent events.

To arrive at an age for the granodiorite and related gneisses requires elimination of samples that have been severely altered by one or more of the mechanisms indicated above. The preliminary results reported by Goldich and Hedge (1974) showed that attention to lithologic characters is helpful; however, careful inspection of samples and thin-section study failed to eliminate anomalous results, and we have resorted to geochemical data to explain these results and to characterize samples. Samples from localities 1, 4, and 6 are plotted in Figure 18. Samples 1 and 3 (from Farhat and Wetherill, 1975), sample 72-74 (the weighted average of three very similar composite samples, Farhat, 1975), and the remaining samples from this study give an age of 3,680 ± 70 m.y. ($R_i$ = 0.7011 ± 0.0014; 95% C.L.) The age is slightly lower than the preliminary age of 3,720 m.y. reported by Goldich and Hedge (1974). Preliminary Sm-Nd model ages on samples of the granodiorite gneiss are in the range of 3,500 to 3,600 m.y. (G. J. Wasserburg and M. T. McCulloch, 1978 written commun.).

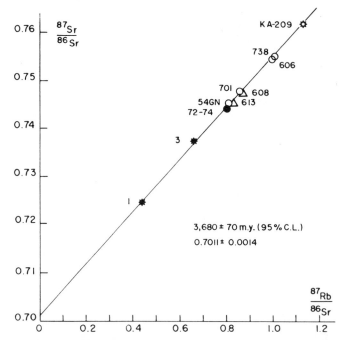

Figure 18. Rb-Sr diagram for ancient granodiorite and adamellite gneiss in the Granite Falls area. Locality symbols as in Figure 1; solid symbols are data from Farhat (1975) and Farhat and Wetherill (1975).

## Review of U-Pb Age Determinations

The U-Pb analyses of zircon samples from the Granite Falls and Morton areas are shown on the concordia diagram of Figure 19 in which some principal rock types are differentiated by symbols. Goldich and Wooden (this volume) have indicated some of the complexities of the U-Pb data. In addition to the igneous and metamorphic events, factors such as relative U contents, lapse of time, and rock composition must be considered. The present discussion is a brief review rather than an analysis of the problems.

A regression of the data for samples of the older gneisses, excluding samples of pegmatite granite and adamellite gneiss, defines the chord (Fig. 19, line I) with upper and lower intercepts of 3,486 ± 101 and 1,789 ± 183 m.y. If samples 1 and 339U are excluded, (Goldich and Wooden, this volume), the intercepts are 3,592 ± 118 and 1,910 ± 173 m.y. These data suggest that the alignment of the samples must be in some way related to the 1,800–m.y. B.P. event, contrary to our earlier treatment of the data (Goldich and others, 1970) wherein we proposed a two-stage model. In that analysis we suggested that an early discordance in the U-Pb ages was caused by the 2,600–m.y. B.P. event (line II, Fig. 19) and that a secondary discordance resulted from the dilatancy effect (Goldich and Mudrey, 1972) in Cretaceous time when the crystalline rocks of southwestern Minnesota were at or near the surface. This relatively recent Pb loss would explain lines E, D, and C in Figure 19. The alignment of the data points for the pegmatite granite gneiss from Morton and the adamellite gneiss from Granite Falls (Fig. 19, line B) suggests a similar relatively recent Pb loss. Although there is a considerable range (809 to 1,437 ppm) in U, the $^{207}Pb$-$^{206}Pb$ ages (Table 19) are similar, and the average for the six samples is 3,045 ± 32 m.y. (95% C.L.). This

probably is the time of emplacement of the pegmatite granite and adamellite. An alternate explanation is that the rocks are older and the U-Pb isotopic systems have been reset during a high-grade metamorphism 3,100 m.y. ago. The age discordance would then be the result of relatively recent Pb loss. The same line of reasoning can be applied to fractions 1, 2, and 3 of zircon from the tonalite gneiss with $^{207}Pb$-$^{206}Pb$ ages of 3,300 m.y. Although the argument can be made that the zircon crystallized 3,300 m.y. ago, the alternate explanation is that the zircon was originally formed 3,600 m.y. ago and either (1) lost some Pb or (2) gained U during the 3,100–m.y. B.P. metamorphic event (see Goldich and Wooden, this volume). U-Pb studies now in progress may provide answers to some of these questions.

## Comparison with the Morton Area

The straight-line distance between Granite Falls and Morton (Fig. 1) is approximately 70 km. Within this distance, there are some marked changes in rock types, but there are noteworthy similarites in the rock sequences. In both areas the field relationships indicate that the old gneisses with enclaves or clasts of amphibolite were folded to form a metamorphic complex before the emplacement of the pegmatitic granite and adamellite. The radiometric ages indicate that the old gneisses date back to approximately 3,600 m.y. ago and the pegmatite and adamellite to approximately 3,050 m.y. ago. Amphibolite is much more abundant in the Morton area than at Granite Falls, but the layered mafic gneisses, prominent and volumetrically important in the Granite Falls area, have not been recognized in the Morton area. The average chemical composition, bulk density, trace-element content, and the ratios K/Rb, Rb/Sr, Sr/Ba, and Ba/Rb for the felsic gneisses of the Granite Falls area are compared with similar data from the Morton area in Table 16.

Tonalite gneiss is not as abundant in the Granite Falls area as in the Morton area and also lacks variety. The four samples (Tables I and II) are all fine- to medium-grained biotite tonalite gneiss. Neither the coarse-grained, dark gray, iron-rich nor the leucocratic iron-poor varieties of the Morton area were found in the Granite Falls area. The average composition is similar to that of fine-grained tonalite gneiss of intermediate composition from the Morton area.

Granodiorite gneiss is much more abundant in the Granite Falls area than in the Morton area. Averages of the high-Ca and low-Ca varieties are given in Table 16 together with the average composition of four samples from the Morton area. The granodiorite in the Morton area is more closely related to the tonalite in composition. The Granite Falls samples are closer to adamellite, and on the average they contain less Fe, MgO, CaO, and Sr, have lower bulk densities, and have somewhat higher $SiO_2$ contents than the average granodiorite from the Morton area. The average of three samples of adamellite from localities 1 and 4 in the Granite Falls area is similar to the average of five samples of adamellite-1 from Morton. CaO and bulk density are in close agreement, but the concentrations of Rb, Sr, and Ba are different, and the average Rb for the adamellites from the Granite Falls area is greater, whereas Sr and Ba are much less than in the Morton averages. A similar relationship can be seen in the values for Rb, Sr, and Ba in the pegmatite from the two areas.

A notable difference is the absence in the Granite Falls area of

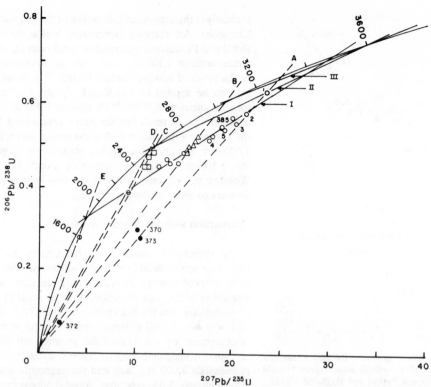

Figure 19. Concordia diagram of U-Pb zircon analyses from the Minnesota River Valley. The data in Figure 17 are shown together with data from the Morton area from Stern and others (1966), Goldich and others (1970), and Goldich and Wooden (this volume). Dashed lines, A through E, as in Figure 17. Note the three discordia lines, which indicate possible times of episodic Pb loss or U gain: I, 1,800 m.y.; II, 2,600 m.y.; and III, 3,050 m.y. ago.

younger adamellite-2 and aplite dikes. Aplite and aplite-pegmatite dikes are abundant throughout the Morton area. Equally striking is the abundance of younger diabase dikes at Granite Falls and their absence at Morton. Diabase dikes, however, are prominent southeast of Morton (Lund, 1956; Goldich and others, 1961; Hanson, 1968).

The amphibolites in the Granite Falls area are similar to those of the Morton area, and two types resembling basaltic komatiite and tholeiite are present. The basaltic komatiite variety is a low-alumina amphibolite with large contents of Cr and Ni compared to the tholeiitic variety. The large-scale interlayering of mafic and felsic gneisses has posed a problem to all investigators. Lund (1956) interpreted the mafic gneisses and amphibolite as a basement complex in which the granitic phases were intruded. Himmelberg (1968) suggested a volcanogenic-sedimentary pile. In this model the granite rocks may have been lavas or pyroclastic material and, in part, intrusive.

Wilson (1976), largely on the basis of trace elements, concluded that the mafic gneisses and amphibolite at Granite Falls and also the amphibolite of the Morton area are metasedimentary and not of igneous origin. The granitic phases and the metagabbro were intruded in a graywacke sedimentary pile. Some of the mafic gneisses are metasedimentary rocks, particularly the biotite-garnet gneiss which gave a large yield of zircon (Goldich and others, 1970, p. 3681). The bulk chemical analyses suggest that some of the hornblende-pyroxene gneiss also may have been derived from graywacke. The rocks classed as amphibolite in their

paper, however, are considered to be of igneous derivation.

The Archean history includes three major igneous-metamorphic events. Regional folding formed a metamorphic complex of tonalite, granodiorite, and adamellite gneisses, mafic gneiss, and amphibolite. The complex was invaded 3,000 to 3,100 m.y. ago by pegmatitic granite and adamellite magmas. The metamorphic complex was fractured and disrupted, but how much of the present structure can be attributed to the 3,000-m.y. B.P. event is not known. In our interpretation the pegmatite and adamellite were generated at depth and came to rest at some moderate depth. The 2,600-m.y. B.P. event involved deformation of both paleosome and granitic neosome. This we regard to have been a high-grade event that was accompanied by the intrusion of a large volume of granitic magma, the Sacred Heart Granite. Adamellite-2 was emplaced in a late stage of the deformation in the Morton area, but has not been found in the Granite Falls area. Similarly, aplite dikes traverse the structure of the Morton Gneiss but appear to be lacking in the Granite Falls area.

The metamorphism in the Granite Falls area has been described by Himmelberg and Phinney (1967) as a hornblende-pyroxene subfacies of the granulite facies. Garnet is abundant in some of the mafic gneisses and occurs in small amounts in the pegmatitic granite and adamellite gneiss in the Granite Falls area. It occurs in small amounts in amphibolite but is rare in the Morton area except in a small area of metasedimentary rocks (see Wooden and others, this volume). The higher grade of metamorphism in the Granite Falls area suggests deeper burial of the rocks during the

2,600–m.y. B.P. event, but this is speculative. Both the 3,100– and the 2,600–m.y. B.P. events produced high-grade metamorphism, and the cumulative effects are seen today.

Shearing occurred at different times. Veins of granoblastic granodiorite were formed at Morton by shearing and recrystallization of layered tonalite gneiss. The pegmatite was granulated and intensely deformed during the 2,600–m.y. B.P. deformation, but the Rb-Sr ages indicate that deformation continued long after this. Shear zones in the Granite Falls area cut a set of diabase dikes and in turn are traversed by a younger set of dikes (Hanson and Himmelberg, 1967). The intense shearing and hydrothermal alteration of the pegmatite and adamellite southeast of Montevideo at locality 1 and northwest of Granite Falls at locality 2 may well be related to the 1,800–m.y. B.P. event. The epidote veins in both the Granite Falls and Morton areas may have been formed at this time.

## ACKNOWLEDGMENTS

We thank G. R. Himmelberg (University of Missouri), R. E. Marvin, Z. E. Peterman, and R. E. Zartman (U.S. Geological Survey) for their constructive criticisms of the manuscript. We thank R. L. Bauer, G. A. Ankenbauer, Jr., T. M. Levy, and R. U. Suda for assistance in the field and laboratory. This study was made possible by financial assistance from the National Science Foundation (Grants GA-40226 and 40266a#1), supplementary grants (41-12122 and 41-23302) from the Graduate School Fund of Northern Illinois University, and logistical support from the Minnesota Geological Survey.

## APPENDIX 1. LOCATION AND DESCRIPTION OF SAMPLES

### Explanation

Sample localities are shown in Figure 1. Key to sample descriptions: mode, modal analysis listed in table of given number; chem., chemical analysis listed in table of given number; $D$ density in grams per cubic centimetre: K, potassium content in weight percent; Ba, barium content in parts per million by weight for data not given in tables.

### Locality 1. Montevideo Vicinity

*A. Inactive Quarries*—2.8 km south of Montevideo, NW cor. sec. 29, T. 117 N., R. 40 W.; lat 44°55′12″N, long 95°42′54″W.

The Montevideo Gneiss of Lund (1956) has four distinctly different rock units: (1) clasts of amphibolite, which are relatively minor, (2) dark gray, fine- to medium-grained, well-foliated gneiss—the granodiorite gneiss of this paper, (3) red, medium- to coarse-grained, "massive" granitic gneiss—the leucoadamellite of this paper, and (4) red, coarse-grained granitic pegmatite gneiss. Phases 2, 3, and 4 are the "granitic gneiss" of Himmelberg (1968).

*613.* Porphyroblastic granodiorite gneiss; fine- to medium-grained; pinkish-gray; plagioclase-quartz-microcline-biotite; average grain size, 1 mm, with microcline porphyroblasts up to 10 mm; alignment of biotite and microcline gives foliation; mode, 2; chem., 9.

*614.* Granodiorite gneiss; pinkish-gray; fine- to medium-grained (1 to 2 mm); thin-layered; plagioclase-quartz-microcline-biotite; mode, 2; chem., 9.

*813.* Pegmatitic granite gneiss; pink to red; coarse to variable grain size; mode, 4; chem., 12.

*B. Outcrops and Cuts*—Along U.S. Highway 212 and railroad tracks, 2.6 km southeast of Montevideo, SE¼ sec. 20, T. 117 N., R. 40 W.; lat 44°55′23″N, long 95°42′07″W.

*KA-368.* Leucoadamellite gneiss; bright red; medium-grained; microcline-oligoclase-quartz; collected by G. R. Himmelberg (Goldich and others, 1970); K, 4.03; Ba, 1,000.

*KA-369.* Granodiorite gneiss; dark gray; fine-grained; foliated; collected by G. R. Himmelberg (Goldich and others, 1970); K, 2.52; Ba, 520.

*KA-384.* Mixed rock; granodiorite-leucoadamellite-pegmatite (Goldich and others, 1970); mode, 4 chem., 10.

*413.* Leucoadamellite gneiss; bright red; medium-grained; oligoclase-microcline-quartz; quartz grains elongated; minor biotite; collected in highway cut; mode, 4; chem., 11.

*431.* Leucoadamellite gneiss; similar to 413; collected along railroad track east of highway; mode, 4; chem., 11.

*KA-608.* Granodiorite gneiss; pinkish-gray; medium- to coarse-grained (0.5 to 1.0 mm); banded with fine-grained biotitic and coarse-grained quartz-feldspar layers; collected by R. L. Bauer (Goldich and Hedge, 1974) from outcrop in SE¼ SW¼ sec. 20 T. 117 N. R. 40 W, southwest of outcrop on U.S. Highway 212; mode, 2; chem., 10.

**Locality 2**—3.6 km northwest of Granite Falls on Peterson farm, SE cor. sec. 30, T. 116 N., R. 39 W.; lat 44°49′32″N, long 95°35′18″W.

Samples were obtained from rock blasted to provide a foundation for a powerline. Rocks are intensely sheared with pervasive oxidation and alteration to a brick-red color similar to that at locality 1 (see Fig. 6).

*KA-605.* Granodiorite gneiss; pinkish-gray; medium- to coarse-grained (2 to 3 mm); oligoclase-quartz-microcline-biotite; similar to KA-608; collected by R. L. Bauer (Goldich and Hedge, 1974); mode, 2; chem., 9.

*790.* Granodiorite gneiss; similar to KA-608; chem., 9.

*791.* Tonalite gneiss; gray; medium-grained (avg 1 mm); plagioclase-quartz-biotite-hornblende; mode, 1; Ba, 240.

*792.* Tonalite gneiss similar to 791; chem., 8.

*793.* Adamellite gneiss; pink; coarse-grained (3 to 5 mm); granoblastic, oligoclase-microcline-quartz-biotite; large sample; mode, 4, chem., 12.

### Locality 3

*A. Outcrops on West Side of River*—3.2 km northwest of Granite Falls, SE cor. sec. 20, T. 116 N., R. 39 W.; lat 44°50′05″N, long 95°34′03″W.

*KA-607.* Tonalite gneiss; grayish-pink; medium-grained (1 to 2 mm); foliated; oligoclase-quartz-biotite; minor idioblastic hornblende; plagioclase commonly antiperthitic; collected by R. L. Bauer (Goldich and Hedge, 1974); mode, 1; chem., 8.

*609.* Granodiorite gneiss; light gray; medium-grained; plagioclase-quartz-microcline; mode, 2; chem., 10.

*620.* Granodiorite gneiss; pinkish-gray; medium-grained (1 to 2 mm); plagioclase-microcline-biotite-quartz; intensely sheared with fabric nearly schistose-lepidoblastic; mode, 2; $D$ 2.680; K, 3.79; Ba, 775.

*B. Outcrops on East Side of River*—Along railroad tracks, 2.5 km north of Granite Falls, NW¼ sec. 28, T. 116 N., R. 39 W.; lat 44°49′3″N, long 95°33′31″W.

*KA-385.* Adamellite gneiss; large sample from Goldich and others (1970); mode, 4; chem., 12

*464M.* Mafic layers, 5 cm thick, from a layered gneiss; dark gray medium- to coarse-grained; sheared; plagioclase-quartz-biotite-hornblende; chem., 13.

*464F.* Felsic layers, 5 cm thick, from the layered gneiss of 464M; light gray; medium- to coarse-grained; plagioclase-quartz-biotite; chem., 13.

*588.* Diabase intruded in Montevideo Gneiss and intruded by adamellite of sec. 28 (Goldich and others, 1961); mode, 7; chem., 14 and 15.

**Locality 4**—Green Company quarry on the northwest edge of Granite Falls, NW cor. sec. 33, T. 116 N., R. 39 W.; lat 44°48′58″N, long 95°33′45″W.

*54CGN.* Layered mafic gneiss with dark gray bands; 0.5 to 2.0 mm thick; andesine-clinopyroxene-orthopyroxene-hornblende-magnetite-ilmenite; light bands 0.5 to 4 mm thick, largely andesine and quartz with minor mafic minerals; mode, 6; chem., 13.

*KA-54GN.* Granodiorite gneiss; pinkish-gray; medium-grained (0.5 to 1.5 mm); thinly banded (2 mm); plagioclase-quartz-microcline-biotite with minor hornblende; mode, 3; chem., 9.

*54GR.* Adamellite gneiss; pink to gray; medium- to coarse-grained (2 to 5 mm) with porphyroblasts of microcline-microperthite; microcline-oligoclase-quartz-biotite; mode, 5; chem., 11.

*KA-606.* Granodiorite gneiss; light to dark reddish-gray; medium-grained (0.5 to 1.5 mm), banded, folded oligoclase-quartz-microcline-biotite; collected by R. L. Bauer (Goldich and Hedge, 1974); mode, 3; chem., 9.

*610.* Granodiorite gneiss; pinkish-gray; medium-grained (1 to 2 mm); banded; with microcline porphyroblasts up to 20 mm; andesine-quartz-microcline-biotite with minor hornblend; mode, 3; *D*, 2.677; K, 2.56; Ba, 610.

*611.* Adamellite gneiss; pink to reddish-brown; coarse-grained (2 to 3 mm) granoblastic, oligoclase-quartz-microcline-biotite; lower level of pit; mode, 3, granodiorite near adamellite; partial chem., 12.

*612.* Adamellite gneiss similar to 611; mode, 5; partial chem., 12.

*624.* Metasedimentary plagioclase-quartz-biotite-hornblende-pyroxene gneiss; dark gray; foliated; thinly layered (2 mm); mode, 6; chem., 13.

*691.* Granodiorite gneiss; greenish-gray; medium-grained (1 mm); granoblastic oligoclase-microcline-microperthite-quartz-biotite; mode, 3; *D*, 2.654; K, 3.24; Ba, 870.

*692.* Adamellite gneiss; red; coarse-grained (3 to 4 mm); granoblastic; K, 3.59; Ba, 945.

*693.* Adamellite gneiss; red; coarse-grained; granoblastic; oligoclase-quartz-microcline-biotite; mode, 3, granodiorite near adamellite; *D*, 2.624; K, 3.90; Ba, 925.

*694.* Granodiorite gneiss; dark green; coarse-grained (2 to 4 mm); granoblastic; oligoclase-microcline-microperthite-quartz-biotite; chem.,9.

*695.* Granodiorite gneiss; red; coarse-grained; granoblastic; mode, 3; chem., 9.

*696.* Adamellite gneiss; pink to red; coarse-grained, similar to 54GR; mode, 5; *D*, 2.632; K, 3.77; Ba, 860.

*700.* Granodiorite gneiss; pinkish-gray; coarse-grained (2 to 4 mm); banded; oligoclase-quartz-microcline-biotite; mode, 3; chem., 10.

*701.* Granodiorite gneiss; green; coarse-grained, similar to 694; chem., 9.

*738.* Adamellite gneiss; pink to gray; medium- to coarse-grained (1 to 3 mm) layered oligoclase-microcline-quartz-biotite; biotite-rich bands 1 to 4 mm thick; mode, 5, chem., 9.

*763.* Low-alumina amphibolite; dark gray; coarse-grained (2 to 7 mm) granoblastic hornblende with small amounts of andesine, pyroxene, and biotite; mode, 7; chem., 14 and 15.

*777.* Amphibolite; dark gray; medium-grained; granoblastic labradorite-pyroxene-hornblende; mode, 7; chem., 14 and 15.

*M-14-16.* Low-alumina amphibolite, similar to 763; collected by R. L. Bauer; chem., 14 and 15.

*775.* Diabase; dark gray; medium-grained diabasic olivine microgabbro porphyry dike cutting the Montevideo Gneiss; mode, 7; chem., 14 and 15.

**Locality 5**

*A. Memorial Park, cut along Highway 67, and two sites in Granite Falls*—
South of Granite Falls, NW cor. sec 3, T. 115 N., R. 39 W.; lat 44°48′00″N, long 95°32′15″W.

*815.* Metagraywacke; dark gray; fine- to medium-grained; massive to weakly foliated andesine-pyroxene-quartz gneiss; mode, 6; chem., 13.

*816.* Metagraywacke; dark gray; fine-grained; thinly layered andesine-quartz-biotite-pyroxene gneiss, similar to 624; mode, 6; chem., 13.

*817.* Metagraywacke; dark gray; fine- to medium-grained (0.2 to 1.0 mm) thinly layered andesine-quartz-biotite-pyroxene-garnet gneiss; mode, 6; chem., 13.

*B. Two Sites in Granite Falls*

*462.* Layered mafic gneiss collected from an excavation for foundation of house at 10th and 11th Streets, Granite Falls, in sec. 33, T. 116 N., R. 39 W.; gray, coarse-grained, layered andesine-quartz-biotite gneiss with minor microcline, clinopyroxene, and orthopyroxene which have been altered to secondary amphiboles, cummingtonite, and so forth; contains some pods or small, boudined masses of pegmatite; small amounts of pyrite in fractures; mode, 6; chem., 13.

*761.* Metagabbro; road cut along U.S. Highway 212, east edge of Granite Falls, just east of the Minnesota River, SW cor. sec. 34, T. 116 N., R. 39 W.; dark gray, sheared, and recrystallized anorthositic gabbro; mode, 7; chem., 14 and 15.

**Locality 6**—Small outcrop on the south bank of the Minnesota River, approximately 4 km southeast of Granite Falls; SW cor. sec. 11, T. 115 N., R. 39 W.; lat 44°46′40″N, long 95°31′54″W.

*KA-209.* Granodiorite gneiss; gray; medium-grained; thin-layered; well-foliated (Goldich and others, 1961); mode, 2; chem., 10.

*819.* Adamellite gneiss; light pink; medium-grained (0.5 to 1.0 mm); granoblastic with weakly developed foliation; oligoclase-quartz-microcline-biotite; mode, 4; partial chem., 11.

*821.* Tonalite gneiss; gray; medium-grained; thin-layered; with fresh biotite and porphyroblasts (2 to 3 mm) of plagioclase; partial chem., 8.

## REFERENCES CITED

Bauer, R. L., 1980, Multiphase deformation in the Granite Falls–Montevideo area, Minnesota River Valley: Geological Society of America Special Paper 182 (this volume).

Catanzaro, E. J., 1963, Zircon ages in southwestern Minnesota: Journal of Geophysical Research, v. 68, p. 2045–2048.

Farhat, J. S., 1975, Geochemical and geochronological investigations of the early Archaean of the Minnesota River Valley, and the effects of metamorphism on Rb-Sr whole rock isochrons [Ph.D. thesis]: Los Angeles, University of California, 174 p.

Farhat, J. S., and Wetherill, G. W., 1975, Interpretation of apparent ages in Minnesota: Nature, v. 257, p. 721–722.

Goldich, S. S., and Hedge, C. E., 1974, 3,800-Myr granitic gneiss in south-western Minnesota: Nature, v. 252, p. 467–468.

Goldich, S. S., and Mudrey, M. G., Jr., 1972, Dilatancy model for discordant U-Pb zircon ages, *in* Tugarinov, A. I., ed., Recent contributions to geochemistry and analytical chemistry: Moscow, Nauka, p. 415–418.

Goldich, S. S., and Wooden, J. L., 1980, Origin of the Morton Gneiss, southwestern Minnesota: Part 3. Geochronology: Geological Society of America Special Paper 182 (this volume).

Goldich, S. S., and others, 1961, The Precambrian geology and geochronology of Minnesota: Minnesota Geological Survey Bulletin 41, 193 p.

Goldich, S. S., Hedge, C. E., and Stern, T. W., 1970, Age of the Morton and Montevideo gneisses and related rocks, southwestern Minnesota: Geological Society of America Bulletin, v. 81, p. 3671-3695.

――――1980, Origin of the Morton Gneiss, southwestern Minnesota: Part 1. Lithology: Geological Society of America Special Paper 182 (this volume).

Hanson, G. N., 1968, K-Ar ages for hornblende from granites and gneisses and for basaltic intrusives in Minnesota: Minnesota Geological Survey Report of Investigations 8, 20 p.

Hanson, G. N., and Himmelberg, G. R., 1967, Ages of mafic dikes near Granite Falls, Minnesota: Geological Society of America Bulletin, v. 78, p. 1429-1432.

Himmelberg, G. R., 1968, Geology of Precambrian rocks, Granite Falls-Montevideo area, southwestern Minnesota: Minnesota Geological Survey Special Publication Series SP-5, 33 p.

Himmelberg, G. R., and Phinney, W. C., 1967, Granulite-facies metamorphism, Granite Falls-Montevideo area, Minnesota: Journal of Petrology, v. 8, p. 325-348.

Johannsen, A., 1939, A descriptive petrography of the igneous rocks, Volume 1 (second edition); Chicago, University of Chicago Press, 318 p.

Lund, E. H., 1950, Igneous and metamorphic rocks of the Minnesota River Valley: [Ph.D. thesis]: Minneapolis, University of Minnesota, 88 p.

――――1956, Igneous and metamorphic rocks of the Minnesota River Valley: Geological Society of America Bulletin, v. 67, p. 1475-1490.

Nielsen, B. V., and Weiblen, P. W., 1980, Mineral and rock compositions of mafic enclaves in the Morton Gneiss: Geological Society of America Special Paper 182 (this volume).

Steiger, R. H., and Jager, E., 1977, Subcommission on Geochronology; Convention and use of decay constants in geo- and cosmochronology: Earth and Planetary Science Letters, v. 37, p. 359-362.

Stern, T. W., Goldich, S. S., and Newell, M. F., 1966, Effects of weathering on the U-Pb ages of zircon from the Morton Gneiss, Minnesota: Earth and Planetary Science Letters, v. 1, p. 369-371.

Wilson, W. E., 1976, Trace element geochemistry and geochronology of early Precambrian granulite facies metamorphic rocks near Granite Falls in the Minnesota River Valley [Ph.D. thesis]: Minneapolis, University of Minnesota, 141 p.

Wilson, W. E., and Murthy, V. R., 1976, Rb-Sr geochronology and trace element geochemistry of granulite facies rocks near Granite Falls, in the Minnesota River Valley, in Proceedings, Twenty-Second Annual Institute on Lake Superior Geology: St. Paul, Minnesota Geological Survey, p. 69.

Wooden, J. L., Goldich, S. S., and Suhr, N. H., 1980, Origin of the Morton Gneiss, southwestern Minnesota: Part 2. Geochemistry: Geological Society of America Special Paper 182 (this volume).

Manuscript Received by the Society April 25, 1979
Revised Manuscript Received November 26, 1979
Manuscript Accepted January 11, 1980

Printed in U.S.A.

Geological Society of America
Special Paper 182
1980

# Origin of the Morton Gneiss, southwestern Minnesota: Part 1. Lithology

**S. S. GOLDICH***
*Northern Illinois University, De Kalb, Illinois 60115*

**J. L. WOODEN**
*Lockheed Engineering and Management Services Company, Inc., NASA–Johnson Space Center, Houston, Texas 77058*

**G. A. ANKENBAUER, JR.***
**T. M. LEVY***
**R. U. SUDA***
*Northern Illinois University, De Kalb, Illinois 60115*

## ABSTRACT

The Morton Gneiss of southwestern Minnesota is a migmatitic hybrid rock. The paleosome consists of tonalitic to granodioritic gneisses and amphibolites; the neosome, of a variety of granitic gneisses. The older rocks are largely biotite-quartz-oligoclase gneisses with minor microcline and hornblende and are tonalitic in composition grading to granodiorite. Locally, shearing and recrystallization of layered tonalitic gneiss developed a granoblastic granodioritic gneiss with idioblastic crystals of hornblende in a granular matrix of microcline, quartz, and oligoclase.

Amphibolite is closely related to the tonalitic gneisses and occurs as pieces or clasts that range in size from small pods or schlieren to large angular blocks. Some of the amphibolite is younger than the tonalitic gneisses and represents fragmented dikes or sill-like masses; however, some may be of the same age or older than the tonalitic gneisses. Occurrence and mineralogy indicate that the amphibolites are all of igneous precursors with compositional variations suggestive of tholeiitic to basaltic komatiite parentage.

The neosome is composed of granitic varieties from oldest to youngest: (1) pegmatite and microcline granite, (2) fine-grained adamellite (adamellite-1), (3) granodiorite and (4) microadamellite porphyry at Morton and gneissic fine-grained adamellite to the northwest (adamellite-2).

*Present addresses: Goldich, Department of Geology, Colorado School of Mines, Golden, Colorado 80401, and U.S. Geological Survey, Box 25046, MS 963, Federal Center, Denver, Colorado 80225; Ankenbauer, Chevron U.S.A., New Orleans, Louisiana 70112; Levy, 922 Jackson Avenue, New Orleans, Louisiana 70130; Suda, Gulf Mineral Resources Co., Albuquerque, New Mexico 87102.

At least two major periods of deformation are recognized. The older involved the tonalitic gneisses and the amphibolites and was developed before the emplacement of the granitic phases. The tonalitic gneiss–amphibolite complex was disrupted at the time of the intrusion of the pegmatitic and adamellitic magmas, but it is not clear how much of the present-day structure of the Morton Gneiss was developed at that time. The second major deformation affected the granitic phases as well as the older tonalitic gneiss–amphibolite complex and was responsible for the highly contorted structure of the Morton Gneiss. The rocks included in adamellite-2 are least deformed, and these appear to have been emplaced in the late stages of the second major deformation.

Straight, relatively undeformed dikes of aplite, aplite-pegmatite, and pegmatite cut the Morton Gneiss. They represent the last Archean igneous activity in the region. A poorly developed foliation in the aplite dikes is attributed to stresses generated during the uplift of the region. In the southeastern part of the Morton area, between Franklin and New Ulm, there are Proterozoic diabase dikes and small plutons containing granitic phases. These rocks were intruded approximately 1,800 m.y. ago.

## INTRODUCTION

Although geologic observations in the Minnesota River Valley date back to Major Stephen H. Long's expedition in 1823 (Winchell, 1884, p. 35.) and numerous studies have been made since then, the Precambrian geology still is not well known. Lund's (1950, 1956) systematic mapping provided a base for the geochronologic investigations of Goldich and others (1961). The Granite Falls area was remapped by Himmelberg (1965, 1968), but the radiometric ages (Goldich and others, 1970) continued to show inconsistencies that reflected problems not recognized in the mapping. The Morton area was remapped by Grant (1972), and here problems also were found.

The Morton area (Fig. 1) is the target of this study. Outcrops are few and scattered, and because it was recognized that the geochronologic problems stem, in large part, from geologic factors, it was decided to carry on three lines of investigation in which (1) field and petrographic and (2) geochemical studies, so far as practicable, were pursued independently of (3) radiometric age measurements. This division of effort proved surprisingly useful, and the principal results are presented in three parts:

Part 1, lithology, presents field observations, modal analyses of rock types, and some aspects of the geologic history deduced from these data. Rock types, previously not recognized, and the recognition of an older period of deformation that formed a tonalitic gneiss–amphibolite complex are significant contributions.

Part 2, geochemistry, is based directly on the findings in Part 1. Major-, minor-, and trace-element data are used to characterize the principal rock types and permit a more detailed and precise approach than the modal analyses. The geochemical data provide new insight into the geologic processes that affect the isotopic decay systems.

Part 3, geochronology, presents the radiometric age determinations and their interpretation, and a summary of the present knowledge of the complex geologic history.

## GEOLOGIC SETTING

The Morton Gneiss is named for outcrops in the vicinity of Morton, Minnesota (Fig. 1). It was first named the Morton quartz monzonite gneiss by Lund (1956, p. 1482):

It is a hybrid rock formed by the strewing out of basic inclusions in a granite magma. It is characterized by a highly contorted structure and displays a great variety of textures and colors.

Lund distinguished three principal rock types: (1) basic or mafic inclusions, (2) gray tonalitic gneiss, and (3) pink granitic gneiss ranging from quartz monzonite to leucogranite. The mafic inclusions of Lund are the amphibolites of this paper. They are closely associated with the tonalitic gneiss.

The principal objective of this study was to determine the major rock types and their relative ages, and the results are largely summarized in Table 1. Modal analyses (volume percent) are given in Tables 2 to 9, and some average compositions in Table 10. The rocks are classified in Johannsen's (1939) system (Fig. 2). The relative ages were determined on the basis of field relationships (Ankenbauer, 1975; Levy, 1975; Suda, 1975), and this phase was facilitated by the earlier work and maps of Lund (1950, 1956) and Grant (1972). A large number of samples were collected during the course of the field studies. Fine-grained rocks presented no special problem, but the coarse-grained gneisses were more difficult to sample. Generally, very large pieces were carried to the laboratory for diamond sawing (see Part 3).

Six localities (Fig. 1) supplied the samples, but the great majority came from Morton where the large active quarry and five smaller inactive quarries and adjacent waste piles afforded fresh rock (Fig. 3). Metasedimentary rocks have not been recognized at the six localities but occur in a small area between localities 5 and 6 where the studies by Grant (1968) are being continued (Wooden and others, 1977). Some amphibolite samples from this area, collected by Nielsen and Weiblen, are discussed in Parts 2 and 3.

Brief descriptions and the locations of individual samples are given in Appendix 1 of Part 3.

## OLDER TONALITIC AND GRANODIORITIC GNEISSES

### Tonalite Gneiss

The tonalitic gneisses are variable in texture, structure, and color. They range from light to dark gray, fine to very coarse grained, thin to thick layered, and straight banded to contorted or convoluted. The contorted structure is generally found in the gneiss containing blocks of amphibolite, which during deformation acted as a relatively competent structural unit.

Because of the coarse grain size and layered structure, modal composition determinations are not precise. Nevertheless, as can be seen in Table 2 and also in Figure 2, most of the samples fall in Johannsen's (1939) restricted field of tonalite in which K-feldspar is limited to 5% of the total feldspar. Modal analyses by Lund (1956, p. 1484) are similar. The tonalitic gneisses are biotite-quartz-oligoclase rocks with minor amounts of microcline and hornblende and can be classed as trondhjemite.

Oligoclase is the abundant mineral in the tonalitic gneisses (Table 2). Microcline is interstitial to the plagioclase and, like the quartz and biotite, in grains considerably smaller than the tabular plagioclase crystals. Some K-feldspar occurs as antiperthite probably formed by exsolution from the oligoclase, but possibly also by replacement. Myrmekitic intergrowths of quartz and oligoclase formed on the margins of plagioclase, usually at or near contacts with microcline. The mafites are biotite and lesser amounts of hornblende and magnetite-ilmenite. Accessory and secondary minerals are apatite, zircon, allanite, sphene, calcite, chlorite, epidote, and sericite. The tonalitic gneisses are relatively unaltered.

At Morton, both fine- to medium-grained, straight-banded gneiss and coarse-grained, convoluted gneiss are common. The latter type generally contains more hornblende (sample 649, Table 2), which in part is a result of recrystallization but largely is contamination from disrupted amphibolite clasts. At locality 3 (Fig. 1), fine- to medium-grained, banded, but not conspicuously layered gneiss is very similar in composition to the gneiss at Morton. At locality 4, fine-grained brownish-gray gneiss shows some mineral alignment but is not layered. The coarse grain size and thick layering of the gneiss at localities 5 and 6 contribute to the variability of the modes (Table 2), but the data support field observations that the coarse-grained layered gneiss is more leucocratic than the tonalitic gneiss to the southeast.

Petrographically, the gneisses at localities 5 and 6 show the same features noted in the rocks at Morton and North Redwood. Textures are more or less recrystallized with abundant mutually sutured grain boundaries but with the oligoclase generally retaining its tabular habit. Hornblende is rare, and minor amounts of microcline occur as small interstitial grains or as antiperthite in

TABLE 1. PRINCIPAL COMPONENTS OF THE MORTON GNEISS

| Rock Type | Description |
|---|---|
| **Neosome** | |
| Adamellite-2 | Least-deformed unit consisting of two rock types: micro-adamellite porphyry similar in composition to the Sacred Heart granite, and gneissic fine-grained adamellite similar to the younger aplite dike rocks |
| Agmatic granodiorite | Dark reddish-brown, fine-grained, brecciated granodiorite with light gray quartz-feldspar veins |
| Adamellite-1 | Deformed lobate masses of gray to pink, fine-grained adamell-ite gneiss. Usually closely associated with the pegmatitic granite gneiss |
| Pegmatitic granite and microcline granite gneiss | Very coarse grained sheared or granulated biotite-oligoclase-microcline pegmatite; the closely associated microcline gneiss may be sheared and recrystallized pegmatite |
| **Paleosome** | |
| Granoblastic granodiori-tic gneiss | Veins of granoblastic gneiss characterized by large idio-blastic hornblende. Hornblende-microcline-quartz-oligoclase gneiss formed by shearing and recrystallization of layered tonalitic gneiss |
| Amphibolite | Granoblastic medium- to coarse-grained; derived from original dike or sill-like masses in the tonalitic gneiss. The pre-cursors were of tholeiitic (plagioclase-hornblende amphibolite) or of basaltic komatiite (hornblende-rich amphibolite) composition |
| Tonalitic and grano-dioritic gneisses | Gray, fine- to coarse-grained, thin- to thick-layered, straight-banded to contorted biotite-quartz-oligoclase gneiss. Principal phase of tonalitic (trondhjemitic) gneiss, minor microcline and hornblende; locally hornblende re-sulted from contamination of the tonalitic gneiss with amphibolite |
| (?) Amphibolite | Raft-like pieces of amphibolite in tonalitic gneiss may be of same age or older than the tonalite |

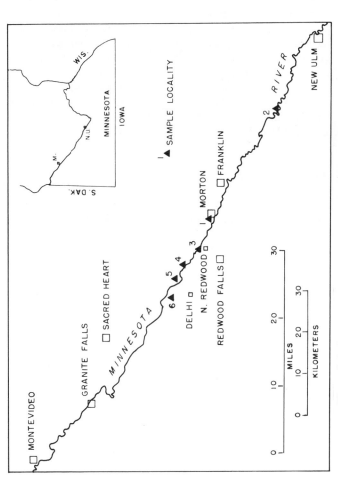

Figure 1. Index map showing sample localities in the Morton area.

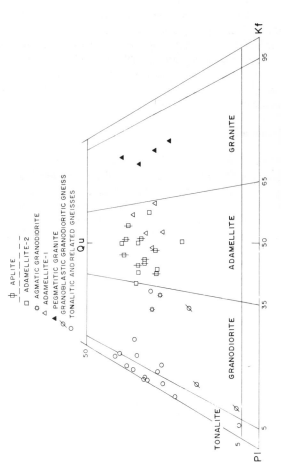

Figure 2. Classification of rocks in Johannsen's (1939) system.

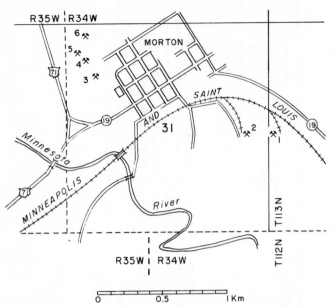

**Figure 3. Map of the Morton vicinity showing quarries.**

block-like intergrowths with the oligoclase. Myrmekite is rare. Sphene is closely associated with biotite and opaque grains and, together with rare rutile, probably is of secondary origin.

The average composition (Table 10), excluding samples 649 and 673, shows the tonalitic gneiss to be quartz-rich (29%), with an average content of biotite of 11% and with less than 2% microcline and hornblende.

## Granodiorite Gneiss

Gradations from tonalitic to granodioritic composition which involve small relative increases in K-feldspar are found both at Morton and at locality 4. These rocks are similar to the tonalitic gneiss and require no special description; however, the granodioritic gneiss at locality 2 southeast of Franklin (Fig. 1) is an unusual rock. In addition to the large amount of microcline, the gneiss contains a small amount of hypersthene (Table 3, sample 659).

The hypersthene is in rounded grains that are difficult to relate to the granoblastic texture of the rock. Both oligoclase and microcline are in large crystals. The K-feldspar is microcline-microperthite with the fine quadrille structure of anorthoclase rather than the plaidlike grating of microcline. Relatively large amounts of albite in string and bead shapes characterize the microperthite. Myrmekitic intergrowths of quartz and oligoclase are abundant and unusually large. In addition, the plagioclase shows antiperthitic intergrowths that range from regular patterns that might be attributed to exsolution of K-feldspar to very irregular blocky intergrowths that suggest a replacement origin.

The rounded hypersthene grains are altered to bastite and a blue-green amphibole. They appear to be xenocrysts corroded by granitic magma and suggest that the original granitic magma was contaminated. The texture of the granodiorite, the microperthite, and the myrmekite suggest that the rock has undergone metamorphism probably during the 2,600–m.y. B.P. event (Goldich

and others, 1970), but it is unlikely that the hypersthene is a metamorphic product of that event. The origin of the granodiorite and its relationship to the granitic phases in the gneiss at Morton are problematical.

The average mode computed for the five samples of granodiorite gneiss (Table 3) is given in Table 10. Exclusion of sample 659 would lower the average microcline content to 5.2% which is still a fivefold increase over the average microcline content of the tonalite gneiss.

## Granoblastic Hornblende Granodioritic Gneiss

Although of minor abundance, the granoblastic granodiorite is a conspicuous rock because of its idioblastic hornblende. It forms veins, 8 cm or more in width, in the straight-banded tonalite gneiss at Morton (Fig. 4). The hornblende megacrysts give the granoblastic rock a speckled appearance; however, there are gneisses of speckled appearance that are unrelated.

The granoblastic granodiorite was formed by shearing and recrystallization of the tonalitic gneiss. Plagioclase grains are rounded and contrast sharply with the tabular oligoclase of the parent rock. Orientation of the twinning shows that the rounded plagioclase grains are variously oriented, whereas microcline, in large grains that enclose both plagioclase and quartz, is in optical continuity. The hornblende megacrysts poikilitically enclose small grains of plagioclase, quartz, and biotite. Shearing and recrystallization upgraded the biotite-quartz-oligoclase banded gneiss to a granoblastic hornblende-microcline-quartz-oligoclase gneiss. Modal analyses are given in Table 4, but comparisons between the original banded and the derived granoblastic gneiss are made in Part 2, utilizing the chemical data. The Rb-Sr system was altered; this directly affected the age determinations, as will be shown in Part 3.

## AMPHIBOLITES

In the Morton area the tonalitic gneisses contain numerous enclaves or fragments of amphibolite. Lund (1956) regarded these to be inclusions or xenoliths derived from older basement rocks. Grant (1972) called them rafts, and in many places the amphibolite occurs in tabular, raftlike masses intercalated with quartz-feldspar gneiss. Some of the amphibolite appears to be broken-up basaltic sills or dikes, but some may have been derived from an older pre-existing terrane and may indeed be inclusions or xenoliths older than the enclosing gneiss. Some of the fine-grained tonalitic gneiss, for example, at localities 3 and 4, may have been derived from flows or pyroclastic material with interlayered basaltic lavas. Thus the amphibolites may have different modes of origin, and because many of the pieces are not raftlike in shape, we use the term "clasts."

### Composition

The modal analyses (Table 5) illustrate two principal types of amphibolite: plagioclase-rich and plagioclase-poor. There is, however, much greater variety in composition than is shown in Table 5. Modes by Lund (1956, p. 1480) of samples from the Morton area show clinopyroxene, 15% to 33%; hornblende, 17%

TABLE 2.  MODAL ANALYSES OF THE OLDER TONALITIC GNEISS SAMPLES

| Sample No. | 649 | 644 | 679D | 739A | 740 | 657 | 673 | 669 | 629B | 339 | 671 |
|---|---|---|---|---|---|---|---|---|---|---|---|
| Locality No. | 1 | 1 | 1 | 3 | 3 | 4 | 5 | 5 | 5 | 6 | 6 |
| Quartz | 26.7 | .31.4 | 31.2 | 23.2 | 27.4 | 28.0 | 4.7 | 21.9 | 27.0 | 34.1 | 39.6 |
| Plagioclase | 52.9 | 50.1 | 53.5 | 58.6 | 54.5 | 59.0 | 78.8 | 67.8 | 64.9 | 48.0 | 53.5 |
| Microcline | tr | 0.7 | 1.0 | 1.6 | 0.6 | 0.4 | 2.7 | 0.5 | 0.5 | 2.8 | 1.3 |
| Biotite | 9.2 | 12.4 | 11.2 | 10.2 | 12.4 | 10.4 | 8.8 | 8.5 | 6.3 | 14.4 | 4.1 |
| Hornblende | 10.6 | 4.5 | 1.6 | 2.2 | 1.6 | 2.2 | 1.8 | -- | 0.1 | -- | -- |
| Opaque | 0.2 | 0.3 | 1.1 | 1.2 | 0.2 | tr | 0.6 | 0.2 | tr | tr | tr |
| Allanite | tr | tr | tr | -- | 0.2 | -- | tr | tr | tr | -- | -- |
| Apatite | 0.1 | 0.2 | 0.2 | 0.2 | 0.5 | tr | 0.7 | 0.1 | 0.2 | tr | 0.1 |
| Sphene | tr | 0.1 | 0.1 | 0.2 | 0.2 | tr | 1.4 | 0.2 | 0.1 | tr | tr |
| Zircon | tr | 0.1 | tr | 0.1 | 0.1 | tr | 0.1 | 0.1 | 0.1 | tr | tr |
| Calcite | tr | tr | -- | 0.1 | tr | -- | -- | tr | -- | tr | -- |
| Chlorite | 0.1 | 0.1 | tr | 1.6 | 1.2 | -- | 0.4 | 0.5 | 0.4 | 0.7 | 1.1 |
| Epidote | 0.1 | 0.1 | 0.1 | 0.4 | 0.6 | tr | tr | 0.1 | tr | tr | tr |
| Rutile | -- | -- | -- | -- | -- | -- | tr | tr | tr | tr | tr |
| Sericite | X | X | X | X | X | X | X | X | X | X | X |
| An content | 24 | 23 | 20 | 25 | 25 | 25 | 23 | 25 | 21 | 18 | 19 |

tr = less than 0.1 percent        X = Included with plagioclase

TABLE· 3.  MODAL ANALYSES OF THE OLDER GRANODIORITIC GNEISS SAMPLES

| Sample No. | 679 | 633 | 659 | 636 | 637 |
|---|---|---|---|---|---|
| Locality No. | 1 | 1 | 2 | 4 | 4 |
| Quartz | 31.9 | 31.1 | 29.0 | 26.3 | 24.8 |
| Plagioclase | 49.2 | 52.2 | 43.0 | 60.2 | 58.0 |
| Microcline | 7.4 | 4.1 | 21.2 | 3.8 | 5.5 |
| Biotite | 8.7 | 9.0 | 4.2 | 8.1 | 8.8 |
| Hornblende | 1.8 | 1.9 | 0.8 | 0.8 | 2.2 |
| Hypersthene | -- | -- | 1.1 | -- | -- |
| Opaque | 0.5 | 0.5 | 0.3 | 0.4 | 0.7 |
| Allanite | tr | tr | -- | -- | -- |
| Apatite | 0.2 | 0.2 | tr | tr | tr |
| Sphene | 0.2 | 0.3 | 0.1 | tr | tr |
| Zircon | tr | 0.1 | tr | tr | tr |
| Calcite | tr | tr | tr | tr | -- |
| Chlorite | 0.1 | 0.2 | 0.2 | 0.2 | -- |
| Epidote | 0.1 | 0.4 | tr | 0.2 | tr |
| Rutile | tr | -- | tr | -- | -- |
| Sericite | X | X | X | X | X |
| An content | 23 | 26 | 27 | 28 | 27 |

tr = less than 0.1 percent;    X = included with plagioclase

to 39%; and plagioclase, 45% to 47%. Lund found a variation in composition with size which was noted also by Nielsen (1976). Clinopyroxene was found only in the larger clasts in which the plagioclase has an An content of 48 or more. The samples in Table 5, on the basis of their composition, all came from relatively small pieces. For further details on compositional variations, see Part 2 and also the paper by Nielsen and Weiblen (this volume).

## Occurrence

The larger clasts (1 to 20 m) are fine-grained (0.5 mm) plagioclase-hornblende amphibolites that have sharp contacts with the enclosing rocks. Fractures are filled with coarse-grained quartz and feldspar. A thin zone of biotite commonly has been developed on the contacts. These larger clasts contain principally plagioclase, hornblende, clinopyroxene, and accessory minerals of basaltic or diabasic rocks of tholeiitic composition.

Small clasts, ranging from 5 to 60 cm in diameter, have larger grain size (1 to 2 mm) and appear to contain larger amounts of hornblende and biotite. This material commonly is found in small pods or lenslike masses, schlieren, screens, or streaks; locally, an appreciable mafic component has been added to the tonalitic gneiss. Some of the smaller clasts are plagioclase-poor amphibolites. These were derived from precursors of basaltic komatiite composition and will be discussed in Part 2.

Brecciated amphibolite healed with coarse-grained quartz-plagioclase veins and interlayered with coarse-grained tonalitic gneiss at locality 5 is shown in Figure 5. Modal analyses of samples 664 and 672 from similar rafts are given in Table 5 and represent the plagioclase-rich amphibolite. A very different type

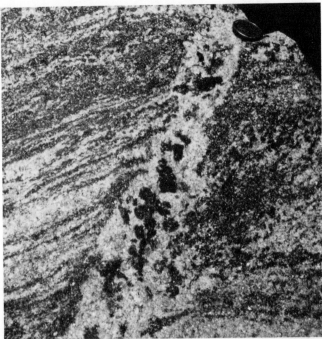

Figure 4. Vein of granoblastic granodioritic gneiss with large idioblastic hornblende in a granular matrix of quartz, microcline, and oligoclase. The granoblastic rock was formed by shearing and recrystallization of the layered tonalitic gneiss. Quarry 4, Morton.

TABLE 4. MODAL ANALYSES OF GRANOBLASTIC HORNBLENDE GRANODIORITIC GNEISS

| Sample No. | 593S | 678S | 667S |
|---|---|---|---|
| Locality No. | 1 | 1 | 5 |
| Quartz | 13.9 | 14.1 | 5.7 |
| Plagioclase | 38.8 | 59.2 | 76.0 |
| Microcline | 16.9 | 5.3 | 5.7 |
| Biotite | 5.4 | 7.6 | 0.4 |
| Hornblende | 24.7 | 12.1 | 11.4 |
| Opaque | 0.2 | 0.8 | tr |
| Allanite | tr | tr | tr |
| Apatite | tr | 0.4 | 0.1 |
| Sphene | 0.1 | 0.2 | 0.1 |
| Zircon | tr | tr | 0.1 |
| Calcite | tr | tr | tr |
| Chlorite | -- | tr | 0.4 |
| Epidote | tr | 0.2 | tr |
| Hematite | -- | tr | tr |
| Sericite | tr | 0.1 | 0.1 |
| An content | 21 | 24 | 21 |

tr = less than 0.1 percent.

of occurrence at locality 5 (Fig. 6) are the thin (20 cm), folded, dikelike masses of plagioclase-poor amphibolites (Table 5, samples 663 and 655). South of Sacred Heart and approximately 12 km northwest of locality 5, plagioclase-poor (basaltic komatiitic) amphibolite is traversed by plagioclase-rich (tholeiitic) amphibolite.

## Relative Ages

The amphibolites are of different ages as is indicated by the modes of occurrence and different compositions (see Part 2). Some of the clasts may indeed by xenoliths in the tonalitic rocks and derived from an older complex as interpreted by Lund (1956). The folded amphibolite and aligned clasts at Morton are interpreted to be fragmented dikes in the tonalitic gneiss. The amphibolites are closely associated with the tonalite, and two periods of deformation are clearly evident. Figure 7, a quarry wall (in quarry 3, Fig. 3), shows a circular section of gray tonalitic gneiss with amphibolite clasts. Surrounding the circular section are light-colored lobate masses of pegmatitic granite. The discordance in structure indicates that the tonalitic gneiss with amphibolite clasts had been deformed before the emplacement of the pegmatitic granite magma. Tonalitic gneiss and intruded pegmatitic granite and adamellite were later involved in deformation approximately 2,600 m.y. ago (Goldich and others, 1970). At localities 4 and 5, pegmatite borders are found on amphibolite clasts, and deformed pegmatite veins occur in the amphibolite.

At Morton, contacts between some of the amphibolite and the

gneiss commonly are sharp. In places the wall rock of the amphibolite is gray layered gneiss with some red granitic layers; this situation suggests that basaltic dikes were intruded after emplacement of pegmatitic granite and adamellite. This interpretation requires a period of deformation after emplacement of the pegmatite and adamellite in the tonalitic gneiss and before the intrusion of the basalt, followed by a third period of deformation at which time the basaltic dikes were broken up. An alternative explanation is that the basalt dikes or sills are older than the pegmatitic and adamellitic phases but were translocated during plastic deformation so as to truncate later structures. It should be noted that in the Franklin area there are diabase dikes, dated at approximately 1,800 m.y. B.P. (Goldich and others, 1970), that were emplaced in the Morton Gneiss. At Morton and localities to the northwest, there are numerous straight and relatively undeformed dikes of aplite and pegmatite, but none of basaltic composition.

## GRANITIC GNEISSES

The neosome of the Morton Gneiss consists of a number of granitic rock types: (1) pegmatite and microcline granite gneiss, (2) fine-grained adamellite (adamellite-1), (3) agmatic granodiorite, and (4) microadamellite porphyry and gneissic leucoadamellite of similar composition, both included in adamellite-2. The order from 1 to 4 is oldest to youngest as determined by field relationships.

TABLE 5.  MODAL ANALYSES OF AMPHIBOLITES

| Sample No. | 594 | 600G | 642 | 664 | 672 | 663 | 665 |
|---|---|---|---|---|---|---|---|
| Locality No. | 1 | 1 | 1 | 5 | 5 | 5 | 5 |
| Plagioclase | 29.3 | 26.2 | 26.5 | 35.0 | 36.1 | 10.7 | 5.8 |
| Hornblende | 65.4 | 62.7 | 72.2 | 63.7 | 62.6 | 88.8 | 92.9 |
| Biotite | 4.2 | X | 1.1 | X | -- | X | X |
| Opaque | X | 3.6 | X | X | X | X | X |
| Apatite | X | X | X | X | X | X | X |
| Calcite | -- | X | -- | -- | X | -- | -- |
| Chlorite | X | 2.5 | X | X | X | -- | X |
| Epidote | X | -- | X | -- | X | X | X |
| Hematite | X | -- | X | X | X | X | X |
| Quartz | -- | 3.0 | -- | -- | -- | -- | -- |
| Sericite | X | X | X | -- | X | X | X |
| Sphene | X | X | X | X | -- | X | X |
| Zircon | X | X | X | -- | X | -- | -- |
| An content | 44 | 32 | 33 | 46 | 44 | 28 | 28 |

X = Usually less than 1 percent.

-- = not detected

## Pegmatite and Microcline Granite

Granitic pegmatite is always in deformed bodies and locally has been granulated and recrystallized to resemble coarse granite. Pegmatite is estimated to constitute at least 20% of the granitic phase. It is composed predominantly of pink to gray perthitic microcline, quartz, and plagioclase with minor amounts of biotite. Locally large crystals, as much as 10 cm in length, are found, and large remnants are abundant throughout the migmatite especially in the Morton vicinity.

In less-deformed areas the pegmatite forms distinct, integral

**Figure 5. Large brecciated amphibolite raft healed with veins of quartz and feldspar. In tonalitic gneiss at locality 5.**

bodies that crosscut the layering in the gray tonalitic gneiss, but more commonly the pegmatite forms irregular pods in the migmatite. These range from 0.3 to 1.0 m in width and are 3 to 4 m in length. Where the pegmatite has been broken up, the fragments are aggregates of microcline and quartz or single large crystals of K-feldspar that have been rounded and apparently displace considerable distances during deformation.

Medium- to coarse-grained microcline granite is closely related to the pegmatite. Generally the color is salmon pink and the grain size is 3 to 4 mm. This granite forms distinct layers concordant with the layering of the gray tonalitic gneiss, but both have been deformed into a highly contorted structure. Some, if not all, of this granite represents granulated pegmatite. Modal compositions are indicated in Table 6.

**Figure 6. Folded dike of hornblende-rich amphibolite in coarse-grained tonalitic gneiss at locality 5.**

**Figure 7. Two periods of deformation. West wall of quarry 3, Morton, showing a circular section of banded tonalitic gneiss with amphibolite clasts surrounded by deformed rock in which lobate masses of granitic gneiss almost completely enclose the tonalite-amphibolite gneiss. The large amphibolite piece is 40 cm in greatest dimension.**

TABLE 6. MODAL ANALYSES OF PEGMATITE AND MICROCLINE GRANITE GNEISS

| Sample No. | 390.L [a] | 674 | 27 [b] | 28 [b] |
|---|---|---|---|---|
| Quartz | 25 | 29.3 | 39 | 34 |
| Plagioclase | 12 | 12.1 | 9 | 14 |
| Microcline | 60 | 56.5 | 50 | 51 |
| Biotite | 2 | 1.5 | X | X |
| Opaque | X [c] | 0.3 | X | X |
| Apatite | X | X | X | X |
| Sphene | X | 0.2 | X | X |
| Zircon | X | X | X | X |
| Chlorite | | 0.2 | | |
| Epidote | | X | X | X |
| Hematite | X | X | | |
| An content | 20 | 14 | 18 | 18 |

a = Estimated mode.

b = Lund (1956, p. 1484).

c = Usually less than 1 percent.

## Adamellite-1

Fine-grained gray to pink leucocratic granite constitutes a major portion of the granitic phase of the migmatite. The texture resembles that of aplite. Pink to grayish-pink microcline, quartz and gray plagioclase (1 to 2 mm) are in roughly equal amounts. A few larger (1 cm) grains of K-feldspar give the rock a porphyritic texture. The adamellite occurs in deformed, irregularly shaped bodies with undulose contacts. There are both discordant and concordant discontinuous bodies, and although deformed, the fine-grained adamellite is in distinct, integral masses that range from 15 to 30 cm in width and can be traced for a few metres.

Modes of adamellite-1 are given in Table 7. The samples from the small quarries northwest of Morton (Fig. 3) are similar; the plagioclase is An$_{20}$ to An$_{22}$. The texture is xenomorphic to granoblastic. Grains have mutually sutured contacts, undulose extinction, and the feldspar shows wedged, discontinuous twinning lamellae. Myrmekite is abundant along oligoclase-microcline contacts. Biotite is altered to chlorite; plagioclase to epidote, sericite, and calcite. The sphene is closely associated with the biotite; magnetite, apatite, and zircon are accessory minerals.

The infolded aplitic adamellite is closely associated with the pegmatite and microcline granite. Locally, masses of the adamellite appear to be younger and crosscut the pegmatite.

## Agmatic Granodiorite

Large tabular masses (6 to 7 m long) of fine-grained, dark reddish-brown granodiorite are exposed in a small quarry (Fig. 3, quarry 3) just west of Morton. The rock is massive with a faint biotite foliation but is broken and veined with pink granitic rock. The brecciaike appearance leads to the name agmatic granodiorite. Modal analyses of two samples of the granodiorite

TABLE 7. MODAL ANALYSES OF ADAMELLITE-1, GRANODIORITE, AND GRANITIC VEIN MATERIAL IN THE AGMATIC GRANODIORITE, MORTON

| Rock Type | Adamellite | | | | | Granodiorite | | Granitic vein |
|---|---|---|---|---|---|---|---|---|
| Sample No. | 600B | 603 | 647 | 651 | 656 | 602 | 646D | 646L |
| Quartz | 28.8 | 36.2 | 32.2 | 26.9 | 29.7 | 25.5 | 27.9 | 28.8 |
| Plagioclase | 24.2 | 24.6 | 31.4 | 32.9 | 35.1 | 43.3 | 46.2 | 45.4 |
| Microcline | 43.0 | 38.2 | 35.7 | 37.9 | 32.9 | 20.8 | 16.8 | 24.4 |
| Biotite | 3.5 | 0.5 | tr | 2.6 | 2.0 | 8.9 | 7.5 | tr |
| Opaques | tr | 0.2 | 0.5 | tr | 0.2 | 0.4 | 0.8 | tr |
| Allanite | tr | -- | -- | tr | tr | -- | -- | -- |
| Apatite | -- | tr | 0.3 | tr | tr | tr | tr | 0.9 |
| Sphene | tr | -- | -- | -- | -- | tr | 0.3 | tr |
| Zircon | tr | tr | tr | -- | -- | tr | tr | tr |
| Calcite | -- | -- | tr | tr | tr | tr | tr | tr |
| Chlorite | 0.6 | tr | tr | 0.5 | tr | 0.7 | 0.3 | 0.2 |
| Epidote | -- | -- | tr | tr | tr | -- | -- | tr |
| Hematite | tr | tr | tr | tr | tr | tr | tr | tr |
| Sericite | tr | tr | tr | tr | tr | tr | tr | tr |
| An content | 22 | 22 | 20 | 22 | 22 | 28 | 28 | 20 |

tr = less than 0.2 percent.

TABLE 8. ADAMELLITE-2, MODAL ANALYSES OF MICROADAMELLITE PORPHYRY FROM MORTON AND GNEISSIC ADAMELLITE FROM LOCALITY 5

| Sample No. | 655 | 680 | 681 | 684 | 631 | 654 |
|---|---|---|---|---|---|---|
| Locality No. | 1 | 1 | 1 | 1 | 5 | 5 |
| Quartz | 20.7 | 31.3 | 33.0 | 28.5 | 37.6 | 33.8 |
| Plagioclase | 36.3 | 26.4 | 39.4 | 40.8 | 28.6 | 39.1 |
| Microcline | 36.6 | 40.8 | 20.7 | 28.0 | 28.4 | 24.9 |
| Biotite | 3.8 | 1.2 | 4.8 | 1.6 | 3.0 | 1.1 |
| Opaques | 0.2 | 0.2 | -- | 0.3 | X | X |
| Allanite | -- | -- | -- | tr | X | X |
| Apatite | 0.3 | tr | tr | -- | X | X |
| Sphene | -- | -- | tr | 0.4 | X | X |
| Zircon | tr | -- | tr | -- | X | X |
| Calcite | tr | tr | -- | tr | -- | -- |
| Chlorite | 1.5 | 0.3 | 0.7 | 0.5 | 1.8 | X |
| Epidote | -- | tr | -- | tr | X | X |
| Hematite | tr | 0.2 | tr | tr | X | X |
| Sericite | tr | tr | tr | X | X | X |
| An content | 20 | 28 | 28 | 26 | 20 | 17 |

X = Less than 1 percent.

tr = Less than 0.2 percent.

TABLE 9. MODAL ANALYSES OF APLITE DIKE ROCKS

| Sample No. | 459 | 648A | 648B | 650 | 652 | 635 | 676 | 587 | 628 | 630 |
|---|---|---|---|---|---|---|---|---|---|---|
| Locality No. | 1 | 1 | 1 | 1 | 1 | 4 | 4 | 5 | 5 | 5 |
| Quartz | 37.0 | 37.4 | 25.1 | 32.7 | 32.4 | 29.1 | 31.5 | 36.8 | 33.1 | 32.9 |
| Plagioclase | 31.4 | 26.7 | 36.6 | 35.6 | 37.6 | 36.0 | 37.0 | 28.5 | 31.1 | 30.2 |
| Microcline | 26.1 | 34.0 | 23.9 | 25.5 | 27.8 | 32.5 | 27.4 | 30.4 | 31.1 | 33.0 |
| Biotite | 3.7 | 0.6 | 8.3 | 5.1 | 1.3 | 1.0 | 1.2 | 1.0 | 1.5 | X |
| Opaque | 0.5 | 0.6 | 3.0 | 0.5 | 0.3 | X | X | X | X | X |
| Allanite | tr | -- | -- | -- | -- | X | X | X | X | X |
| Apatite | 0.4 | -- | 0.4 | tr | -- | X | X | X | X | X |
| Sphene | 0.5 | 0.5 | 0.2 | tr | tr | X | -- | -- | -- | -- |
| Zircon | -- | -- | -- | tr | -- | X | X | X | X | X |
| Calcite | tr | -- | tr | -- | -- | X | X | -- | -- | -- |
| Chlorite | 0.4 | tr | 3.0 | 0.8 | 0.6 | X | 1.4 | 1.7 | 2.0 | 1.9 |
| Epidote | -- | -- | tr | tr | -- | X | X | X | X | X |
| Hematite | tr | tr | tr | tr | tr | X | X | X | -- | X |
| Sericite | X | X | X | X | X | X | X | X | X | X |
| An content | 22 | 18 | 22 | 22 | 22 | 18 | 18 | 18 | 19 | 18 |

X= less than 1 percent.          tr = less than 0.2 percent.

are given in Table 7. The granodiorite is composed of equigranular (1 to 2 mm) microcline, plagioclase, and biotite. Its composition is distinctly different from that of the infolded adamellite-1, notably in the more-calcic plagioclase, the plagioclase/microcline ratio, and the larger content of biotite. A mode of the quartz-feldspar vein (Table 7, sample 646-L) is similar but contains only a trace of biotite.

The agmatic granodiorite in exposures looks like an inclusion; however, it is more likely a dike and probably had considerable original extent since a search of the rock debris piles revealed a large number of pieces. Its age is somewhat speculative, but on the basis of its occurrence, it is younger than the tonalitic gneisses and of the same age or somewhat older than the infolded adamellite-1.

### Adamellite-2

In the quarries at Morton, a distinctive porphyritic, fine- to medium-grained, dark gray to red rock contains abundant phenocrysts of feldspar and quartz (2 to 3 mm) in a granular matrix of quartz, microcline, oligoclase, and biotite (1 to 2 mm). The texture suggests a high-level granitic intrusive, and the modal composition (Table 8) is similar to that of the Sacred Heart Granite (Lund, 1956, p. 1486; Goldich and others, 1961, p. 139). Because of its texture this rock is called microadamellite porphyry to distinguish it from the older adamellite-1. The microadamellite porphyry is not well exposed and does not apear to be an abundant rock type. Where contacts with other rock types could be seen, they are irregular. The porphyry is relatively little deformed and must be a late phase that was involved in the very late stage of deformation. Fractures in the rock contain well-crystallized epidote.

The microadamellite porphyry has not been recognized outside of the vicinity of Morton; however, at locality 5, pink fine-grained gneiss forms small irregular pods (0.6 to 1.5 m) in the gray tonalitic gneiss. The structure in the gneissic leucoadamellite (Table 8, samples 631 and 654) is apparent from the alignment of the biotite. Similar rocks occur at localities 3 and 4. Neither the microadamellite porphyry nor the gneissic adamellite resemble the infolded adamellite (adamellite-1) at Morton either in structure or in composition. Modally the gneissic adamellite of locality 5 resembles the late straight aplite dike rocks (Table 9), but their occurrence appears to be different; hence, they are tentatively grouped with the microadamellite porphyry in adamellite-2.

### APLITE DIKES

Aplite dikes are common and locally abundant in the Morton area, and some of the larger ones were mapped by Lund. There are also pegmatite and pegmatite-aplite composite dikes, but they are not as abundant as the aplite, and the pegmatite was not sampled in the present study. The aplites are straight, relatively undeformed dikes that traverse both the tonalitic and granitic phases; hence they set a limiting time on the deformation of the Morton Gneiss. They range from a few centimetres to metres in width and in places can be traced for many metres. They are vertical or stand at high angles.

The aplites have the typical sugary-grained texture. The xenomorphic to granoblastic quartz and feldspar grains range from 1 to 3 mm. The mineralogy is uniform and consists of quartz, oligoclase, and microcline in nearly equal amounts, with minor amounts of biotite and accessory magnetite, allanite, apatite, sphene, and zircon. Modes of samples from localities 1, 4, and 5 are given in Table 9. There are local variations, for example, a large (1 m) pegmatite-aplite dike at Morton has a middle pegmatite zone (20 cm wide) grading to pink aplite (sample 648A, Table 9) with a marginal gray aplitic phase (sample 648B) that contains an unusually high content of biotite and magnetite. Much of the biotite has been altered to chlorite.

Although the dikes are straight and not deformed, they are offset on minor faults, and in nearly all there is some evidence of secondary structure and foliation that trends off from the strike of the dikes. This foliation is well-developed in the large dike (6 m) at locality 5. The aplite has undergone some shearing, and alteration of the plagioclase to sericite is extensive. Chlorite, epidote, hematite, and calcite also are fairly common as secondary minerals in the aplites.

The similarity in composition of the gneissic adamellite included in adamellite-2 (Table 10) to the aplites should be noted. It appears likely that the rocks are closely related.

## DISCUSSION

The relative time sequence of rock types that compose the Morton Gneiss is given in Table 1. We have accepted by definition (Lund, 1956; Goldich and others, 1970) that the Morton Gneiss is a hybrid unit. The paleosome consists of amphibolite and tonalitic to granodioritic gneisses. The neosome consists of a number of granitic types. In places the amphibolite occurs in masses large enough to be mapped separately by Lund (1950, 1956), but much of the amphibolite is so intimately mixed and structurally integrated that it is a part of the Morton Gneiss. The amphibolites are closely associated with the tonalitic gneisses. Raftlike intercalations may represent original sills or possibly contemporaneous lavas. The folded dikelike masses and the aligned clasts are disrupted dikes in the tonalitic gneisses.

The original basaltic rocks are the best metamorphic indicators and show that the amphibolite grade was attained throughout the Morton area. In none of the rocks from the six samples localities (Fig. 1) was garnet or other high-grade minerals recognized in hand specimens. Nielsen (1976), however, reported small amounts of garnet in the thin sections of amphibolites and orthopyroxene associated with clinopyroxene in thin sections of samples from two of the larger amphibolite enclaves at Morton. A small amount of hypersthene was found in the gneiss at locality 2.

The tonalitic gneisses also exhibit textural, structural, and compositional differences which suggest that they are of different origins and probably of somewhat different ages. At localities 3 and 4 (Fig. 1), the tonalitic gneiss is fine grained and has distinct foliation but is not well layered. At Morton, fine- to medium-grained, straight-banded, thin-layered gneiss is abundant; but at localities 5 and 6, the gneiss is light gray, medium to coarse grained, and thick layered. The fine-grained gneiss may represent surface flows or pyroclastic material, and the coarse-grained varieties may be intrusive. The structure in the tonalitic gneiss was developed in two or more periods of deformation. An early deformation formed an amphibolite–tonalitic gneiss complex. This was followed by a period of shearing and recrystallization that formed the veins of granoblastic granodioritic gneiss before the emplacement of the pegmatite, the oldest phase of the granitic neosome of the Morton Gneiss.

Pegmatite and associated microcline granite gneiss are closely related. Granulation and recrystallization of the pegmatite form-

TABLE 10.  AVERAGE MODES OF QUARTZ-BEARING COMPONENTS OF THE MORTON GNEISS AND YOUNGER APLITES

| Composite No. | 1 | 2 | 3 | 4 | 5 | 6 | 7 | 8 | 9 |
|---|---|---|---|---|---|---|---|---|---|
| No. Samples | (9) | (5) | (3) | (4) | (5) | (2) | (4) | (2) | (10) |
| Quartz | 29.3 | 28.6 | 11.2 | 32 | 30.8 | 26.7 | 28.4 | 35.7 | 32.8 |
| Plagioclase | 56.7 | 52.5 | 58.0 | 12 | 29.6 | 44.8 | 35.7 | 33.9 | 33.1 |
| Microcline | 1.0 | 8.4 | 9.3 | 54 | 37.5 | 18.8 | 31.5 | 26.7 | 29.2 |
| Biotite | 10.0 | 7.8 | 4.5 | 1 | 1.7 | 8.2 | 2.9 | 2.1 | 3.6[a] |
| Hornblende | 1.4 | 1.5 | 16.1 | | | | | | |

1.  Tonalitic gneiss (Table 2, samples 649 and 673 omitted).

2.  Granodioritic gneiss (Table 3).

3.  Granoblastic hornblende granodioritic gneiss (Table 4).

4.  Pegmatite and microcline granite gneiss (Table 6).

5.  Adamellite-1 (Table 7).

6.  Agmatic granodiorite (Table 7).

7.  Adamellite-2, microadamellite porphyry, Morton (Table 8).

8.  Adamellite-2, gneissic adamellite, Loc. 5 (Table 8).

9.  Aplite dike rocks (Table 9).

[a]  Includes chlorite

ed the microcline granite gneiss. Fine-grained leucogranitic rock—adamellite-1—is also closely associated with the pegmatite and microcline granite gneiss. The relative ages of these rocks could not be determined precisely, but the adamellite may be younger than the pegmatite. Lobate masses of pegmatite and microcline granite crosscut the structure of the amphibolite–tonalitic gneiss, but commonly the rock units appear to be concordant as a result of the severe deformation that affected the granitic phases as well as the older amphibolite and tonalitic phases. Pegmatite borders on amphibolite clasts are fairly common at Morton and also at other localities, but nowhere have we found irrefutable evidence of amphibolite being younger than the pegmatitic and adamellite-1 gneisses.

The complex of folded amphibolite and tonalitic gneiss was ruptured and invaded by the pegmatitic and adamellitic magmas, but it is not possible to determine how much of the present-day structure dates back to the time of this event. Granitic material infiltrated cracks and crevices and reacted with the older rocks to locally form hornblende and biotite. The veins of granoblastic granodioritic gneiss were similarly altered.

Large blocks of granodiorite are fine-grained gray to red rock that has been brecciated and infilled with coarse-grained quartz-oligoclase veins to form the agmatic granodiorite. This rock is exposed in the inactive quarry 3 on the northwest side of Morton and was also observed in the active quarry. The large blocks are interpreted to be remnants of disrupted dikes. The massive fine-grained granodiorite fractured but resisted plastic deformation. The age can only be surmised to be essentially that of adamellite-1.

The youngest phase of the Morton Gneiss at Morton is a microadamellite porphyry with abundant phenocrysts of microcline and quartz in a groundmass of quartz-microcline-oligoclase-biotite. This microadamellite porphyry is the least deformed of the granitic phases. Modally, it resembles the Sacred Heart Granite, which in the area south of Sacred Heart intrudes gneiss mapped by Lund (1956) as the Morton Gneiss. The Sacred Heart Granite obviously cannot be included as a component of the Morton Gneiss, but it is not likely that the microadamellite porphyry can be separated from the Morton Gneiss at Morton. Gneissic adamellite occurring in small pods in the tonalitic gneiss at localities 3, 4, and 5 appear to be late-stage emplacements and are tentatively grouped with the microadamellite porphyry at Morton as adamellite-2. Whether or not these rocks—adamellite-2—should be considered a component of the Morton Gneiss cannot be resolved at this time, but they appear to be somewhat older than the straight aplite and pegmatite dikes that cut the Morton Gneiss.

The younger straight aplite, aplite-pegmatite, and pegmatite dikes are relatively undeformed. They are locally displaced on minor faults and have a weak but distinct secondary foliation. These features may have been introduced during the uplift of the region. The aplites represent the youngest Archean igneous activity in the area and set a limiting age on the last major deformation that affected the Morton Gneiss.

The present investigation was directed primarily to field and laboratory studies that might have immediate applications to the problems of the radiometric dating of the ancient gneisses and related rocks in the Minnesota River Valley. Detailed metamorphic and structural studies could not be pursued; for a broader view of these aspects of the regional geology, the reader is referred to the review by Grant (1972). Our field studies, however, show that the relationships and the geologic history in the Morton area are much more complex than envisioned by Grant (1972), Goldich and others (1970), and earlier investigators. The origin of migmatitic gneisses or migmatites has been a controversial subject. In the Morton area the principal process involved a number of periods of deformation and magmatic intrusions of granitic composition into pre-existing metamorphic assemblages.

Multiple thermal events resulted in overprinting and retrogressive metamorphism, and we are unable to relate a grade of metamorphism with a time of deformation. The one area with a distinctive mineral assemblage is the locality east of the Minnesota river between localities 5 and 6 that was not included in the present investigation. The metasedimentary rocks in this area have been studied by Grant (1968) and Grant and Weiblen (1971), who defined a number of mineral assemblages in which the principal minerals, in addition to quartz, plagioclase, and biotite are garnet, anthophyllite, cordierite, sillimanite, and microcline. These minerals characterize the high-temperature, low-pressure Abukuma-type of metamorphism (Miyashiro, 1958; Shido, 1958). The development of the mineral assemblages in the metasedimentary rocks, however, may have been influenced by the water content. Goldich and others (1970) suggested that the high-grade metamorphism was related to the 2,600-m.y. B.P. event, but we now feel that this will prove to be incorrect (see Part 3).

The diversity of closely intermingled rock types, all of which were subjected to one or more thermal events ranging from low to high amphibolite grade, accounts in large part for the difficulties that have been encountered in determining the original age of the various components of the Morton Gneiss. The tonalitic gneisses are contaminated by (1) the incorporation of mafic material from the amphibolites, which probably are not of a single age, and (2) infiltration of granitic material from the younger pegmatite, adamellite-1, and adamellite-2. There has been movement and redistribution of radiogenic $^{87}Sr$ (Goldich and others, 1970), and mixed-rock samples (Goldich and Hedge, 1974) further complicate the radiometric dating.

Much of the deformation in the Morton area involved cataclasis and recrystallization, which occurred at different times. Even the late aplite dikes show effects of shearing. Epidote is a ubiquitous minerals in all the rock types. Veins and well-crystallized encrustations of epidote on fracture surfaces occur in all the rock types, including adamellite-2. This epidote may have been formed by hydrothermal activity associated with the regional low-temperature event approximately 1,800 m.y. ago.

## ACKNOWLEDGMENTS

We are grateful for financial support from the National Science Foundation (Grants GA-40226 and 40226a#1) and from the Graduate School Fund of Northern Illinois University (Grants 41-12122 and 41-23302). We are indebted to the Minnesota Geological Survey for logistical support, to the Cold Spring Granite Company, Cold Spring, Minnesota, and to property owners who graciously permitted access and collection of samples. We thank G. R. Himmelberg, University of Missouri, and C. E. Hedge and Z. E. Peterman, U.S. Geological Survey, for their constructive criticisms of the manuscript.

## REFERENCES CITED

Ankenbauer, G. A., Jr., 1975, Early Precambrian tonalites, Morton area, Minnesota [M.Sc. thesis]: De Kalb, Northern Illinois University, 41 p.

Goldich, S. S., and Hedge, C. E., 1974, 3,800-Myr granitic gneiss in south-western Minnesota: Nature, v. 252, p. 467–468.

Goldich, S. S., and others, 1961, The Precambrian geology and geochronology of Minnesota: Minnesota Geological Survey Bulletin 41, 193 p.

Goldich, S. S., Hedge, C. E., and Stern, T. W., 1970, Age of the Morton and Montevideo gneisses and related rocks, southwestern Minnesota: Geological Society of America Bulletin, v. 81, p. 3671–3695.

Grant, J. A., 1968, Partial melting of common rocks as a possible source of cordierite-anthophyllite–bearing assemblages: American Journal of Science, v. 266, p. 908–931.

———1972, The Precambrian geology of the Minnesota River Valley, in Sims, P. K., and Morey, G. B., eds., Geology of Minnesota: A centennial volume: Minneapolis, Minnesota Geological Survey, p. 177–196.

Grant, J. A., and Weiblen, P. W., 1971, Retrograde zoning in garnet near the second sillimanite isograd: American Journal of Science, v. 270, p. 281–296.

Himmelberg, G. R., 1965, Precambrian geology of the Granite Falls-Montevideo area, Minnesota [Ph.D. thesis]: Minneapolis, University of Minnesota, 101 p.

———1968, Geology of Precambrian rocks, Granite Falls-Montevideo area, southwestern Minnesota: Minnesota Geological Survey Special Publication Series SP-5, 33 p.

Johannsen, A., 1939, Petrography, Volume 1 (second editon): Chicago, University of Chicago Press, 318 p.

Levy, T. M., 1975, An age sequence in ancient rocks near Delhi, Minnesota [M.Sc. thesis]: De Kalb, Northern Illinois University, 40 p.

Lund, E. H., 1950, Igneous and metamorphic rocks of the Minnesota River Valley [Ph.D. thesis]: Minneapolis, University of Minnesota, 88 p.

——1956, Igneous and metamorphic rocks of the Minnesota River Valley: Geological Society of America Bulletin, v. 67, p. 1475–1490.

Miyashiro, A., 1958, Regional metamorphism of the Gosaisyo-Takanuki district in the central Abukuma plateau: Tokyo University Faculty of Science, Journal, v. 11, pt. 2, p. 217–272.

Nielsen, B. V., 1976, The mafic enclaves of the Morton Gneiss, Morton, Minnesota [M.Sc. thesis]: Minneapolis, University of Minnesota, 121 p.

Nielsen, B. V., and Weiblen, P. W., 1980, Mineral and rock compositions of mafic enclaves in the Morton Gneiss: Geological Society of America Special Paper 182 (this volume).

Shido, F., 1958, Plutonic and metamorphic rocks of the Nakoso and Iritono districts of the central Abukuma plateau: Tokyo University Faculty of Science, Journal, v. 11, pt. 2, p. 131–217.

Suda, R. U., 1975, The Morton Gneiss, Morton, Minnesota [M.Sc. thesis]: De Kalb, Northern Illinois Universtiy, 36 p.

Winchell, N. H., 1884, Historical sketch: Minnesota Geological and Natural History Survey, v. 1, p. 1–110.

Wooden, J. L., Grant, J. A., and Nyquist, L. E., 1977, A metasedimentary-amphibolitic-trondhjemitic complex within the Morton Gneiss, Minnesota River Valley, Minnesota: Geological Society of America Abstracts with Programs, v. 9, p. 1234.

MANUSCRIPT RECEIVED BY THE SOCIETY APRIL 25, 1979
REVISED MANUSCRIPT RECEIVED NOVEMBER 26, 1979
MANUSCRIPT ACCEPTED JANUARY 11, 1980

Geological Society of America
Special Paper 182
1980

# Origin of the Morton Gneiss, southwestern Minnesota: Part 2. Geochemistry

**J. L. WOODEN**
*Lockheed Engineering and Management Services Company, Inc., NASA–Johnson Space Center, Houston, Texas 77058*

**S. S. GOLDICH***
*Northern Illinois University, De Kalb, Illinois 60115*

**N. H. SUHR**
*Pennsylvania State University, University Park, Pennsylvania 16802*

## ABSTRACT

The paleosome of the hybrid Morton Gneiss consists of tonalitic and granodioritic gneisses with enclaves or clasts of amphibolite. The neosome contains granitic gneisses ranging from granodiorite to K-rich granite. The major rock groups are differentiated in plots of the normative orthoclase:plagioclase:quartz ratios and of $K_2O$ versus $Na_2O$. These groups are further subdivided on the basis of major, minor, and trace elements in discrimination diagrams in which both concentrations and ratios of elements are considered. All the principal rock types were originally of igneous origin.

The tonalite gneisses are subdivided into Fe-rich and Fe-poor varieties; both are biotite-quartz-oligoclase assemblages with minor microcline and hornblende and are classed as trondhjemitic gneiss. With increase in microcline, the tonalite gneiss grades to granodiorite gneiss that cannot be differentiated in the field but is a distinct petrographic and chemical type. Locally, shearing and recrystallization of layered tonalite gneiss formed veins of granoblastic granodioritic gneiss. Relatively large amounts of Rb were lost during the shearing and recrystallization.

The basic chemical patterns of the amphibolites indicate that they were derived from basaltic precursors of two different types. A hornblende-plagioclase amphibolite is tholeiitic, and a hornblende-rich, low-alumina variety is komatiitic in composition. Both are Fe-rich, and within each group there are variations, which are assigned to magmatic processes, but also some that are related to secondary metasomatic processes. Relatively large amounts of K and Rb (and, in some samples, also Ba and Sr) were introduced. The amphibolite also must be regarded as the source

of some contamination by physical mixing during deformation of the closely associated tonalite gneiss.

Microcline pegmatitic granite, granodiorite, and adamellite are recognized in the granitic gneiss phase of the Morton Gneiss. Two groups of adamellite are distinguished: an older gneiss called adamellite-1 and a younger gneiss called adamellite-2. Adamellite-2 is further subdivided into microadamellite porphyry that occurs at Morton and a gneissic fine-grained adamellite at localities northwest of Morton. The microadamellite porphyry at Morton chemically resembles a phase of the Sacred Heart Granite, but the gneissic adamellite to the northwest resembles in composition some of the aplite dikes that traverse the structure of the Morton Gneiss. The aplites, however, show the effects of strong fractionation, and the average aplite composition differs from that of adamellite-2 in the relative enrichment in Rb and impoverishment in Sr and Ba.

## INTRODUCTION

As outlined in Part 1 (Goldich and others, 1980b, this volume), the Morton Gneiss is a complex migmatitic gneiss composed of a number of rock types. The paleosome consists of tonalitic and granodioritic gneisses and amphibolites; the neosome includes granitic types ranging from granodiorite to K-rich pegmatite and granite. Relatively undeformed dikes of aplite and pegmatite that cut the gneiss and date the last major deformation are included in this study.

Major, minor, and trace elements have been determined on samples of the principal rock types composing the Morton Gneiss and of the late aplite dikes from six localities (Fig. 1). The results, together with some averaged compositions, are given in Tables 1 through 13. Figure 2 is a chemical classification in which the ratios of normative orthoclase:plagioclase:quartz are plotted.

---

*Present address: Department of Geology, Colorado School of Mines, Golden, Colorado 80401, and U.S. Geological Survey, Box 25046, MS 963, Federal Center, Denver, Colorado 80225.

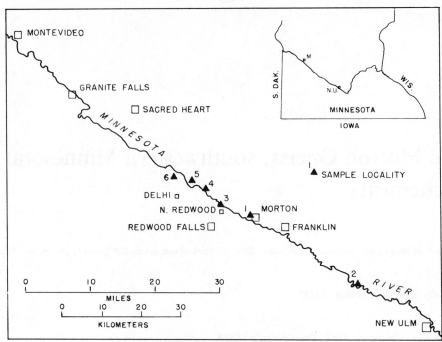

**Figure 1. Index map showing sample localities.**

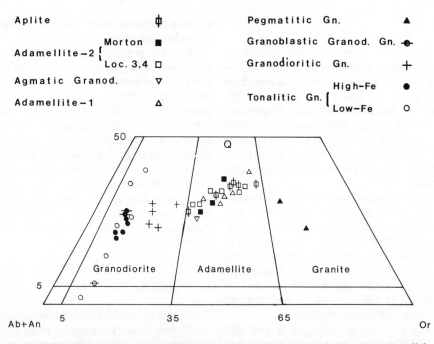

**Figure 2. Classification of rocks, excluding amphibolites, on the basis of normative ratios; divisions based on Johannsen's (1939) classification. Granoblastic granodioritic gneiss: solid slashed circles = high Fe, open slashed circles = low Fe.**

Because K₂O is calculated as orthoclase with no provision for biotite, rocks that are classified modally as tonalite (see Fig. 2 in Part 1) are classed normatively as granodiorites. Figure 2, however, serves to show the distribution and the range of the samples, including the range in the late aplite dikes. The amphibolite samples are not included in Figure 2, but are shown in Figure 3, a plot of K₂O versus Na₂O.

## ANALYTICAL METHODS

The new chemical analyses in this paper were made by combination of methods in the Mineral Constitution Laboratories of Pennsylvania State University (PSU) and in the Department of Geology of Northern Illinois University (NIU). SiO₂, Al₂O₃, TiO₂, total Fe, MnO, CaO, Na₂O, and K₂O were determined

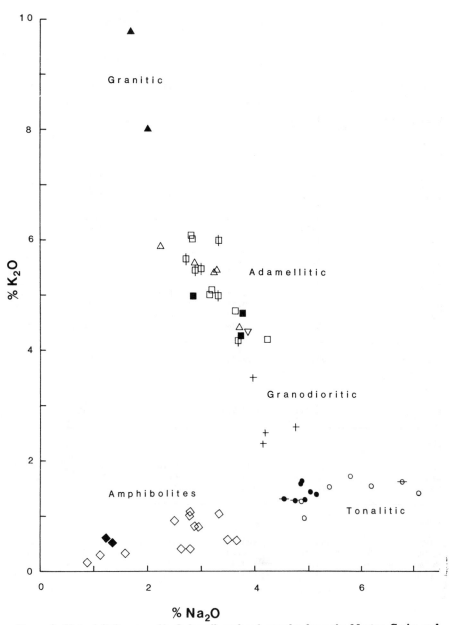

Figure 3. Plot of $K_2O$ versus $Na_2O$ for all analyzed samples from the Morton Gneiss and associated rocks. Amphibolites are shown by diamonds; solid diamonds are samples from Granite Falls. Other symbols as in Figure 2.

(PSU) by atomic absorption spectrophotometry following dissolution of 200-mg samples by the lithium metaborate–nitric acid technique (Suhr and Ingamells, 1966). $P_2O_5$ and F were determined (PSU) on separate samples by colorimetry (Bodkin, 1976) and by specific ion electrode (Bodkin, 1977), respectively, following $LiBO_2$ fusion.

Sample preparation (NIU) involved slabbing of large samples with a diamond saw in water, crushing, and grinding to less than 60 mesh for the FeO analyses [determined (NIU) titrimetrically with $KMnO_4$], and to less than 200 mesh for all other work. $H_2O$ and $CO_2$ were determined (NIU) by a modified Penfield method and by absorption train (3- to 5-g samples), respectively. A few complete analyses were made (NIU) by conventional methods. These can be identified in that $SiO_2$ and $Al_2O_3$ are reported to two decimals. A large number of duplicate determinations were made (NIU) of total Fe with $K_2Cr_2O_7$, of $Na_2O$ and $K_2O$ by flame photometry with Li as internal indicator following acid dissolution, and of $P_2O_5$ by gravimetry.

The trace elements Rb, Sr, and Ba were determined on an additional 21 samples that were not analyzed for the major constituents, and these data, given in Part 3 (Goldich and Wooden, this volume), are used in a number of diagrams. Rb and Sr were determined by isotope dilution and Ba by X-ray fluorescence (NIU). Determinations of Cr, Ni, and other trace elements were made on

TABLE 1.   CHEMICAL ANALYSES OF TONALITE GNEISS
(Iron-rich type)

| Sample No. | 649B | 799 | 644 | 389D | 739A |
|---|---|---|---|---|---|
| Locality No. | 1 | 1 | 1 | 1 | 3 |
| **wt %** | | | | | |
| $SiO_2$ | 65.0 | 66.2 | 68.2 | 69.1 | 65.1 |
| $Al_2O_3$ | 15.9 | 16.0 | 14.9 | 15.6 | 16.0 |
| $TiO_2$ | 0.52 | 0.50 | 0.39 | 0.39 | 0.63 |
| $Fe_2O_3$ | 1.61 | 0.85 | 0.27 | 0.90 | 1.56 |
| FeO | 3.12 | 3.14 | 3.72 | 3.42 | 3.30 |
| MnO | 0.09 | 0.07 | 0.08 | 0.04 | 0.11 |
| MgO | 1.92 | 1.73 | 1.56 | 0.86 | 1.63 |
| CaO | 4.63 | 4.38 | 3.53 | 3.14 | 4.26 |
| $Na_2O$ | 4.93 | 4.88 | 4.86 | 5.17 | 5.04 |
| $K_2O$ | 1.28 | 1.63 | 1.57 | 1.38 | 1.42 |
| $P_2O_5$ | 0.21 | 0.16 | 0.15 | 0.09 | 0.29 |
| $H_2O+$ | 0.44 | 0.25 | 0.38 | 0.27 | 0.56 |
| $H_2O-$ | 0.07 | 0.05 | 0.08 | 0.04 | 0.07 |
| $CO_2$ | 0.03 | 0.02 | 0.08 | n.d. | 0.04 |
| F | 0.07 | 0.08 | 0.07 | n.d. | 0.09 |
| Total* | $99.7_9$ | $99.9_1$ | $99.8_1$ | $100.4_0$ | $100.0_6$ |
| **$g/cm^3$** | | | | | |
| D | 2.731 | 2.725 | 2.725 | 2.712 | 2.735 |
| **ppm** | | | | | |
| Rb | 67.0 | 80.6 | 92.6 | 72.9 | 81.7 |
| Sr | 362 | 379 | 362 | 364 | 380 |
| Ba | 240 | 315 | 335 | 290 | 250 |
| **ratios** | | | | | |
| K/Rb | 159 | 168 | 141 | 156 | 144 |
| Rb/Sr | 0.19 | 0.21 | 0.26 | 0.20 | 0.22 |
| Sr/Ba | 1.5 | 1.2 | 1.1 | 1.3 | 1.5 |
| Ba/Rb | 3.6 | 3.9 | 3.6 | 4.0 | 3.1 |

* *Totals corrected for F.*

a number of amphibolite samples by emission spectroscopy (PSU).

Bulk density (grams per cubic centimetre) was determined on most of the samples (NIU) by determining the volume by water displacement. Replicate determinations give a precision of ±0.007 g/cm³.

## OLDER TONALITIC AND GRANODIORITIC GNEISSES

The older, generally gray gneisses are divisible into two major groups: (1) tonalite and (2) granodiorite. They are intimately associated and generally cannot be distinguished in the field, but as we will show, the granodioritic gneisses form a distinct chemical group. The tonalitic gneisses are further divided into Fe-rich and Fe-poor types.

## Tonalite Gneiss

**Fe-rich Type.** Chemical analyses of five samples of Fe-rich tonalite gneiss (Table 1) show a consistent composition. The uniformity is shown in K₂O (1.28% to 1.63%), Na₂O (4.86% to 5.17%), and CaO (3.14% to 4.63%). Similar small ranges are shown by the trace elements Rb, Sr, and Ba and in the ratios K/Rb, Rb/Sr, Sr/Ba, and Ba/Rb (Table 1). The average bulk density is 2.73 g/cm³ with a limited range from 2.712 to 2.735 g/cm³.

**Fe-poor Type.** In contrast to the Fe-rich tonalite gneiss at Morton and North Redwood, the Fe-poor samples from localities 4, 5, and 6 (Table 2) show considerable variability in major and trace elements. The average total Fe and MgO contents are less than half that of the Fe-rich variety (Table 5), and the bulk densities (Tables 2 and 5) are uniformly lower. Samples 673, 669, and

TABLE 2.   CHEMICAL ANALYSES OF TONALITE GNEISS
(Iron-poor type)

| Sample No. | 657 | 673 | 669 | 629B | 339 | 671 |
|---|---|---|---|---|---|---|
| Locality No. | 4 | 5 | 5 | 5 | 6 | 6 |
| **wt %** | | | | | | |
| $SiO_2$ | 68.7 | 60.6 | 65.8 | 74.2 | 72.0 | 76.5 |
| $Al_2O_3$ | 16.8 | 21.1 | 18.2 | 14.4 | 15.4 | 13.0 |
| $TiO_2$ | 0.36 | 0.68 | 0.54 | 0.24 | 0.21 | 0.20 |
| $Fe_2O_3$ | 0.66 | 0.77 | 0.39 | 0.33 | 0.38 | 0.32 |
| FeO | 1.75 | 1.53 | 1.74 | 1.10 | 1.44 | 1.11 |
| MnO | 0.04 | 0.04 | 0.02 | 0.01 | 0.02 | 0.01 |
| MgO | 0.86 | 0.79 | 0.82 | 0.58 | 0.64 | 0.67 |
| CaO | 3.72 | 5.03 | 3.82 | 3.07 | 2.22 | 1.70 |
| $Na_2O$ | 5.39 | 7.10 | 6.19 | 4.92 | 5.80 | 4.87 |
| $K_2O$ | 1.52 | 1.40 | 1.53 | 0.95 | 1.71 | 1.25 |
| $P_2O_5$ | 0.09 | 0.31 | 0.13 | 0.07 | 0.01 | 0.03 |
| $H_2O+$ | 0.26 | 0.32 | 0.37 | 0.20 | 0.32 | 0.36 |
| $H_2O-$ | 0.08 | 0.06 | 0.02 | 0.07 | 0.05 | 0.05 |
| $CO_2$ | 0.00 | 0.03 | 0.10 | 0.00 | n.d. | 0.04 |
| F | 0.05 | 0.05 | 0.04 | 0.03 | n.d. | 0.04 |
| Total* | $100.2_6$ | $99.7_9$ | $99.6_9$ | $100.1_6$ | 100.2 | $100.1_3$ |
| **$g/cm^3$** | | | | | | |
| D | 2.693 | 2.659 | 2.698 | 2.672 | 2.668 | 2.627 |
| **ppm** | | | | | | |
| Rb | 57.2 | 61.4 | 71.0 | 46.3 | 38.8 | 35.1 |
| Sr | 478 | 1448 | 1061 | 836 | 588 | 494 |
| Ba | 400 | 330 | 585 | 360 | 553 | 380 |
| **ratios** | | | | | | |
| K/Rb | 221 | 189 | 179 | 194 | 366 | 296 |
| Rb/Sr | 0.12 | 0.042 | 0.067 | 0.055 | 0.066 | 0.071 |
| Sr/Ba | 1.2 | 4.4 | 1.8 | 2.3 | 1.1 | 1.3 |
| Ba/Rb | 7.0 | 5.4 | 8.2 | 7.8 | 14 | 11 |

*Totals Corrected for F.*

629B (Table 2), from locality 5, illustrate the variations in composition from a small sampling area. The change from 60.6% to 74.2% $SiO_2$ is marked by an increase in modal quartz from 5% to 27% (Part 1). Similarly, the changes in $Na_2O$ from 7.10% to 4.92% and in CaO from 5.03% to 3.07% are reflected in a decrease in modal plagioclase from 79% to 65%. Comparable modal and chemical variations do not exist in the Fe-rich tonalites.

## Granodiorite Gneiss

Chemical analyses of four samples of biotite-microcline-quartz-oligoclase gneiss (Table 3) are different from the tonalite gneiss (Figs. 2 and 3). The average concentrations of Rb and Sr (Table 5) are similar to the Fe-rich tonalite gneiss, but Ba is greater by a factor of 2. Sample 659 from locality 2 is different petrographically

(see Part 1) from the granodiorite gneiss at localities 1 and 4 and is characterized by its relatively high $K_2O$. This rock is similar in composition to the older granodiorite and adamellite gneisses in the Granite Falls area (Goldich and others, 1980a, this volume).

## Granoblastic Granodioritic Gneiss

Although it is a minor rock type, the granoblastic granodioritic gneiss is of special interest because of the effects of the shearing and recrystallization on the Rb-Sr isotopic system. Chemical analyses of two samples from Morton and one from locality 5 are given in Table 4. The bulk chemical analyses show no apparent major differences; samples 649S and 678S resemble the tonalite gneiss at Morton (Table 1), and sample 667S resembles the gneiss at locality 5 (Table 2).

The overall similarity in chemical composition between

TABLE 3.  CHEMICAL ANALYSES OF GRANODIORITE GNEISS

| Sample No. | 633 | 679 | 659 | 636 |
|---|---|---|---|---|
| Locality No. | 1 | 1 | 2 | 4 |
| wt % | | | | |
| $SiO_2$ | 69.2 | 69.9 | 71.7 | 69.1 |
| $Al_2O_3$ | 15.0 | 14.6 | 14.8 | 15.7 |
| $TiO_2$ | 0.40 | 0.40 | 0.30 | 0.39 |
| $Fe_2O_3$ | 1.01 | 0.96 | 0.13 | 1.11 |
| FeO | 2.27 | 2.22 | 2.12 | 1.99 |
| MnO | 0.06 | 0.05 | 0.02 | 0.05 |
| MgO | 1.33 | 1.26 | 0.74 | 1.00 |
| CaO | 3.35 | 3.35 | 2.49 | 2.98 |
| $Na_2O$ | 4.20 | 4.17 | 3.96 | 4.75 |
| $K_2O$ | 2.51 | 2.30 | 3.50 | 2.60 |
| $P_2O_5$ | 0.16 | 0.16 | 0.10 | 0.09 |
| $H_2O+$ | 0.31 | 0.27 | 0.24 | 0.26 |
| $H_2O-$ | 0.07 | 0.07 | 0.09 | 0.09 |
| $CO_2$ | 0.15 | 0.02 | 0.06 | 0.03 |
| F | 0.06 | 0.06 | 0.06 | 0.06 |
| Total* | $100.0_5$ | $99.7_6$ | $100.2_8$ | $100.1_7$ |
| $g/cm^3$ | | | | |
| D | 2.707 | 2.702 | 2.659 | 2.684 |
| ppm | | | | |
| Rb | 86.0 | 74.3 | 86.4 | 77.4 |
| Sr | 342 | 360 | 283 | 513 |
| Ba | 540 | 600 | 750 | 755 |
| ratios | | | | |
| K/Rb | 242 | 257 | 336 | 279 |
| Rb/Sr | 0.25 | 0.21 | 0.31 | 0.15 |
| Sr/Ba | 0.63 | 0.60 | 0.38 | 0.68 |
| Ba/Rb | 6.3 | 8.1 | 8.7 | 9.8 |

*Totals corrected for F.*

granoblastic granodioritic gneiss and the layered tonalite gneiss supports the conclusion arrived at in the field and in the petrographic study (Part 1) that the granoblastic gneiss was formed by local shearing and recrystallization of layered tonalite gneiss. Because of the difficulty in sampling the gneisses, a precise chemical balance between the granoblastic and the original gneiss has not been attempted; however, a comparison of the two rock types on the basis of the samples from Morton is shown in Figure 4. Mn, Mg, and Fe show relative gains; Ti, P, and Rb show large relative losses.

Prograde metamorphism with the development of hornblende and microcline at the expense of biotite and plagioclase has been noted by Mehnert (1968) and Winkler (1974). The chemical and mineralogic data suggest that recrystallization was favored by a fluid, that metamorphism was not strictly isochemical, and that the following reaction took place: Plagioclase + quartz + biotite → plagioclase + quartz + hornblende + microcline + Rb + Ti + P + fluid. Destruction of biotite, the principal host for Rb and Ti, released these elements for which sites in the new minerals hornblende and microcline were inadequate; hence, the elements were partially removed. K and Ba were more or less retained in the microcline, and Sr, in the plagioclase. The destruction of biotite would contribute to a fluid phase, but it seems likely that a high-temperature fluid or vapor was largely generated elsewhere and moved along the shear zones. Important to our discussion, however, is the loss of Rb and the retention of Sr, as this phenomenon has direct bearing on Rb-Sr radiometric age determination; further reference to this subject is made in Part 3.

TABLE 4. CHEMICAL ANALYSES OF GRANOBLASTIC GRANODIORITIC GNEISS

| Sample No. | 649S | 678S | 667S |
|---|---|---|---|
| Locality No. | 1 | 1 | 5 |
| **wt %** | | | |
| $SiO_2$ | 67.5 | 67.6 | 62.2 |
| $Al_2O_3$ | 14.6 | 15.0 | 19.7 |
| $TiO_2$ | 0.37 | 0.28 | 0.22 |
| $Fe_2O_3$ | 1.25 | 1.08 | 0.74 |
| $FeO$ | 3.57 | 3.36 | 1.91 |
| $MnO$ | 0.10 | 0.10 | 0.05 |
| $MgO$ | 2.00 | 1.78 | 1.00 |
| $CaO$ | 3.91 | 3.80 | 4.96 |
| $Na_2O$ | 4.54 | 4.75 | 6.78 |
| $K_2O$ | 1.30 | 1.25 | 1.62 |
| $P_2O_5$ | 0.12 | 0.13 | 0.17 |
| $H_2O+$ | 0.45 | 0.37 | 0.22 |
| $H_2O-$ | 0.05 | 0.03 | 0.04 |
| $CO_2$ | 0.12 | 0.12 | 0.02 |
| $F$ | 0.06 | 0.06 | 0.04 |
| Total* | $99.9_1$ | $99.6_8$ | $99.6_5$ |
| **$g/cm^3$** | | | |
| D | 2.739 | 2.734 | 2.700 |
| **ppm** | | | |
| Rb | 52.9 | 50.0 | 29.9 |
| Sr | 337 | 370 | 1010 |
| Ba | 265 | 265 | 655 |
| **ratios** | | | |
| K/Rb | 204 | 207 | 450 |
| Rb/Sr | 0.16 | 0.14 | 0.030 |
| Sr/Ba | 1.3 | 1.4 | 1.5 |
| Ba/Rb | 5.0 | 5.3 | 22 |

*Totals corrected for F.*

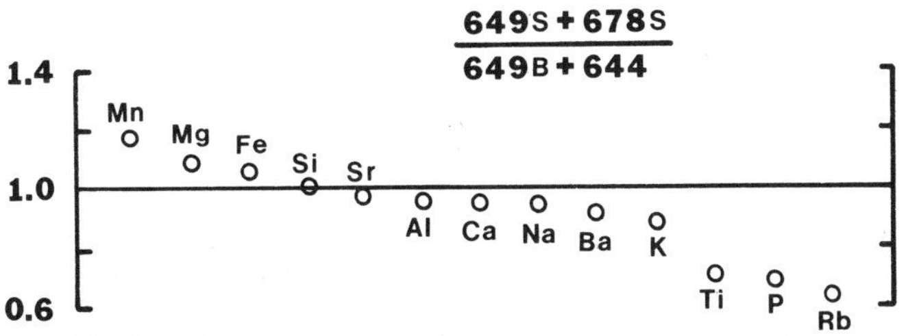

**Figure 4. Comparison of the composition of the granoblastic granodioritic gneiss with (samples 649S and 687S) that of the layered tonalite gneiss at Morton (samples 649B and 644).**

## AMPHIBOLITES

During the course of this study, 24 chemical analyses of amphibolites and 2 of younger diabase dikes were made. Bulk chemical analyses of eight amphibolite samples are reported in Table 6, and the trace-element data are given in Table 7. Trace-element analyses on a few additional samples are included in Part 3. Other analyses are reported by Goldich and others (1980a, this volume), and Nielsen and Weiblen (this volume). For the discussion of the chemical variations within the amphibolites, reference is made to the analyses reported in other papers. A total of 18 chemical analyses of amphibolites is shown in Figure 5 in which the amphibolites of tholeiitic composition are divided on the basis of a plot of $TiO_2$ versus $Na_2O/CaO$ into four subgroups (A through D). The amphibolites of komatiitic composition make up the fifth subgroup (E).

TABLE 5.  AVERAGE COMPOSITION OF TONALITIC AND GRANODIORITIC GNEISSES, MINNESOTA RIVER VALLEY
AND THE NORTHERN LIGHT LAKE GNEISS, ONTARIO

| | Tonalitic Gneiss | | Granodioritic Gneiss | | Northern Light Lake Gneiss ** |
|---|---|---|---|---|---|
| | Iron-rich* | Iron-poor+ | Biotitic Layered$ | Hornblende Granoblastic# | |
| No. Samples | (5) | (6) | (4) | (2) | (3) |
| **wt %** | | | | | |
| $SiO_2$ | 66.7 | 69.8 | 70.0 | 67.6 | 69.1 |
| $Al_2O_3$ | 15.7 | 16.5 | 15.0 | 14.8 | 16.6 |
| $TiO_2$ | 0.49 | 0.37 | 0.37 | 0.33 | 0.26 |
| $Fe_2O_3$ | 1.04 | 0.48 | 0.80 | 1.17 | 0.71 |
| FeO | 3.34 | 1.45 | 2.15 | 3.47 | 1.56 |
| MnO | 0.08 | 0.02 | 0.05 | 0.10 | 0.03 |
| MgO | 1.54 | 0.73 | 1.08 | 1.89 | 0.85 |
| CaO | 3.99 | 3.26 | 3.04 | 3.86 | 3.42 |
| $Na_2O$ | 4.98 | 5.71 | 4.27 | 4.65 | 5.42 |
| $K_2O$ | 1.46 | 1.39 | 2.73 | 1.28 | 1.14 |
| $P_2O_5$ | 0.18 | 0.11 | 0.13 | 0.13 | 0.15 |
| $H_2O+$ | 0.38 | 0.31 | 0.27 | 0.41 | 0.46 |
| $H_2O-$ | 0.06 | 0.06 | 0.08 | 0.04 | 0.03 |
| $CO_2$ | 0.04 | 0.03 | 0.07 | 0.12 | 0.04 |
| F | 0.08 | 0.04 | 0.06 | 0.06 | |
| **$g/cm^3$** | | | | | |
| D | 2.726 | 2.670 | 2.668 | 2.736 | |
| **ppm** | | | | | |
| Rb | 79 | 52 | 81 | 51 | 28 |
| Sr | 369 | 818 | 375 | 354 | 500 |
| Ba | 286 | 451 | 660 | 265 | 300 |
| **ratios** | | | | | |
| K/Rb | 153 | 224 | 280 | 208 | 338 |
| Rb/Sr | 0.21 | 0.06 | 0.22 | 0.14 | 0.056 |
| Sr/Ba | 1.3 | 1.8 | 0.57 | 1.3 | 1.7 |
| Ba/Rb | 3.6 | 8.7 | 8.1 | 5.2 | 10.7 |

*Table 1.   +Table 2.   $Table 3.   #Table 4, Nos. 649S and 678S.
**Goldich and others (1972, p. 159); Hanson and Goldich (1972, p. 181).

## Tholeiitic Amphibolite

Although additional samples may appreciably alter the groups in Figure 5, they are useful for purposes of discussion. The ratio $Na_2O/CaO$ might be used as an index of magmatic differentiation, and the trend within group A of increasing Ti and Fe with increasing values for the ratio $Na_2O/CaO$ then suggests fractionation of common or similar parent magmas. It is more difficult, however, to establish genetic relationships between the groups. Group B, for example, with the highest average content of $SiO_2$ (51.4%), has the lowest content of $Na_2O$ (1.21%). Group D differs from the others in the high concentrations of Sr and Ba (Table 8). The most definitive trait of the amphibolites is the strong enrichment in Fe and Ti with decreasing Mg (Tables 6 and

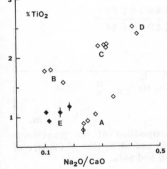

Figure 5. Chemical groups of amphibolites differentiated in a plot of $TiO_2$ versus $Na_2O/CaO$. Samples from Granite Falls denoted by vertical bar; letter designations explained in text.

TABLE 6. CHEMICAL ANALYSES OF AMPHIBOLITES FROM THE MORTON AREA

| Sample No. | 801 | 600G | 594 | 731 | 664 | 672 | 665 | 663 |
|---|---|---|---|---|---|---|---|---|
| Locality No. | 1 | 1 | 1 | 2 | 5 | 5 | 5 | 5 |

wt %

| | | | | | | | | |
|---|---|---|---|---|---|---|---|---|
| $SiO_2$ | 48.2 | 48.7 | 49.1 | 47.6 | 48.70 | 49.1 | 48.1 | 51.69 |
| $Al_2O_3$ | 12.6 | 12.5 | 12.6 | 12.9 | 15.29 | 15.1 | 8.14 | 7.10 |
| $TiO_2$ | 2.21 | 2.19 | 2.20 | 2.21 | 0.95 | 0.89 | 1.10 | 0.95 |
| $Fe_2O_3$ | 5.55 | 5.25 | 4.72 | 5.25 | 2.28 | 2.22 | 2.61 | 2.47 |
| FeO | 11.84 | 12.01 | 12.00 | 13.77 | 9.49 | 9.18 | 11.09 | 10.14 |
| MnO | 0.25 | 0.34 | 0.25 | 0.29 | 0.21 | 0.20 | 0.20 | 0.22 |
| MgO | 5.05 | 4.87 | 4.90 | 4.58 | 7.10 | 7.48 | 12.80 | 12.14 |
| CaO | 9.30 | 8.90 | 8.90 | 9.13 | 11.10 | 10.60 | 11.70 | 11.45 |
| $Na_2O$ | 2.89 | 2.80 | 2.79 | 2.64 | 2.80 | 2.52 | 1.25 | 1.35 |
| $K_2O$ | 0.83 | 1.02 | 1.06 | 0.41 | 0.40 | 0.91 | 0.62 | 0.53 |
| $P_2O_5$ | 0.22 | 0.25 | 0.24 | 0.27 | 0.10 | 0.08 | 0.11 | 0.12 |
| $H_2O+$ | 1.09 | 1.05 | 1.07 | 0.79 | 1.14 | 1.75 | 1.86 | 1.45 |
| $H_2O-$ | 0.08 | 0.05 | 0.08 | 0.06 | 0.05 | 0.08 | 0.07 | 0.06 |
| $CO_2$ | 0.04 | 0.10 | 0.05 | 0.04 | 0.00 | 0.07 | 0.03 | 0.01 |
| F | 0.09 | 0.11 | 0.11 | 0.04 | 0.02 | 0.05 | 0.15 | 0.07 |
| Total* | $100.2_0$ | $100.0_9$ | $100.0_2$ | $99.9_6$ | 99.62 | $100.2_1$ | $99.7_7$ | 99.72 |

$g/cm^3$

| | | | | | | | | |
|---|---|---|---|---|---|---|---|---|
| D | 3.080 | 3.085 | 3.027 | 3.126 | 3.015 | 3.012 | 3.109 | 3.066 |

* Totals corrected for F.

TABLE 7. TRACE ELEMENT DETERMINATIONS ON AMPHIBOLITE FROM THE MORTON AREA

| Sample No. | 810 | 600G | 594 | 731 | 664 | 672 | 665 | 663 |
|---|---|---|---|---|---|---|---|---|
| ppm | | | | | | | | |
| Rb | 15.7 | 13.4 | 21.4 | 13.1 | 6.4 | 36.7 | 7.0 | 8.8 |
| Sr | 121.0 | 137.0 | 122.8 | 109.3 | 111.6 | 163.0 | 170.1 | 210.3 |
| Ba | 70 | 65 | 80 | 60 | 45 | 110 | 35 | 35 |
| Cr | 84 | 65 | 65 | 21 | 200 | 200 | 800 | 940 |
| Ni | 76 | 65 | 50 | 50 | 110 | 150 | 320 | 350 |
| V | 360 | 400 | 380 | 480 | 260 | 240 | 200 | 210 |
| Cu | 140 | 130 | 120 | 200 | 150 | 170 | <10 | <10 |
| Y | 50 | 55 | 60 | 70 | 25 | 20 | 25 | 25 |
| Zr | 170 | 170 | 170 | 230 | 85 | 75 | 120 | 120 |
| ratios | | | | | | | | |
| K/Rb | 440 | 630 | 410 | 260 | 520 | 205 | 735 | 500 |
| Rb/Sr | 0.13 | 0.10 | 0.17 | 0.12 | 0.057 | 0.23 | 0.041 | 0.042 |
| Sr/Ba | 1.7 | 2.1 | 1.5 | 1.8 | 2.5 | 1.5 | 4.9 | 6.0 |
| Ba/Rb | 4.5 | 4.9 | 3.7 | 4.6 | 7.0 | 3.0 | 5.0 | 4.0 |

TABLE 8.   CHEMICAL TYPES OF THE AMPHIBOLITE AND AVERAGE COMPOSITION OF A
LARGE AMPHIBOLITE BLOCK AT MORTON

| Type | A | B | C | D | E | 74-23 Av. |
|------|-----|-----|-----|-----|-----|-----------|
| No. Samples | (4) | (3) | (4) | (2) | (4) | (5) |
| wt % | | | | | | |
| $SiO_2$ | 48.6 | 51.4 | 48.4 | 49.2 | 48.4 | 48.0 |
| $Al_2O_3$ | 14.6 | 14.3 | 12.7 | 13.5 | 8.42 | 14.4 |
| $TiO_2$ | 1.06 | 1.72 | 2.20 | 2.48 | 1.08 | 1.05 |
| $Fe_2O_3$ | 2.82 | 1.15 | 5.19 | 4.22 | 2.57 | 3.24 |
| FeO | 9.71 | 13.45 | 12.41 | 12.43 | 10.43 | 9.91 |
| MnO | 0.22 | 0.22 | 0.28 | 0.23 | 0.22 | 0.24 |
| MgO | 6.95 | 5.55 | 4.85 | 4.31 | 13.33 | 7.16 |
| CaO | 10.51 | 9.19 | 9.06 | 8.57 | 10.99 | 10.47 |
| $Na_2O$ | 2.91 | 1.21 | 2.78 | 3.58 | 1.56 | 2.95 |
| $K_2O$ | 0.80 | 0.26 | 0.83 | 0.57 | 0.67 | 0.84 |
| $P_2O_5$ | 0.11 | 0.25 | 0.25 | 0.23 | 0.10 | 0.10 |
| $H_2O+$ | 1.32 | 1.07 | 1.00 | 0.32 | 1.52 | 1.29 |
| $H_2O-$ | 0.06 | 0.09 | 0.07 | 0.06 | 0.09 | 0.05 |
| $CO_2$ | 0.04 | 0.02 | 0.06 | 0.04 | 0.14 | 0.07 |
| F | 0.07 | 0.04 | 0.09 | 0.06 | 0.08 | 0.07 |
| $(FeO)_t$ | (12.24) | (14.48) | (17.08) | (16.23) | (12.74) | (12.83) |
| $g/cm^3$ D | 3.013 | 3.017 | 3.080 | 3.059 | 3.103 | |
| ppm | | | | | | |
| Rb | 20 | 3.3 | 16 | 6.5 | 12.5 | 25 |
| Sr | 126 | 146 | 123 | 260 | 145 | 119 |
| Ba | 65 | 52 | 70 | 200 | 44 | 62 |
| Cr | 174 | 95 | 59 | ~20 | 1085 | 166 |
| Ni | 121 | 83 | 60 | 55 | 465 | 124 |
| V | 274 | 357 | 405 | 400 | 220 | 274 |
| Cu | 150 | 120 | 148 | 148 | ~100 | 90 |
| Y | 27 | 42 | 59 | 38 | ~20 | 27 |
| Zr | 87 | 163 | 185 | 175 | 105 | 78 |

8). The low concentrations of Sr and Ba, except for group D, also are characteristic of tholeiitic basalt. The variable amounts of $K_2O$ and Rb, however, are not solely the result of primary compositional differences or of igneous processes and are more adequately explained by secondary processes, which may have also altered the concentration of Sr, Ba, and other constituents, as we will discuss below.

### Komatiitic Amphibolite

In contrast with the amphibolites of tholeiitic composition, two samples (665 and 663 in Table 6) differ considerably in their low contents of $Al_2O_3$ and high contents of MgO, Cr, and Ni. Chemically, the samples are similar to the basaltic komatiite of South Africa (Viljoen and Viljoen, 1969). These samples, together with two samples of similar compostion from Granite Falls, are plotted in Figure 5. The hornblende-rich, low-alumina amphibolites occur in narrow (20-cm) dikelike bodies (see Fig. 6 in Part 1). Their occurrence and composition rule out the possibility of their being hornblendites or crystal cumulates

related to the tonalite gneiss. The close association of the low-alumina amphibolites with the tholeiitic types also argues that they were intruded as dikes and sills in the tonalite gneiss.

### YOUNGER GRANITIC PHASES

The neosome of the Morton Gneiss is composed of a number of granitic rock types for which chemical analyses are given in Tables 9, 10, 11. These samples are also plotted in Figures 2 and 3. Six analyses of aplite from the relatively undeformed dikes in the Morton Gneiss are given in Table 12. Averages for the principal rock types are compared in Table 13.

### Potassic Pegmatitic Granite

A new chemical analysis of coarse pegmatite granite (674) is given in Table 9 together with the earlier analysis (Goldich and others, 1970). The pegmatite is characterized by its high K and Ba and low Ca and Na contents (Table 9).

TABLE 9.   CHEMICAL ANALYSES OF GRANITIC PHASES OF MORTON GNEISS, MORTON

| | Pegmatite | | Adamellite-1 | | | | |
|---|---|---|---|---|---|---|---|
| | 390L | 674 | 647 | 656 | 600B | 651 | 603 |
| **wt %** | | | | | | | |
| $SiO_2$ | 71.2 | 74.2 | 73.5 | 73.8 | 74.3 | 75.1 | 77.3 |
| $Al_2O_3$ | 14.5 | 13.1 | 14.5 | 14.5 | 13.3 | 13.8 | 11.9 |
| $TiO_2$ | 0.03 | 0.13 | 0.09 | 0.08 | 0.10 | 0.03 | 0.08 |
| $Fe_2O_3$ | 0.82 | 0.35 | 0.66 | 0.19 | 0.53 | 0.09 | 0.21 |
| FeO | 1.07 | 0.77 | 0.40 | 0.49 | 0.68 | 0.22 | 0.44 |
| MnO | 0.01 | 0.02 | 0.01 | 0.00 | ..01 | 0.00 | 0.01 |
| MgO | 0.10 | 0.26 | 0.09 | 0.15 | 0.22 | 0.08 | 0.16 |
| CaO | 0.42 | 0.43 | 1.92 | 1.54 | 1.34 | 1.37 | 0.93 |
| $Na_2O$ | 1.70 | 2.02 | 3.72 | 3.28 | 2.88 | 3.30 | 2.27 |
| $K_2O$ | 9.78 | 8.01 | 4.43 | 5.43 | 5.60 | 5.44 | 5.88 |
| $P_2O_5$ | | 0.03 | 0.07 | 0.03 | 0.02 | 0.01 | 0.01 |
| $H_2O+$ | 0.13 | 0.19 | 0.10 | 0.13 | 0.22 | 0.07 | 0.16 |
| $H_2O-$ | 0.03 | 0.03 | 0.04 | 0.01 | 0.00 | 0.06 | 0.04 |
| $CO_2$ | | 0.06 | 0.14 | 0.10 | 0.10 | 0.08 | 0.01 |
| F | | 0.02 | 0.02 | 0.02 | 0.01 | 0.01 | 0.01 |
| Total* | $99.7_9$ | $99.6_1$ | $99.6_8$ | $99.7_4$ | $99.3_1$ | $99.6_6$ | $99.4_1$ |
| **$g/cm^3$** | | | | | | | |
| D | 2.582 | 2.611 | 2.680 | 2.629 | 2.612 | 2.618 | 2.619 |
| **ppm** | | | | | | | |
| Rb | 216 | 189.6 | 90.8 | 114.7 | 112.9 | 106.2 | 134.3 |
| Sr | 399 | 312.0 | 377.0 | 384.0 | 350.0 | 363.4 | 310.1 |
| Ba | 3630 | 3280 | 1420 | 1920 | 1900 | 1840 | 1800 |
| **ratios** | | | | | | | |
| K/Rb | 376 | 350 | 405 | 393 | 412 | 426 | 363 |
| Rb/Sr | 0.54 | 0.61 | 0.24 | 0.30 | 0.32 | 0.29 | 0.43 |
| Sr/Ba | 0.11 | 0.10 | 0.27 | 0.20 | 0.18 | 0.20 | 0.17 |
| Ba/Rb | 17 | 17 | 16 | 17 | 17 | 17 | 13 |

*Totals corrected for F.*

## Adamellite-1

Two rocks of adamellite composition of different age are found in the gneiss at Morton. The older and more highly deformed variety is called adamellite-1 and is a sugary-grained rock that was involved with the pegmatite in the folding and deformation that developed the highly contorted structure. Five chemical analyses (Table 9) of adamellite-1 are similar. These adamellites are distinguished from adamellite-2 by their low Fe and Mg contents and their high K/Rb ratios (Table 13 and Fig. 6).

## Agmatic Granodiorite

A chemical analysis of the granodiorite from quarry 3 at Morton is included in Table 10 (sample 602). The sample is from a large block in the Morton Gneiss and is considered to be a piece of a dike fragmented in the last major deformation.

## Adamellite-2

At Morton, small intrusive masses have a distinctive texture and were described (Part 1) as microadamellite porphyry. Sample 655, both in bulk composition and in trace elements, resembles an analyzed sample of the Sacred Heart Granite (Goldich and others, 1970, Table 3, No. 388). The adamellite intrusives in the tonalite gneisses at localities 3, 4, and 5 are fine grained with poorly developed foliation. In outcrops of limited exposure, the gneissic adamellite is easily confused with aplite. Six samples (Table 11) show some variations, but the group is fairly coherent and is characterized by higher Rb, and lower Ba, Sr, and K/Rb than adamellite-1 (Figs. 6 through 8 and Table 13).

## Mixed-Rock Samples

Two chemical analyses of mixed rock are included in Table 10. Sample 76 is a composite from the active quarry at Morton and

TABLE 10.   CHEMICAL ANALYSES OF GRANODIORITE AND MICROADAMELLITE PORPHYRY
AND MIXED-ROCK SAMPLES FROM MORTON

| | Grano-diorite | Microadamellite Porphyry (Adamellite-2) | | | Mixed-Rock Samples | |
|---|---|---|---|---|---|---|
| Sample No. | 602 | 655 | 681 | 680 | 338 | 76 |
| **wt %** | | | | | | |
| $SiO_2$ | 69.0 | 68.9 | 71.5 | 75.7 | 70.8 | 71.54 |
| $Al_2O_3$ | 14.9 | 14.6 | 14.3 | 13.1 | 14.2 | 14.62 |
| $TiO_2$ | 0.57 | 0.46 | 0.28 | 0.11 | 0.25 | 0.26 |
| $Fe_2O_3$ | 0.82 | 1.92 | 0.93 | 0.50 | 0.74 | 0.69 |
| FeO | 2.38 | 1.39 | 1.57 | 0.62 | 2.03 | 1.64 |
| MnO | 0.04 | 0.03 | 0.04 | 0.01 | 0.04 | 0.04 |
| MgO | 0.76 | 0.76 | 0.56 | 0.21 | 0.74 | 0.77 |
| CaO | 2.09 | 1.98 | 1.85 | 1.31 | 2.32 | 2.08 |
| $Na_2O$ | 3.86 | 3.75 | 3.74 | 2.85 | 4.07 | 3.84 |
| $K_2O$ | 4.36 | 4.70 | 4.29 | 4.98 | 3.40 | 3.92 |
| $P_2O_5$ | 0.16 | 0.18 | 0.09 | 0.02 | 0.08 | 0.10 |
| $H_2O+$ | 0.44 | 0.35 | 0.30 | 0.21 | 0.26 | 0.30 |
| $H_2O-$ | 0.05 | 0.04 | 0.06 | 0.03 | 0.05 | 0.26 |
| $CO_2$ | 0.12 | 0.24 | 0.13 | 0.09 | | 0.14 |
| F | 0.04 | 0.06 | 0.03 | 0.02 | | |
| Total* | $99.5_7$ | $99.3_3$ | $99.6_6$ | $99.7_5$ | 99.0 | 99.96 |
| **$g/cm^3$** | | | | | | |
| D | 2.677 | 2.684 | 2.638 | 2.626 | | |
| **ppm** | | | | | | |
| Rb | 129.5 | 158.1 | 110.4 | 108.6 | 112 | 132 |
| Sr | 296.2 | 530.1 | 356.3 | 320.5 | 287 | 282 |
| Ba | 1200 | 2370 | 1300 | 1570 | 785 | 790 |
| **ratios** | | | | | | |
| K/Rb | 280 | 247 | 322 | 380 | 252 | 247 |
| Rb/Sr | 0.44 | 0.30 | 0.31 | 0.34 | 0.39 | 0.47 |
| Sr/Ba | 0.25 | 0.22 | 0.27 | 0.20 | 0.37 | 0.36 |
| Ba/Rb | 9.3 | 15 | 12 | 14 | 7.0 | 6.0 |

*Totals corrected for F.*

was used by Goldich (1938) as an average composition of fresh rock to which residual clay samples, developed on the Morton Gneiss, were compared. Samples 76 and 338 were included by Goldich and others (1970) in an earlier geochronologic investigation which led to the recognition that the tonalitic and granitic phases of the Morton Gneiss are not of the same age. Much of the geochemical work reported here was done to determine and characterize the individual rock types, and sampling of mixed rock was avoided as much as possible. Samples 76 and 338 have a composition similar to the granodiorite (Table 10, sample 602); they are of no further interest to this discussion.

## APLITE DIKE ROCKS

Aplite, composite aplite-pegmatite, and pegmatite dikes were intruded following the last major deformation in the Morton area. The pegmatites are difficult to sample, and only the aplitic rock was anaylzed (Table 12). Samples 648A and 648B are phases of a zoned aplite-pegmatite dike. Dark gray rock, with conspicuous crystals of magnetite, forms the outer or wall phase of aplite and is represented in 648B, whereas 648A is pink aplite that grades to an inner zone of pegmatite. Sample 650 also is the aplitic phase from a composite aplite-pegmatite dike. The average composition of the six aplite samples is given in Table 13.

## DISCUSSION

### Tonalitic and Granodioritic Gneisses

The Fe-rich and Fe-poor tonalites are separable on the basis of both major (Fe, Mg, Na, Al) and trace (Sr, Rb, Ba) elements

TABLE 11.  CHEMICAL ANALYSES OF ADAMELLITE-2, NORTHWEST OF MORTON

| Sample No. | 783 | 772 | 784 | 745 | 654 | 729 |
|---|---|---|---|---|---|---|
| Locality No. | 3 | 3 | 3 | 4 | 5 | 5 |
| **wt %** | | | | | | |
| $SiO_2$ | 71.5 | 74.3 | 74.4 | 72.1 | 72.9 | 74.1 |
| $Al_2O_3$ | 13.8 | 13.4 | 13.4 | 13.7 | 14.2 | 13.7 |
| $TiO_2$ | 0.40 | 0.26 | 0.23 | 0.34 | 0.21 | 0.21 |
| $Fe_2O_3$ | 1.15 | 0.98 | 0.70 | 1.10 | 0.96 | 1.00 |
| FeO | 2.18 | 0.78 | 1.15 | 1.74 | 0.92 | 0.88 |
| MnO | 0.04 | 0.02 | 0.02 | 0.03 | 0.01 | 0.02 |
| MgO | 0.53 | 0.29 | 0.29 | 0.61 | 0.26 | 0.26 |
| CaO | 1.56 | 0.74 | 0.93 | 1.35 | 1.31 | 1.10 |
| $Na_2O$ | 3.20 | 2.82 | 2.82 | 3.20 | 4.24 | 3.65 |
| $K_2O$ | 5.00 | 6.10 | 6.02 | 5.11 | 4.20 | 4.72 |
| $P_2O_5$ | 0.10 | 0.06 | 0.03 | 0.08 | 0.08 | 0.04 |
| $H_2O+$ | 0.42 | 0.24 | 0.26 | 0.45 | 0.32 | 0.34 |
| $H_2O-$ | 0.05 | 0.06 | 0.07 | 0.06 | 0.06 | 0.05 |
| $CO_2$ | 0.12 | 0.06 | 0.09 | 0.16 | 0.07 | 0.10 |
| F | 0.10 | 0.03 | 0.04 | 0.04 | 0.02 | 0.02 |
| Total* | $100.1_1$ | $100.1_3$ | $100.4_3$ | $100.0_5$ | $99.7_5$ | $100.1_8$ |
| **$g/cm^3$** | | | | | | |
| D | 2.663 | 2.621 | 2.628 | 2.660 | 2.637 | 2.640 |
| **ppm** | | | | | | |
| Rb | 169.9 | 189.7 | 181.0 | 178.7 | 100.6 | 187.6 |
| Sr | 228.6 | 156.4 | 167.4 | 346.2 | 430.7 | 229.4 |
| Ba | 1490 | 1450 | 1480 | 1100 | 1750 | 920 |
| **ratios** | | | | | | |
| K/Rb | 244 | 267 | 276 | 237 | 347 | 209 |
| Rb/Sr | 0.74 | 1.21 | 1.08 | 0.52 | 0.23 | 0.82 |
| Sr/Ba | 0.15 | 0.11 | 0.11 | 0.31 | 0.25 | 0.25 |
| Ba/Rb | 8.8 | 7.6 | 8.2 | 6.2 | 17 | 4.9 |

*Totals corrected for F.*

(Figs. 6 through 8, Table 5). The Fe-poor gneisses are found at localities 5 and 6 and are coarse-grained, light-colored rocks. The Fe-rich gneisses are found at localities 1 and 3 and are generally medium grained and darker colored. Samples from locality 4 fall between the two groups and do not fit either group very well with respect to composition, perhaps being best described as transitional. These rocks are texturally distinct because they are massive, well foliated, and finer grained than samples from other locations.

All the tonalitic samples are considered to be metaigneous rocks. Compositional variations within groups, especially the Fe-poor group, probably represent igneous differentiation. The strong compositional differences between groups are assumed to result from original differences in the parental tonalitic magmas. The regular textural and geographic variations of the groups are related to different magmas being emplaced as discrete bodies, as

flows, and as shallow to medium-depth intrusions.

Trondhjemite (Goldschmidt, 1916) is an unusual rock; hence, it is advantageous to adhere to a rigid definition. Johannsen (1932, p. 387) studied the description and modes of Goldschmidt and placed the rock in groups 288P in his classification. The principal minerals are Na-rich plagioclase (oligoclase or andesine) and quartz. K-feldspar is minor or lacking, and biotite is the important mafite. The tonalitic gneisses described above have these features and may properly be classified as trondhjemite.

Tonalitic and trondhjemitic gneisses are abundant in terranes of early Precambrian age as well as in younger orogenic belts. The average composition of the trondhjemitic Northern Light Lake Gneiss (Hanson and others, 1971; Goldich and others, 1972; Hanson and Goldich, 1972) is similar to the averages for the tonalitic gneisses in the Morton area (Table 5). Total Fe as FeO is 2.20% for the Northern Light Lake Gneiss, and a number of major

TABLE 12.  CHEMICAL ANALYSES OF APLITIC DIKE ROCKS

| Sample No. | 648B | 648A | 650 | 459 | 652 | 587 |
|---|---|---|---|---|---|---|
| Locality No. | 1 | 1 | 1 | 1 | 1 | 5 |
| **wt %** | | | | | | |
| $SiO_2$ | 69.9 | 75.4 | 72.4 | 74.24 | 74.7 | 75.0 |
| $Al_2O_3$ | 13.8 | 12.9 | 13.4 | 13.09 | 12.6 | 13.2 |
| $TiO_2$ | 0.45 | 0.05 | 0.31 | 0.14 | 0.14 | 0.18 |
| $Fe_2O_3$ | 2.00 | 0.56 | 1.47 | 1.05 | 0.88 | 0.35 |
| $FeO$ | 2.57 | 0.19 | 1.39 | 0.89 | 0.87 | 0.95 |
| $MnO$ | 0.05 | 0.01 | 0.03 | 0.01 | 0.01 | 0.02 |
| $MgO$ | 0.60 | 0.02 | 0.44 | 0.25 | 0.24 | 0.27 |
| $CaO$ | 1.73 | 0.70 | 1.28 | 1.07 | 1.18 | 0.80 |
| $Na_2O$ | 3.71 | 3.33 | 3.29 | 2.73 | 2.90 | 3.00 |
| $K_2O$ | 4.19 | 5.99 | 4.98 | 5.66 | 5.49 | 5.55 |
| $P_2O_5$ | 0.11 | 0.00 | 0.05 | 0.03 | 0.03 | 0.04 |
| $H_2O+$ | 0.33 | 0.09 | 0.26 | 0.23 | 0.21 | 0.34 |
| $H_2O-$ | 0.06 | 0.03 | 0.09 | 0.06 | 0.04 | 0.07 |
| $CO_2$ | 0.17 | 0.07 | 0.10 | 0.11 | 0.19 | 0.02 |
| F | 0.05 | 0.01 | 0.04 | | 0.02 | 0.02 |
| Total* | $99.7_0$ | $99.3_5$ | $99.5_1$ | 99.56 | $99.4_9$ | $99.8_0$ |
| **$g/cm^3$** | | | | | | |
| D | 2.681 | 2.623 | 2.645 | 2.654 | 2.625 | 2.598 |
| **ppm** | | | | | | |
| Rb | 181.5 | 291.9 | 199.8 | 150.3 | 135.3 | 229.4 |
| Sr | 140.1 | 35.2 | 132.4 | 192.6 | 199.8 | 97.1 |
| Ba | 490 | 90 | 630 | 810 | 805 | 805 |
| **ratios** | | | | | | |
| K/Rb | 192 | 170 | 207 | 313 | 337 | 201 |
| Rb/Sr | 1.30 | 8.29 | 1.51 | 0.78 | 0.68 | 2.36 |
| Sr/Ba | 0.29 | 0.39 | 0.21 | 0.24 | 0.25 | 0.12 |
| Ba/Rb | 2.7 | 0.31 | 3 2 | 5.4 | 5.9 | 3.5 |

* *Totals corrected for F.*

(MgO, CaO, $Na_2O$), minor (MnO, $P_2O_5$), and trace elements (Sr, Ba) fall between the values for the Fe-rich and Fe-poor varieties. $K_2O$ and Rb are notably lower in the Northern Light Lake Gneiss, and the average K/Rb ratio of 338 is appreciably greater than the ratios for Morton area rocks (Table 5). The average Rb content (29 ppm) is much less than that of the granoblastic granodioritic gneiss (51 ppm, Table 5). Caution is urged to avoid the erroneous conclusion that low Rb contents indicate loss of Rb.

The granodiorite gneisses, exclusive of the granoblastic gneisses, are grouped on the basis of their K and Ba contents. With the exception of samples 633 and 679, the granodioritic gneisses are probably not genetically related to each other. Samples 633 and 679 are from locality 1 and are interlayered with Fe-rich tonalite gneiss. The simplest explanation of these samples is that they represent more-fractionated members of the Fe-rich group found at locations 1 and 3. Sample 636 is closely related to Fe-poor tonalite sample 657; they are both from locality 4. Sample 659 is unusually K-rich compared to other granodiorite samples and comes from locality 2, which has not been sampled in detail. The preferred origin for the majority of the granodiorite gneisses is that they are the result of magmatic differentiation of the two or three tonalitic magma types discussed above; however, some of the samples may have been produced by the physical mixing of tonalitic material with granitic material during intrusion or deformation or by infiltration of the tonalite by K-, Rb-, and Ba-rich fluids associated with intrusive or metamorphic events.

## Amphibolites

The amphibolites from the Morton area and Granite Falls are shown in Figure 9, a diagram used by Naldrett and Arndt (1975) to show the relationships between the komatiites and Fe-rich

TABLE 13.  AVERAGE COMPOSITION OF GRANITIC ROCK TYPES AND OF APLITIC DIKE ROCKS

| Rock Type | Pegmatitic Granite | Adamellite-1 | Agmatic Grano. | Adamellite-2 | | Aplitic Rocks |
| | | | | Morton | Loc. 3-4-5 | |
|---|---|---|---|---|---|---|
| No. Samples | (2) | (5) | (1) | (3) | (6) | (6) |
| **wt %** | | | | | | |
| $SiO_2$ | 72.7 | 74.8 | 69.0 | 72.0 | 73.2 | 73.6 |
| $Al_2O_3$ | 13.8 | 13.6 | 14.9 | 14.0 | 13.7 | 13.2 |
| $TiO_2$ | 0.08 | 0.08 | 0.57 | 0.28 | 0.28 | 0.21 |
| $Fe_2O_3$ | 0.59 | 0.34 | 0.82 | 1.12 | 0.98 | 1.05 |
| $FeO$ | 0.92 | 0.45 | 2.38 | 1.19 | 1.28 | 1.14 |
| $MnO$ | 0.02 | 0.01 | 0.04 | 0.03 | 0.02 | 0.02 |
| $MgO$ | 0.18 | 0.14 | 0.76 | 0.51 | 0.37 | 0.30 |
| $CaO$ | 0.43 | 1.42 | 2.09 | 1.71 | 1.17 | 1.12 |
| $Na_2O$ | 1.86 | 3.09 | 3.86 | 3.45 | 3.32 | 3.16 |
| $K_2O$ | 8.90 | 5.36 | 4.36 | 4.66 | 5.19 | 5.31 |
| $P_2O_5$ | 0.03 | 0.03 | 0.16 | 0.10 | 0.07 | 0.04 |
| $H_2O+$ | 0.16 | 0.14 | 0.44 | 0.29 | 0.34 | 0.24 |
| $H_2O-$ | 0.03 | 0.03 | 0.05 | 0.04 | 0.06 | 0.06 |
| $CO_2$ | 0.06 | 0.09 | 0.12 | 0.14 | 0.10 | 0.11 |
| F | 0.02 | 0.01 | 0.04 | 0.04 | 0.04 | 0.03 |
| **$g/cm^3$** | | | | | | |
| D | 2.597 | 2.632 | 2.677 | 2.649 | 2.642 | 2.638 |
| **ppm** | | | | | | |
| Rb | 203 | 112 | 130 | 126 | 168 | 198 |
| Sr | 356 | 357 | 296 | 402 | 260 | 133 |
| Ba | 3455 | 1770 | 1200 | 1745 | 1365 | 605 |
| **ratios** | | | | | | |
| K/Rb | 364 | 397 | 278 | 307 | 256 | 223 |
| Rb/Sr | 0.57 | 0.31 | 0.44 | 0.31 | 0.65 | 1.49 |
| Sr/Ba | 0.10 | 0.20 | 0.25 | 0.23 | 0.19 | 0.22 |
| Ba/Rb | 17 | 16 | 9.2 | 14 | 8.1 | 3.1 |

tholeiites in the Munro Township area, Ontario. Figure 9 also shows averages for peridotitic and basaltic komatiites from South Africa (Viljoen and Viljoen, 1969) and from Western Australia (Nesbitt and Sun, 1976), and the approximate field for the komatiite of the Abitibi belt, Ontario. The Minnesota River Valley amphibolites form two distinct groups. The low-alumina samples plot near the average Barberton basaltic komatiite. The high-Al samples plot in the area and along a trend associated with tholeiitic basalts.

Green and Schulz (1977) published chemical analyses of basaltic komatiites from the Vermilion district in northeastern Minnesota; in a plot similar to Figure 9, their samples with less than 10% $Al_2O_3$ fall to the right of the line in the field of the amphibolites and the average Barberton basaltic komatiite. McGregor and Mason (1977) showed that metabasaltic enclaves in the Amitsoq gneisses in the Godthåb region, West Greenland, have komatiitic affinities. In a diagram similar to Figure 9, their samples, with one exception, plot to the right of the line. It would appear that the chemical characterization of peridotitic and

basaltic komatiite is a difficult problem, as has been shown by Prinz (1967) and more recently by Pearce and Cann (1973) for basaltic rocks in general.

The extent of chemical change during metamorphism is uncertain. It should be noted, however, that the $CaO/Al_2O_3$ ratios in the four analyzed samples are greater than 1, $TiO_2$ ranges from 0.95% to 1.18%, Fe as FeO ranges from 12.4% to 13.4%, and MgO from 12.1% to 14.5%. These features are considered to be very close to those of the original igneous rocks. K, Na, Rb, Sr, and Ba concentrations in these rocks are complicated by metasomatic processes; we will discuss this below. Since the natural variations in komatiite chemistry are uncertain at this time, the composition of the plagioclase-poor amphibolites of the Minnesota River Valley is best described as resembling Fe-Ti–rich basaltic komatiites. It is perhaps significant that these Fe-Ti–rich komatiiticlike rocks are associated with tholeiitic amphibolites that also show strong Fe-Ti enrichment. There is considerable range in the contents of K (0.14% to 0.88%), Rb (2.2 to 37.4 ppm), Sr (18.6 to 270 ppm), and Ba (25 to 220 ppm) in the am-

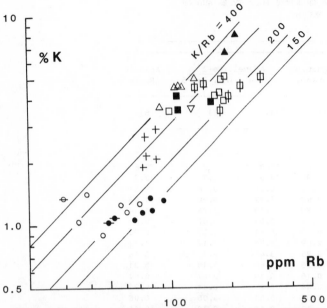

Figure 6. Plot of K versus Rb for all samples except the amphibolites. Major groupings are essentially the same as in Figure 3. Adamellite-1 samples are distinguishable from the aplites, but adamellite-2 samples overlap both groups. Symbols as in Figure 2.

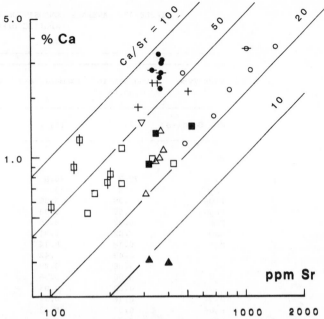

Figure 7. Plot of Ca versus Sr for all samples except the amphibolites. The Fe-rich (solid circles) and Fe-poor (open circles) tonalite gneisses are separated on this diagram. Symbols as in Figure 2.

phibolites. The K/Rb ratio ranges from 1,770 to 210, and the Ba/Rb ratio from 32 to 2. Sr concentrations, in general, are much less variable than Rb, and most samples fall within the range from 100 to 125 ppm. The two samples of amphibolite in group D (Table 8) are unusual with average values for Sr of 260 ppm and for Ba of 200 ppm. Some of the variations are primary differences, but the wide ranges indicate that secondary processes also were involved.

That the variability of K, Rb, Sr, and Ba in part resulted from secondary metasomatic processes is demonstrated by five chemical analyses made of samples from a single large block of amphibolite exposed in quarry 3 at Morton. These analyses are reported by Nielsen and Weiblen (this volume), who made detailed petrographic and chemical studies. The average composition calculated from the five analyses is given in Table 8 (column labeled 74-23 Av.). The variations in $Na_2O$ and $TiO_2$ with K (Fig. 10A) are small; $H_2O$, $CO_2$ and F, however, increase with increasing content of K, as do Sr, Ba, and Rb in a regular manner (Fig. 10B). The percentage increase for Rb, however, is considerably greater than for Sr or Ba. The slope for the K/Rb line is sharply negative, but the slopes for K/Ba and K/Sr are flat or slightly positive (Fig. 10C).

Lund (1956) noted a selvage of biotite on the margins of amphibolite blocks which he interpreted as a reaction product between basaltic xenoliths and the tonalitic and granitic magmas. Nielsen and Weiblen (this volume) found that the abundance of biotite in large amphibolite enclaves at Morton decreases from the margins to the center. Because relatively large amounts of Rb were introduced into the block from which the five 74-23 samples were taken, it is possible to determine a minimum time for this event, as will be shown in Part 3. The chemical data also show that Sr and Ba were mobile elements in the amphibolites.

## Granitic Phases

The K-rich pegmatite or pegmatitic granite is chemically the most distinctive granitic phase of the Morton Gneiss. Field relationships indicate that the pegmatite was emplaced in the tonalitic gneiss–amphibolite complex. The degree and style of deformation of adamellite-1 is similar to that of the K-rich pegmatite so that the relative ages of the two are thought to be close. It is difficult, however, to determine if there is a genetic relationship between the two rock types. There are some chemical similarities between the two groups (low mafic contents, high Ba, high K/Rb ratio, approximately equal Sr); however, the chemistry of the two groups is not consistent with the K-rich pegmatite being a simple fractionation product of adamellite-1. Although some of the compositional differences are consistent with fractionation of feldspar (Ca and Na are lower and K and Rb are higher in the K-rich pegmatite), others are not (the K-rich pegmatite has higher Fe, Mg, Ba, about the same Sr, and lower Si contents than adamellite-1). These two phases do represent the oldest K-rich phases of Morton Gneiss, and they are distinctly younger than the Fe-rich tonalitic and related granodioritic rocks with which they are found at Morton.

Adamellite-2 samples represent a rather heterogeneous group. There are general chemical differences between the microadamellite porphyry samples and the rest of adamellite-2 as Tables 10 and 11 and Figures 6 through 8 show; however, there are also compositional overlaps between the groups. Field, chemical, and isotopic data are compatible with the interpretation that adamellite-2 samples represent a series of small intrusive bodies emplaced in the Morton Gneiss during the later stages of the last major deformation to affect the area. The fact that these samples are from several distinct bodies explains the composi-

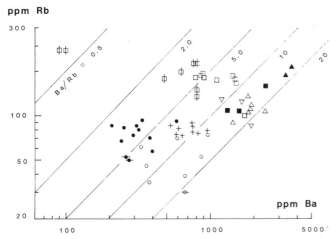

**Figure 8. Plot of Rb versus Ba for all samples except the amphibolites. More samples are shown in this diagram than in Figures 6 and 7 because the major elements were not determined on all samples. Fe-poor and Fe-rich tonalite gneisses are separated, as are the granodiorite gneisses. Note the similarity in the Ba/Rb ratios for adamellite-1, pegmatite, agmatic granodiorite, and adamellite-2 from Morton (loc. 1) compared to the adamellite-2 samples (locs. 3 and 4). Symbols as in Figure 2.**

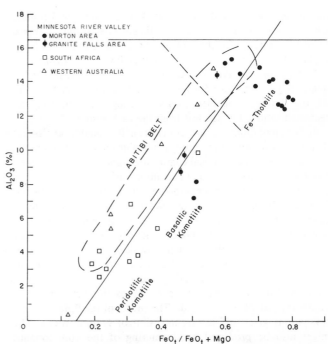

**Figure 9. Basaltic komatiite and tholeiite fields on a plot of $Al_2O_3$ versus $FeO_t/(FeO_t + MgO)$ (after Naldrett and Arndt, 1975). Amphibolites from the Minnesota River Valley fall into two separate fields. One corresponds to an Fe-enrichment trend for normal tholeiites; the other to some South African basaltic komatiites. Open symbols are average compositons. The field of komatiite analyses from the Abitibi belt is encircled.**

tional heterogeneity and the geographic subgroups. The overall compositional similarity of the group and of some samples in adamellite-2 to the Sacred Heart Granite may result from all these intrusions being related to the same magmatic event at approximately 2,600 m.y. B.P. (see Part 3). As illustrated in Figures 6 through 8, the general chemical features of this group might be explained by feldspar fractionation where Sr, Ba, and Ca are depleted and Rb and K are enriched. Relationships among the various intrusions certainly may be more complicated than simple fractionation, and compositional variation as a result of intrusion of discrete partial melts over a range of time should also be considered.

The trace-element plots (Figs. 6 though 8) were used as discrimination diagrams (in the manner of Pearce and Cann, 1973), in conjunction with the bulk chemical and petrographic data, so that the various granitic phases could be reasonably well characterized for geochronologic purposes. In other words, the geochemical data, particularly the trace-element data, supply strong evidence that the assignment of a sample to a particular group is correct and that isochrons based on samples from a single group will at least not reflect mixing lines between two or more groups with possibly different ages (see Part 3). The various major- and trace-element plots used in this paper also show a clear distinction between the tonalitic-granodioritic gneiss complex and all the younger granitic phases. This was helpful in characterizing the oldest phase of the Morton Gneiss and in selecting samples for the isotopic studies.

## Aplite Dike Rocks

The bulk chemistry of the aplite dike rocks is similar to that of adamellite-2 (Table 13). On the average, however, the aplite samples are relatively enriched in Rb and impoverished in Sr and Ba. Variations within the group suggest that the differences

reflect fractionation of a parent or similar parental magmas. Samples 648A and 648B, two phases from the same dike, differ appreciably in trace-element concentrations. Sample 587 is still different in its relatively high Rb and Ba but low Sr contents. Differences in the aplite samples are shown in the patterns developed in Figures 6, 7, and 8. In spite of internal differences, the similarity of the aplite samples to the gneissic adamellite-2 from localities 3, 4, and 5 is apparent, and the two groups may be genetically related. Adamellite-2 is considered to be late kinematic, that is, emplaced in the late stages of the major deformation 2,600 m.y. ago.

## CONCLUSIONS

1. On the basis of texture, structure, mineralogy, and chemistry, the older tonalite and granodiorite gneisses in the Morton area are considered to represent original igneous rocks. Fe-rich and Fe-poor tonalitic varieties are geographically distinct, and differences in texture and composition suggest original differences in mode of origin and emplacement. The coarse-grained, thick-layered gneisses at localities 5 and 6 are regarded to be intrusives. The fine-grained, weakly foliated gneiss at locality 4 and the fine- to medium-grained thin-layered varieties at Morton may represent surface materials or near-surface intrusions. Additional work may alter in some details the present classification of the tonalitic and granodioritic gneisses.

2. The granodiorite gneiss at Morton is closely related to the Fe-

rich tonalite gneiss and, at locality 4, to a transitional tonalitic gneiss sample (657). The granodioritic gneiss at locality 2 appears to be different, but for the present we are unable to fully define or explain the differences in texture, mineralogy, and chemistry.

3. A minor rock type is a granoblastic hornblende-microcline-quartz-oligoclase gneiss formed by shearing and recrystallization of the layered tonalitic gneiss. During the shearing and recrystallization, Rb was lost from the rock system.

4. The amphibolite enclaves or clasts in the tonalitic gneisses are divisable into two major groups: hornblende-rich, low-alumina amphibolite with a chemical composition similar to an Fe-Ti–rich basaltic komatiite and a more common variety of tholeiitic composition. The tholeiitic amphibolites can be further divided into four tentative groups. Both the tholeiitic and the komatiitic amphiboles are Fe-rich. The amphibolites on the basis of structure, mineralogy, and chemistry are considered to be of igneous origin.

5. Within amphibolites of either tholeiitic or komatiitic composition, large variations of K, Rb, Sr, and Ba are in part secondary and result from metasomatic processes. Biotite selvages on the margins of amphibolite blocks represent reaction between the amphibolites and alkali-rich fluids. The source of the fluids might be emanations given off during intrusion of granitic rocks. They might also be produced by dewatering of the bulk tonalitic-granodioritic-granitic complex during high-grade amphibolite metamorphism. Regardless of the origin of the fluids, it must be remembered that the basaltic rocks would be alkali sinks compared to any other component of the Morton Gneiss. The data indicate that significant amounts of K and Rb were added to most amphibolites, but that Sr and Ba concentrations were not affected as much as the K and Rb and are significantly changed only where unusually large amounts of K and Rb were added.

6. The tonalitic rocks and some of the granodioritic and basaltic rocks underwent metamorphism of a regional nature before the emplacement of the K-rich granite and other granitic types composing the neosome (Part 1). Chemically, the pegmatitic granite is distinctive. It is closely associated with adamellite-1, and the rock types may be related in some manner.

7. The agmatic granodiorite is a chemically distinctive rock. It is regarded to have been emplaced at the time of, or shortly after, adamellite-1.

8. The youngest granitic phase of the Morton Gneiss includes two varieties, a microadamellite porphyry at Morton and a gneissic fine-grained adamellite at localities 3, 4, and 5. Both types are included in adamallite-2. The microadamellite porphyry at Morton resembles a phase of the Sacred Heart Granite (sample KA-388, Goldich and other, 1961) that has been dated at approximately 2,600 m.y. B.P. (Goldich and others, 1970). Adamellite-2 was emplaced in the Morton Gneiss during a late stage of the last deformation that involved both the paleosome and the neosome.

9. The straight and relatively undeformed aplite and aplite-pegmatite dikes are postkinematic intrusives. The gneissic adamellite subgroup closely resembles the aplite dike rocks in composition. Both are somewhat similar to a leucocratic phase of the Sacred Heart Granite exposed near Echo, Minnesota (sample KA-24, Goldich and others, 1961).

10. Bulk chemistry and trace elements supplemented with petrographic data and field observations serve to characterize the various rock types that compose the Morton Gneiss and afford a basis of sample selection for geochronologic investigations.

11. Two important aspects of the present study also have direct application to geogchronology: documentation of Rb loss as a result of shearing and recrystallization of tonalitic gneiss and of Rb gain in amphibolites as a result of metasomatic processes. Further reference to these findings is made in Part 3.

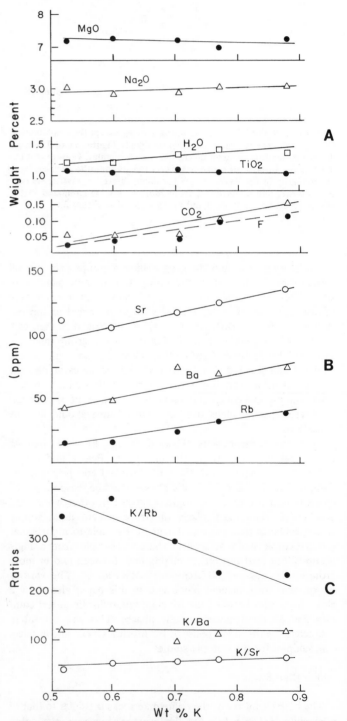

Figure 10. Variations of some major, minor, and trace elements and of ratios with K content in five samples from a single block of amphibolite at Morton. See text for discussion.

## ACKNOWLEDGMENTS

This study was made possible by National Science Foundation Grants GA-40226 and 40266a #1 and supplementary grants 41-12122 and 41-23302 from the Graduate School Fund of Northern Illinois University. We thank R. U. Suda, T. M. Levy, and G. A. Ankenbauer, Jr., for assistance with the sample preparation, J. B. Bodkin and R. C. Wamser for assistance with the analytical work, and L. Romero for help in preparing the manuscript. We are indebted to G. R. Himmelberg, University of Missouri, and Z. E. Peterman, U.S. Geological Survey, for constructive criticism of the manuscript.

## REFERENCES CITED

Bodkin, J. B., 1976, Colorimetric determination of phosphorus in silicates following fusion with lithium metaborate: Analyst, v. 101, p. 44–48.

_____1977, Determination of fluorine in silicates by use of an ion-selective electrode following fusion with lithium metaborate: Analyst, v. 102, p. 409–413.

Goldich, S. S., 1938, A study in rock-weathering: Journal of Geology, v. 46, p. 17–58.

Goldich, S. S., and Wooden, J. L., 1980, Origin of the Morton gneiss, southwestern Minnesota: Part 3. Geochronology: Geological Society of America Special Paper 182 (this volume).

Goldich, S. S., and others, 1961, The Precambrian geology and geochronology of Minnesota: Minnesota Geological Survey Bulletin 41, 193 p.

Goldich, S. S., Hedge, C. E., and Stern, T. W., 1970, Age of the Morton and Montevideo gneisses and related rocks, southwestern Minnesota: Geological Society of America Bulletin, v. 81, p. 3671–3695.

Goldich, S. S., and others, 1961, The Precambrian geology and geochronology of Minnesota: Minnesota Geological Survey Bulletin 41, 193 p.
Studies in mineralogy and Precambrian geology: Geological Society of America Memoir 135, p. 151–177.

_____1980a, Archean rocks of the Granite Falls area, southwestern Minnesota: Geological Society of America Special Paper 182 (this volume).

_____1980b, Origin of the Morton Gneiss, southwestern Minnesota: Part 1. Lithology: Geological Society of America Special Paper 182 (this volume).

Goldschmidt, V. M., 1916, Geologisch-petrographische Studien im Hochgebirge des sudlichen Norwegens. IV Ubersicht der Eruptivgesteine im kaledonischen Gebirge zwichen Stavanger und Trondhjem: Videnskabers Selskabs Skrifter, Kristiania, v. 2, 76 p.

Green, J. C., and Schulz, K. J., 1977, Iron-rich basaltic komatiites in the early Precambrian Vermilion district, Minnesota: Canadian Journal of Earth Sciences, v. 14, p. 2181–2192.

Hanson, G. N., and Goldich, S. S., 1972, Early Precambrian rocks in the Saganaga Lake–Northern Light Lake area, Minnesota-Ontario, Part II, Petrogenesis, in Doe, B. R., and Smith, D. K., eds., Studies in mineralogy and Precambrian geology: Geological Society of America Memoir 135, p. 179–192.

Hanson, G. N., and others, 1971, Age of the early Precambrian rocks of the Saganaga Lake–Northern Light Lake area, Ontario-Minnesota: Canadian Journal of Earth Sciences, v. 8, p. 1110–1124.

Johannsen, A., 1932, A descriptive petrography of the igneous rocks, Volume 2: Chicago, University of Chicago Press, p. 387.

_____1939, A descriptive petrography of the igneous rocks, Volume 1 (second edition): Chicago, University of Chicago Press, 318 p.

Lund, E. H., 1956, Igneous and metamorphic rocks of the Minnesota River Valley: Geological Society of America Bulletin, v. 67, p. 1475–1490.

McGregor, V. R., and Mason, B., 1977, Petrogenesis and geochemistry of metabasaltic and metasedimentary enclaves in the Amitsoq gneisses, West Greenland: American Mineralogist, v. 62, p. 887–904.

Mehnert, K. R., 1968, Migmatites and the origin of granitic rocks: New York, Elsevier, 393 p.

Naldrett, A. J., and Arndt, N. T., 1975, Volcanogenic nickel deposits with some guides for exploration: Transactions of the American Institute of Mining, Metallurgical, and Petroleum Engineers, v. 260, p. 13–15.

Nesbitt, R. W., and Sun, S.-S., 1976, Geochemistry of Archean spinifex-textured peridotites and magnesian and low-magnesian tholeiites: Earth and Planetary Science Letters, v. 31, p. 433–453.

Nielsen, B. V., and Weiblen, P. W., 1980, Mineral and rock compositions of mafic enclaves in the Morton Gneiss: Geological Society of America Special Paper 182 (this volume).

Pearce, J. A., and Cann, J. R., 1973, Tectonic setting of basic volcanic rocks determined using trace element analyses: Earth and Planetary Science Letters, v. 19, p. 290–300.

Prinz, M., 1967, Geochemistry of basaltic rocks: Trace elements, in Hess, H. H., and Poldervaart, A., eds., Basalts: New York, John Wiley & Sons, p. 271–323.

Suhr, N. H., and Ingamells, C. O., 1966, Solution technique for analysis of silicates: Analytical Chemistry, v. 38, p. 730–734.

Viljoen, M. J., and Viljoen, R. P., 1969, The geology and geochemistry of the lower ultramafic unit of the Onverwacht Group and a proposed new class of igneous rocks: Geological Society of South Africa Special Publication 2, p. 55–85.

Winkler, H.G.F., 1974, Petrogenesis of metamorphic rocks: New York, Springer-Verlag, 320 p.

MANUSCRIPT RECEIVED BY THE SOCIETY APRIL 25, 1979
REVISED MANUSCRIPT RECEIVED NOVEMBER 26, 1979
MANUSCRIPT ACCEPTED JANUARY 11, 1980

Geological Society of America
Special Paper 182
1980

# Origin of the Morton Gneiss, southwestern Minnesota: Part 3. Geochronology

**S. S. GOLDICH***
*Northern Illinois University, De Kalb, Illinois 60115*

**J. L. WOODEN**
*Lockheed Engineering and Management Company, Inc. NASA-Johnson Space Center, Houston, Texas 77058*

## ABSTRACT

The Morton Gneiss in the Minnesota River Valley, southwestern Minnesota, consists of an older metamorphic complex of tonalitic gneisses and amphibolites (the paleosome) and younger granitic gneisses (the neosome). Discordant U-Pb ages on zircon from the tonalitic gneisses give a minimum age of 3,300 m.y. with indications of an older age. An age of approximately 3,500 m.y. is indicated by Rb-Sr data for the tonalitic and related granodioritic gneisses; however, subsequent events have disturbed the Rb-Sr as well as the U-Pb systems in the Archean rocks. In addition to the early folding that formed the metamorphic complex, these subsequent events include high-grade metamorphic events 3,050 and 2,600 m.y. ago and a thermal event in Proterozoic time, 1,800 m.y. ago.

The Rb-Sr isotopic system has been variously affected in different rock units as a result of loss and gain of both Rb and Sr. Rb loss from veins of granoblastic granodioritic gneiss formed by shearing and recrystallization of layered tonalitic gneiss results in ages that are too old. Secondary isochrons for the amphibolites and for the tonalitic and granodioritic gneisses in the Morton vicinity reflect Rb gain approximately 3,000 m.y. ago. The geologic relationships suggest that the Rb, K, and other elements were introduced at the time of invasion of the tonalitic gneiss-amphibolite complex by the pegmatitic granite and adamellite-1 magmas. The average $^{207}Pb$-$^{206}Pb$ age of four samples of zircon from the pegmatitic granite gneiss is 3,043 ± 26 m.y., but because of discordance in the U-Pb ages, a somewhat older original age cannot be precluded.

The pegmatitic granite gneiss, adamellite-1 gneiss, and agmatic granodiorite, the principal units of the neosome, all give Rb-Sr ages that are too young because of loss of radiogenic $^{87}Sr$. Loss of radiogenic Sr from some samples of the aplite dikes is clearly shown in model ages that range from 2,060 to 2,465 m.y.; whereas the Rb-Sr isochron age of less-disturbed samples is 2,590 ± 40 m.y. [$R_i$, 0.7036 ± 0.0020 (95% C.L.].). Samples of adamellite-2, which is chemically similar to the Sacred Heart Granite (2,600 m.y.) and to the aplite dike rocks, give an isochron age of 2,555 ± 55 m.y. ($R_i$, 0.7029 ± 0.0013). These rocks probably were emplaced in a late stage of the major deformation 2,600 m.y. ago that affected both the neosome and paleosome and developed the structure of the Morton Gneiss. Veinlets of well-crystallized epidote are common in adamellite-2 and in older rock units. The Rb-poor, Sr-rich epidote has a high $^{87}Sr/^{86}Sr$ ratio, and the veinlets were formed by hydrothermal activity probably associated with the 1,800-m.y. B.P. event.

Redistribution of radiogenic $^{87}Sr$ among the mineral phases and between intimately mixed rock units has occurred in the Morton Gneiss—hence the need for careful sampling and interpretation of the Rb-Sr results. A relatively good isochron [2,640 ± 115 m.y.; $R_i$, 0.7048 ± 0.0014 (95% C.L.] for samples of adamellite-1 and the secondary isochrons for the amphibolites and the tonalitic and granodioritic gneisses (about 3,000 m.y.) are essentially mixing lines.

## INTRODUCTION

Earlier efforts to date the Morton Gneiss encountered many problems. Not all of these have been resolved, but considerable progress has been made in the recognition of the geologic events, in large part, as a result of the lithologic and geochemical studies reported in Parts 1 (Goldich and others, 1980b, this volume) and 2 (Wooden and others, this volume). Relative ages and some absolute ages are now assigned to the principal rock types and to the geologic events. The confidence limits, however, range very considerably, and some major problems remain. Each successive deformation and thermal event has had some effect, particularly as registered in the disturbance of the isotopic systems. Some new

---

*Present address: Department of Geology, Colorado School of Mines, Golden, Colorado 80401, and U.S. Geological Survey, Box 25046, MS 963, Federal Center, Denver, Colorado 80225.

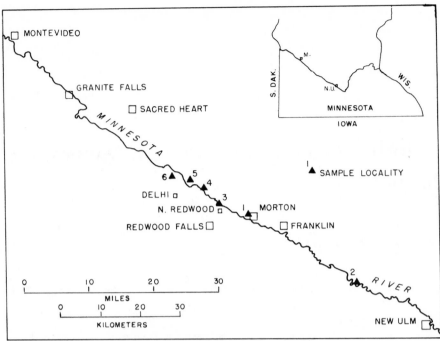

**Figure 1. Index map showing sample localities.**

U-Pb analyses of zircon are reported, and this phase of the geochronologic study is being continued. In addition, Sm-Nd studies with G. J. Wasserburg and M. T. McCulloch of the California Institute of Technology are in progress.

**Previous Work**

The first K-Ar age determinations from the Morton area were reported by Goldich and others (1956). The results published since then are summarized by Goldich and others (1970). Farhat and Wetherill (1974) reported an estimated age of 3.28 ± 0.05 b.y. for the "Morton gneiss formation." Michard-Vitrac and others (1977) reported U-Pb ages on single zircon grains from the Morton Gneiss supplied by G. W. Wetherill and accepted 3,280 m.y. as the probable age. These data and some of Farhat's determinations pertinent to the discussion are considered.

**Sample Localities**

The samples used in the present study include a few from earlier investigations (Goldich and others, 1961, 1970). Most of the new samples are from the vicinity of Morton and from localities northwest of Morton (Fig. 1). Localities 2, southeast of Morton, and 6, the most northwesterly locality, are represented by only a few samples. The old samples are fully identified, and the new ones are briefly described in Appendix 1.

**The Sampling Problem**

The sampling problem is universal and inescapable in any and all forms of analytical chemistry. In the laboratory, the problem commonly is a small-scale one—how large a sample is needed for the determination of elements in different concentrations by dif-

ferent analytical techniques or instruments. In the field, texture and structure pose additional problems, particularly in layered metamorphic rocks. The requisites for a Rb-Sr isochron further complicate the sampling. Not only must the samples be representative, but they must represent facies that are sufficiently different in Rb/Sr to permit construction of an isochron. If one is using the isochron to determine an age, it follows that the facies samples should be genetically related and of the same original age.

**Earlier Sampling Procedures.** In the sampling for the earlier work (Goldich and others, 1970), the Morton Gneiss was considered to be a two-phase system consisting of a gray tonalitic gneiss with inclusions of amphibolite and of a red granitic gneiss. The two phases were thought to be of similar age. Large samples (60 to 80 kg) were collected to avoid the effects of redistribution of radiogenic $^{87}$Sr which had been demonstrated in mineral analyses. Large samples also provided the relatively large zircon concentrates needed for the analytical methods of the time. The large samples, after washing and drying, were crushed and mixed in stages. In an early stage a number of pieces were removed for hand specimens and thin sections. In a later stage, when a considerable reduction in size of pieces had been achieved, a representative fraction was removed by coning and quartering. This sample was further crushed by hand in a hardened steel mortar and ground in a large agate mortor to pass a 200-mesh screen. Portions of the final mixed powder were used for the chemical and isotopic analyses.

The large samples made some problems in limiting the material to a single rock type. For example, a re-examination of the hand specimens reserved from the 80-kg sample of 389D showed small amounts of amphibolitic and granitic contaminants. Our present work also shows that the distinction between paleosome and neosome on the basis of color is unreliable; locally, the pegmatitic granite is light gray.

TABLE 1.  CONSTANTS USED IN CALCULATING ISOTOPIC AGES

| Parent | Daughter | Decay Constant |
|--------|----------|----------------|
| $^{238}U$ | $^{206}Pb$ | $1.551 \times 10^{-10} yr^{-1}$ |
| $^{235}U$ | $^{207}Pb$ | $9.848 \times 10^{-10} yr^{-1}$ |
| $^{232}Th$ | $^{208}Pb$ | $4.947 \times 10^{-11} yr^{-1}$ |
| $^{87}Rb$ | $^{87}Sr$ | $1.42 \times 10^{-11} yr^{-1}$ |

Atomic ratios:

$^{238}U/^{235}U = 137.88$

$^{85}Rb/^{87}Rb = 2.593$

*Constants recommended by the IUGS Subcommission on Geochronology (Steiger and Jäger, 1977).*

**Later Methods.** The radiometric age determinations (Goldich and others, 1970) on samples of the Montevideo gneiss of Lund (1956) showed that the two phases—foliated granodioritic gneiss and more massive leucocratic adamellite gneiss—are of different ages. This information required a radical change in the methods of sampling the gneisses and stimulated the field and laboratory studies that are reported in Parts 1 and 2.

The size of the sample collected was varied with the rock type. Fine-grained rocks of uniform texture, such as the aplites and the adamellites, are much less of a problem than the coarse-grained, thick-layered tonalitic gneiss. Samples of fine-grained rocks generally ranged from 3 to 5 kg, but very large pieces of the coarse-grained layered gneisses were collected. These were cut normal to the foliation with a 60-cm diamond saw in water so that the pieces could be inspected. They were then further cut into slices with a diamond saw, and representative samples were prepared by hand as described above.

## Analytical Methods

Rb-Sr dissolutions and chemistry followed the methods of Goldich and others (1961) with minor modifications, such as spiking for both Rb and Sr after weighting of 0.25- to 0.50-g samples in 50-ml platinum evaporating dishes. Spiking was accomplished by syringes, weighed before and after delivery of spike solutions to the platinum dishes, a technique developed at the National Bureau of Standards.

Mass spectrometric measurements were made on a 15-cm instrument with close adherence to the techniques developed at the National Bureau of Standards (Moore and others, 1973). These methods lead to good precision and permit corrections of both Sr and Rb for instrumental fractionation. We used NBS SRM 988 for our $^{84}Sr$ spike, SRM 984 to calibrate our $^{87}Rb$ spike solution, SRM 607 to monitor Sr and Rb concentration determinations, and SRM 987 to monitor isotopic composition. Our long-term average for $^{87}Sr/^{86}Sr$ for SRM 987 is 0.71051 ± 0.00021, normalized to $^{86}Sr/^{88}Sr$ of 0.1194. From replicate determinations, we estimate that our precision in determining $^{87}Sr/^{86}Sr$ is ± 0.0003 and in determining concentrations of Sr and Rb, ±1% at the 95% confidence level. All our ages and those from the literature are based on the new constants (Table 1). Isochron ages and initial ratios ($R_i$) were calculated using the program of York (1966) and are reported at the 95% confidence level.

The U-Pb analyses of zircon were made in the Department of Earth and Space Sciences, State University of New York at Stony Brook; and technique of Krogh (1973) and a mixed U and Pb spike were used. Samples ranged from 1 to 5 mg; the errors are $^{207}Pb/^{206}Pb$, ~0.1%; $^{206}Pb/^{204}Pb$, ~0.3%); and the concentrations, ~1%.

## Rb-Sr AGE DETERMINATIONS

Earlier work with the Rb-Sr dating technique clearly showed that the isotopic system in the Minnesota River Valley gneisses of early Precambrian age was disturbed as a result of the younger geologic events (Goldich and others, 1970). A second difficulty in the early work results from mixing rock types of different ages (Goldich and Hedge, 1974). Farhat and Wetherill (1975) used a method of "block sampling" in the Granite Falls area, and samples obtained by this method gave a good isochron which they attributed to homogenization of $^{87}Sr$ through metamorphic re-equilibration. The hypothesis of large-scale homogenization was also advanced for the Morton area (Farhat, 1975).

Metamorphic re-equilibration on a small scale (as, for example, among the mineral phases in the Morton Gneiss) was demonstrated by Goldich and others (1970), and they considered the possibility "that the Morton Gneiss, approximately 3,300 m.y. old, was metamorphosed 2,650 m.y. ago with redistribution of radiogenic Sr[87]" (Goldich and others, 1970, p. 3685, Figs. 9

TABLE 2.  Rb-Sr ANALYTICAL DATA FOR TONALITIC AND RELATED GNEISSES

| Sample No. | Rb (ppm) | Sr (ppm) | $^{87}Rb/^{87}Sr$ (atomic) | $^{87}Sr/^{86}Sr$ |
|---|---|---|---|---|
| **Locality 1. Morton Vicinity** | | | | |
| 389D | 72.9 | 364 | 0.580 | 0.7280 |
| 633 | 86.0 | 342 | 0.730 | 0.7360 |
| 633B | 81.8 | 345 | 0.686 | 0.7332 |
| 644 | 92.6 | 362 | 0.742 | 0.7358 |
| 649S | 52.9 | 337 | 0.456 | 0.7241 |
| 649B | 67.0 | 362 | 0.537 | 0.7279 |
| 678S | 50.0 | 370 | 0.392 | 0.7222 |
| 679 | 74.3 | 360 | 0.598 | 0.7308 |
| 679D | 85.5 | 512 | 0.484 | 0.7258 |
| 682 | 69.5 | 407 | 0.495 | 0.7254 |
| 713 | 73.4 | 403 | 0.531 | 0.7265 |
| 799 | 80.6 | 379 | 0.617 | 0.7308 |
| **Locality 2.  SE of Franklin** | | | | |
| 337D | 75.7 | 300 | 0.732 | 0.7382 |
| 659 | 86.4 | 283 | 0.888 | 0.7459 |
| 659C | 89.5 | 275 | 0.946 | 0.7465 |
| **Locality 3.  North Redwood** | | | | |
| 739A | 81.7 | 380 | 0.624 | 0.7302 |
| 740 | 85.5 | 373 | 0.665 | 0.7328 |
| 785 | 90.0 | 378 | 0.691 | 0.7339 |
| **Locality 4.  NW of North Redwood** | | | | |
| 636 | 77.4 | 513 | 0.438 | 0.7223 |
| 637 | 76.8 | 505 | 0.441 | 0.7221 |
| 657 | 57.2 | 478 | 0.346 | 0.7162 |
| **Locality 5.  NE of Delhi** | | | | |
| 629B | 46.3 | 836 | 0.160 | 0.7073 |
| 667S | 29.9 | 1010 | 0.086 | 0.7063 |
| 669 | 71.0 | 1061 | 0.194 | 0.7093 |
| 669C | 75.7 | 701 | 0.313 | 0.7139 |
| 673 | 61.4 | 1448 | 0.123 | 0.7059 |
| **Locality 6.  N of Delhi** | | | | |
| 339 | 38.8 | 588 | 0.191 | 0.7097 |
| 671 | 35.1 | 494 | 0.206 | 0.7103 |
| 671C | 52.6 | 566 | 0.269 | 0.7125 |

and 11). The block sampling method of Farhat and Wetherill (1975) reproduced the earlier results, but as all the samples were obtained within a limited area, the data points show less scatter. This aspect of our differences with Farhat and Wetherill (1975; Goldich and Hedge, 1975), is discussed in this paper and the paper on the Granite Falls area (Goldich and others 1980a, this volume).

The homogenization of $^{87}Sr$ between tonalitic and granitic phases in rocks in which the two phases are intimately mixed is considered to be real, and the elimination of such samples is imperative. The intense deformation that the Morton Gneiss has

undergone must also be considered. As a result of shearing and flowage, phases of different composition and age have been mixed; this produced mechanical homogenization further altered by movements of fluids.

Various types of disturbances of the Rb-Sr isotopic system in minerals and rocks have been noted by many writers. An early example is the work by Fairbairn and others (1961). Theoretical analyses also have been presented, for example, by Riley and Compston (1962). References to additional papers can be found in Faure (1977). In the present discussion we supplement earlier findings in the Minnesota River Valley and demonstrate that the

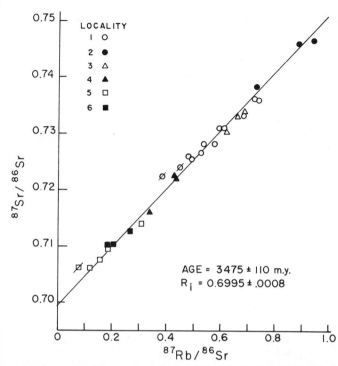

Figure 2. Rb-Sr diagram for samples of tonalitic and related gneisses. Granoblastic granodioritic gneiss samples (slash symbols) are not included in the regression.

Rb-Sr isotopic system has been variously affected in different rock types as a result of Rb loss, Rb gain, Sr loss, and Sr gain.

## Tonalitic Gneiss–Amphibolite Complex

The paleosome of the Morton Gneiss is a metamorphic complex of tonalite and minor granodiorite gneisses and amphibolites that was formed before the emplacement of the magmas of the pegmatitic granite and adamellite-1 (see Parts 1 and 2).

**Tonalite and Related Gneisses.** Twenty-nine samples of tonalite and granodiorite gneisses were analyzed (Table 2). A regression of the data for 26 samples (granoblastic granodioritic gneiss excluded) gives an age of $3,475 \pm 110$ m.y. and $R_i$ of $0.6995 \pm 0.0008$ (Fig. 2). Samples of the granoblastic gneiss give ages that are too old because of Rb loss and are discussed in a later section. In addition, two features of the isochron that require further explanation are the fine structure that is apparent in the en echelon pattern of the data points and the low initial ratio; both are attributed to Rb gain. The chemical data for the amphibolites (Part 2) clearly show that these rocks have gained Rb as well as K, Ba, Sr, $H_2O$, and F; hence, the Rb-Sr data for the amphibolites are first considered in the following section.

**Amphibolites.** K metasomatism during the alteration of basaltic rocks long has been recognized, and the geochemical coherence of K and Rb is well known. Fratta and Shaw (1974) have shown that diabase dikes are contaminated with K, Rb, Li, and Tl derived from the intruded host rock. As noted in Part 1, the occurrence of the amphibolite masses in the tonalitic gneisses indicates that some of the amphibolite was originally emplaced as basaltic dikes or sills, and these may have gained Rb from the tonalitic gneisses. The older gneisses, however, have low Rb con-

tents, and Rb gained at the time of intrusion will not alter the Rb-Sr age, provided that the rock system has remained closed. A logical conclusion is to associate Rb gain with the time of intrusion of the pegmatitic granite for which U-Pb ages on zircon give a minimum age of approximately 3,045 m.y., as will be shown in a later section.

The analytical data for samples of amphibolite are given in Table 3. Eighteen of the samples are plotted in Figure 3 which does not include four samples from locality 5, two of which are of the low-alumina variety. The 18 samples fall along two lines (Fig. 3, lines A and B). The enrichment in Rb and the resulting range in $^{87}Rb/^{86}Sr$ can be compared with that of the tonalite and granodiorite gneisses from the Morton vicinity (Fig. 3, line C). Amphibolite samples along line B are low in Rb, ranging from 2 to 13 ppm, whereas the samples along line A range from 15 to 37 ppm, except for sample 338.I (143 ppm) and 667 (8.4 ppm). The exceptionally low Sr content (18.4 ppm) accounts for the high $^{87}Rb/^{86}$ Sr ratio for sample 677 (Table 3). Obviously the amphibolites have undergone additions of varying amounts of Rb. Sr, if added as appears to be the case in sample 388.I (675 ppm, Table 3), was not appreciably enriched in $^{87}Sr$.

Samples A to E (line A, Fig. 3) are from a single block of amphibolite sampled by Nielsen (Nielsen and Weiblen, this volume). Starting with these samples, regression analyses give ages and initial ratios ($R_i$) as follows: 74-23 A through E, $2,730 \pm 275$ m.y. ($0.7003 \pm 0.0024$); 74-23 + 801, $2,800 \pm 250$ m.y. ($0.6996 \pm 0.0020$); 74-23 + 801 + 677, $2,870 \pm 150$ m.y. ($0.6991 \pm 0.0014$); and 74-23 + 801 + 677 + 388.I, $2,870 \pm 145$ m.y. ($0.6990 \pm 0.0014$). For eight samples, the age is approximately 2,870 m.y., and $R_i$ is 0.6990. We consider this to be the approximate time of amphibolitization. Large additions of Rb would cause a shift in the Rb/Sr ratios and a lowering of the initial ratio. Sr probably was added in varying amounts, but the principal factor was Rb again.

Samples with a lesser degree of enrichment in Rb (line B, Fig. 3) give an age of $2,950 \pm 245$ m.y. and $R_i$ of $0.7021 \pm 0.0007$. The weighted averages for seven samples with 7.0 ppm or less of Rb (Table 3) are 0.7055 and 0.093 for $^{87}Sr/^{86}Sr$ and $^{87}Rb/^{86}Sr$, respectively. If these rocks were emplaced 3,500 m.y. ago, the initial ratio is calculated at 0.7007.

Rb-Sr analyses of two samples from Morton and U-Pb analyses of zircon from these amphibolites by Farhat (1975) provide valuable additional information. The Rb-Sr data plot on line A (Fig. 3), reduce the regressions error, and give an age of $2,900 \pm 125$ m.y. ($R_i$, $0.6988 \pm 0.0012$). Farhat (1975) obtained essentially concordant U-Pb ages on one of the zircon samples with a $^{207}Pb$-$^{206}Pb$ age of 3,070 m.y. The second zircon sample gave discordant U-Pb ages which reflect the 2,600-m.y. B.P. event. In Figure 3, samples 594 and 74-23E plot below the isochron, and these samples may have gained some Rb during the 2,600-m.y. B.P. event. The Rb-Sr data do not give a precise age; however, all the data considered above indicate that the amphibolites were involved in a high-grade metamorphic event approximately 3,000 m.y. ago as well as in the 2,600-m.y. B.P. event.

**Fine Structure in Isochron for Old Gneisses.** A regression of the data for 10 samples of tonalite and granodiorite gneisses from the Morton vicinity gives an age of $2,920 \pm 325$ m.y. and $R_i$ of $0.7047 \pm 0.0028$ (Fig. 3, line C). The line is parallel to the lines for the amphibolites, and our interpretation is that the gneisses, like the

TABLE 3.  Rb-Sr ANALYTICAL DATA FOR AMPHIBOLITES

| Sample No. | Rb (ppm) | Sr (ppm) | $^{87}Rb/^{86}Sr$ (atomic) | $^{87}Sr/^{86}Sr$ |
|---|---|---|---|---|
| Locality 1. | | | | |
| BN-74-11 | 13.1 | 111.4 | 0.340 | 0.7175 |
| BN-74-23A | 23.3 | 117.5 | 0.575 | 0.7235 |
| -23B | 15.6 | 105.7 | 0.428 | 0.7170 |
| -23C | 15.0 | 111.9 | 0.390 | 0.7159 |
| -23D | 37.4 | 135 | 0.802 | 0.7329 |
| -23E | 32.8 | 123.7 | 0.769 | 0.7295 |
| BN-74-31 | 7.22 | 269.8 | 0.077 | 0.7052 |
| -32 | 5.84 | 250.1 | 0.068 | 0.7048 |
| 338.I | 143.0 | 674.7 | 0.614 | 0.7241 |
| 594 | 21.4 | 122.8 | 0.503 | 0.7181 |
| 600G | 13.4 | 137.0 | 0.283 | 0.7146 |
| 677 | 8.42 | 18.6 | 1.311 | 0.7541 |
| 801 | 15.7 | 121.0 | 0.376 | 0.7143 |
| Locality 2. | | | | |
| 731 | 13.1 | 109.3 | 0.347 | 0.7163 |
| 733 | 6.21 | 115.5 | 0.155 | 0.7083 |
| Locality 5. | | | | |
| 663 | 8.79 | 210 3 | 0.121 | 0.7061 |
| 664 | 6.39 | 111.6 | 0.165 | 0.7069 |
| 665 | 7.04 | 170.1 | 0.120 | 0.7055 |
| 672 | 36.7 | 163.0 | 0.653 | 0.7224 |
| Between Localities 5 and 6 (See Appendix) | | | | |
| BN-74-25 | 4.98 | 129.8 | 0.111 | 0.7061 |
| -26 | 2.64 | 98.2 | 0.078 | 0.7061 |
| -28 | 2.18 | 208.7 | 0.030 | 0.7041 |

amphibolites, have gained Rb during the high-grade metamorphism approximately 3,000 m.y. ago. The tonalitic gneiss–amphibolite complex was formed by folding early in the history of the Morton Gneiss. The basaltic precursors of the amphibolites behaved as competent rock and were fragmented to form the enclaves or clasts in the tonalitic gneiss. The structure was developed before the emplacement of the pegmatitic granite and adamellite-1, and field relationships (Part 1) also suggest that the veins of granoblastic gneiss were formed before the invasion of the complex by the granitic magmas.

The veins of coarse-grained granoblastic granodioritic gneiss were formed by shearing and recrystallization of the layered tonalite gneiss (Part 1, Fig. 4). The formation of hornblende and microcline at the expense of biotite resulted in a net loss of Rb (Part 2). This type of prograde metamorphism has been cited as the cause of depletion of the large cations K, Cs, Rb, U, and Th in granulites (Heier, 1965). Comparative data for rocks of differing metamorphic grade appear to support this thesis (Lambert and Heier, 1968; Tarney and others, 1972). Biotite is the principal or only mafic silicate in the tonalite and granodiorite gneisses, and we have been able to demonstrate Rb loss only from the granoblastic veins, which are of limited extent.

The fine structure in the isochron for the tonalitic and related gneisses is shown in Figure 4. We cannot dismiss the possibility that some of the samples plotted may have lost Rb and for this reason plot to the left of the line. The introduction of Rb at the time of the invasion of the tonalitic gneiss–amphibolite complex, however, appears to be the major cause for the fine structure and low initial ratio of the isochron (Fig. 4). Locally, some infiltration and reaction with the granoblastic gneiss occurred, but less well-healed fractures were the avenues for infiltration of granitic magma that contaminated the gneiss and reacted with the amphibolites.

## Granitic Gneisses and Aplite Dikes

The neosome of the Morton Gneiss is composed of a number of granitic rock types. Rb-Sr analytical data for the pegmatitic granite, adamellite-1, and the agmatic granodiorite are given in Table 4. The microadamellite porphyry at Morton and the gneissic adamellite at localities 3, 4, and 5 appear to be similar in composition to the Sacred Heart Granite (Part 2). Locally, the rock can be recognized at Morton, but it cannot be mapped separately. Although we consider the microadamellite to be a

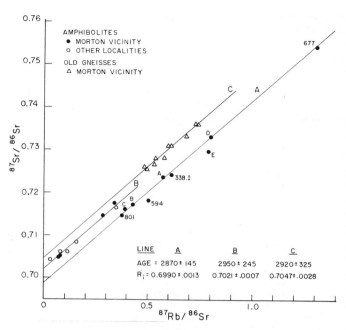

Figure 3. Rb-Sr diagram for samples of amphibolites from Morton (line A) and from Morton and other localities (line B) and samples of tonalitic gneiss from Morton (line C).

minor rock type at Morton, additional study may show it to be intermingled with the granitic phases. Analytical data for samples of adamellite-2 are given in Table 5. Chemically, the samples of adamellite-2 are similar to the aplite dike rocks, and because of the difficulties encountered in dating the various granitic phases, the aplite dikes are included in this discussion, although they cut the structure of the Morton Gneiss and are not considered part of the neosome. Analytical data for the aplite dike rocks are given in Table 6.

Rb-Sr isochron diagrams have been prepared for adamellite-1 (Fig. 5), adamellite-2 (Fig. 6), and the aplite dike rocks (Fig. 7). The results for these rocks, the pegmatitic granite gneiss, and the agmatic granodiorite are summarized in Table 7. The rock units are arranged from oldest to youngest (1 to 5) on the basis of field relationships (Part 1), and this order is retained in the Rb-Sr ages. Samples of the pegmatitic granite gneiss give model ages of 2,735 and 2,655 m.y. with an assumed $R_i$ of 0.701. In view of U-Pb ages of 3,000 m.y. or more on zircons, the Rb-Sr ages cannot be accepted. Six samples of adamellite-1 give a good alignment (Fig. 5) with an apparent age of 2,640 ± 115 m.y. and $R_i$ or 0.7048 ± 0.0014. Three samples of the agmatic granodiorite would plot near the line for adamellite-1, and the calculated age, if the initial ratio for adamellite-1 is used, is 2,670 m.y.

The Rb-Sr plot (Fig. 6) for 10 samples of adamellite-2 gives an age of 2,555 ± 55 m.y. and $R_i$ of 0.7029 ± 0.0013. The analytical error is less than that for adamellite-1; nevertheless, the two isochron ages are not resolved at the 95% confidence level, and on the basis of the field relationships, the apparent ages for adamellite-1 and the agmatic granodiorite appear to be low.

The age given by the aplite dike rock samples (Fig. 7) is 2,590 ± 40 m.y. with $R_i$ of 0.7036 ± 0.0020 and supports the conclusion from field and geochemical studies that adamellite-2 and the aplites are closely related. A regression of the adamellite-2 and the aplite dike rocks gives an age of 2,590 ± 40 m.y. and $R_i$ of 0.7024

± 0.0011 (Table 7). The 2,590-m.y. age for the aplites is based on six samples. Actually nine samples were analyzed (Table 6), but three were eliminated from the regression. If $R_i$ = 0.7036 (Fig. 7), the calculated ages for the three samples follow: sample 676, 2,080 m.y.; sample 648A, 2,420 m.y.; and sample 741, 2,465 m.y. The low ages are attributed to loss of radiogenic $^{87}Sr$.

Granitic rocks with large amounts of K-feldspar are particularly susceptible to loss of radiogenic $^{87}Sr$, probably in large part because of the many cleavage and parting planes of the brittle K-feldspar. Sr, like Ca, is readily leached, but Rb, like K, may be retained in alteration productions. Brooks (1968) called attention to changes in the Rb-Sr istopic system in a granite as the result of alteration of the feldspar. Studies in the Rainy Lake district, Ontario (Hart and Davis, 1969; Peterman and others, 1972), also are instructive. Under conditions of low-grade metamorphism, granitic rocks commonly undergo a loss of radiogenic $^{87}Sr$, and the Rb-Sr isochron age may be appreciably lower than the U-Pb zircon age. In the Saganaga Lake–Northern Light Lake area, Ontario-Minnesota, Hanson and others (1971) found that the sheared samples of tonalite had lost radiogenic $^{87}Sr$.

Goldich and others (1970) concluded that the granitic rocks with abundant K-feldspar were more or less open systems during and following the 2,600-m.y. B.P. event, and radiogenic $^{87}Sr$ was lost. The argument may be made that the 2,640-m.y. isochron indicates redistribution of $^{87}Sr$ and homogenization on a large scale. Farhat (1975) made this argument with reference to a precise isochron he obtained for six samples by his block-sampling technique. The whole-rock samples, which Farhat described as "gray type of the Morton Gneiss," and a plagioclase separate give an age of 2,410 ± 20 m.y. [$R_i$ 0.7100 ± 0.0001 (2σ)]. In view of

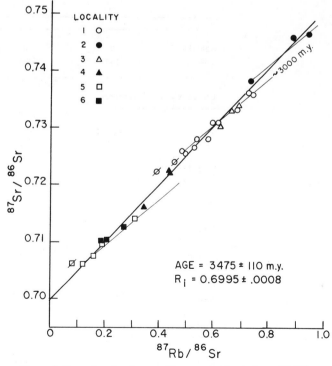

Figure 4. Secondary isochrons with an approximate age of 3,000 m.y. result from samples of contaminated rock which have gained Rb, probably at the time of emplacement of the pegmatitic granite and adamellite-1.

TABLE 4.  Rb-Sr ANALYTICAL DATA FOR PEGMATITIC GRANITE, ADAMELLITE-1,
AND AGMATIC GRANODIORITE, MORTON

| Sample No. | Rb (ppm) | Sr (ppm) | $^{87}Rb/^{86}Sr$ (atomic) | $^{87}Sr/^{86}Sr$ |
|---|---|---|---|---|
| **Pegmatitic Granite** | | | | |
| 290L | 216.0 | 401.4 | 1.567 | 0.7612 |
| 674 | 189.6 | 312.0 | 1.771 | 0.7711 |
| **Adamellite-1** | | | | |
| 600B | 112.9 | 350.0 | 0.937 | 0.7402 |
| 603 | 134.3 | 310.1 | 1.259 | 0.7533 |
| 647 | 90.8 | 377.0 | 0.699 | 0.7314 |
| 651 | 106.2 | 363.4 | 0.849 | 0.7371 |
| 656 | 114.7 | 384.0 | 0.867 | 0.7375 |
| 720A | 105.9 | 472.2 | 0.651 | 0.7301 |
| **Agmatic Granodiorite** | | | | |
| 602 | 129.5 | 296.2 | 1.271 | 0.7540 |
| 782 | 124.4 | 358.0 | 1.009 | 0.7433 |
| 646(vein) | 86.1 | 358.2 | 0.698 | 0.7322 |

TABLE 5.  Rb/Sr ANALYTICAL DATA FOR ADAMELLITE-2

| Sample No. | Rb (ppm) | Sr (ppm) | $^{87}Rb/^{86}Sr$ (atomic) | $^{87}Sr/^{86}Sr$ |
|---|---|---|---|---|
| **Locality 1.   Morton** | | | | |
| 655 | 158.1 | 530.1 | 0.865 | 0.7345 |
| 680 | 108.6 | 320.5 | 0.984 | 0.7403 |
| **Locality 3.   North Redwood** | | | | |
| 772 | 189.7 | 156.4 | 3.555 | 0.8339 |
| 783 | 169.9 | 228.6 | 2.168 | 0.7854 |
| 784 | 181.0 | 167.4 | 3.163 | 0.8197 |
| **Locality 4.   NW of North Redwood** | | | | |
| 743 | 183.8 | 143.5 | 3.756 | 0.8383 |
| 745 | 178.7 | 346.2 | 1.501 | 0.7582 |
| **Locality 5.   NE of Delhi** | | | | |
| 654 | 100.6 | 430.7 | 0.678 | 0.7273 |
| 729 | 187.6 | 229.4 | 2.385 | 0.7898 |
| 730 | 196.9 | 230.7 | 2.491 | 0.7954 |

TABLE 6.  Rb-Sr ANALYTICAL DATA FOR APLITIC DIKE ROCKS

| Sample No. | Rb (ppm) | Sr (ppm) | $^{87}Rb/^{86}Sr$ (atomic) | $^{87}Sr/^{86}Sr$ |
|---|---|---|---|---|
| **Locality 1.    Morton** | | | | |
| 459 | 150.3 | 192.6 | 2.276 | 0.7889 |
| 648A | 291.9 | 35.2 | 26.17 | 1.620 |
| 648B | 181.5 | 140.1 | 3.802 | 0.8479 |
| 650 | 199.8 | 132.4 | 4.440 | 0.8709 |
| 652 | 135.3 | 199.8 | 1.974 | 0.7770 |
| **Locality 3.    North Redwood** | | | | |
| 741 | 283.1 | 26.6 | 34.52 | 1.933 |
| **Locality 4.    NW of North Redwood** | | | | |
| 676 | 197.7 | 173.9 | 3.320 | 0.8036 |
| **Locality 5.    NE of Delhi** | | | | |
| 587 | 229.4 | 97.1 | 7.010 | 0.9644 |
| 628 | 228.6 | 99.5 | 6.813 | 0.9568 |

the 2,640-m.y. isochron for adamellite-1, 2,555-m.y. isochron for adamellite-2, and 2,590 m.y. isochron for the aplite dikes, all with initial ratios lower than 0.710, we cannot accept a high-grade metamorphism at 2,410 m.y. ago in the Morton area. A more reasonable interpretation is that Farhat's isochron is a mixing line.

## U-Pb AGE DETERMINATIONS

U-Pb zircon ages commonly are greater than corresponding mineral ages by Rb-Sr and K-Ar methods as was early shown by Aldrich and others (1965). In the recent literature, a number of examples are given in which the $^{207}Pb$-$^{206}Pb$ age is appreciably greater than the Rb-Sr isochron age. The conclusion first apparent is that zircon is a stable mineral; however, the sharp discordancy in U-Pb ages does not support the stability concept; rather, it is the systematics of decay systems of the two U isotopes that permit interpretations, and for this purpose Wetherill's (1956) concordia diagram is most useful.

## Tonalite Gneiss

Zircon was concentrated from sample 673 of coarse-grained, layered tonalitic gneiss from locality 5. An approximate mode is given in Part 1 (Table 2) and a chemical analysis in Part 2 (Table 2). The zircon grains are well-formed crystals with a large range in size, and analyses were made of size fractions, a method introduced by Silver and Deutsch (1961). The fractions and the calculated

ages are given in Table 8. The $^{207}Pb$-$^{206}Pb$ ages range from 3,170 to 3,305 m.y. The large clear crystals (Table 9, sample 673.1) contain the least U (195 ppm) and give the oldest age. Dark crystals (sample 673.4) in the smallest handpicked size give the youngest age. The fraction with the smallest crystals (Sample 673.5) was not handpicked and contains both clear and dark grains; the mix-

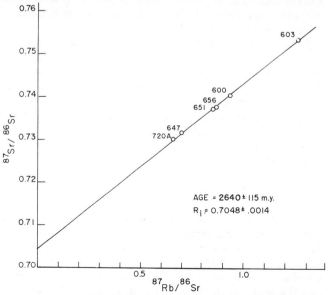

Figure 5. Rb-Sr diagram for samples of adamellite-1.

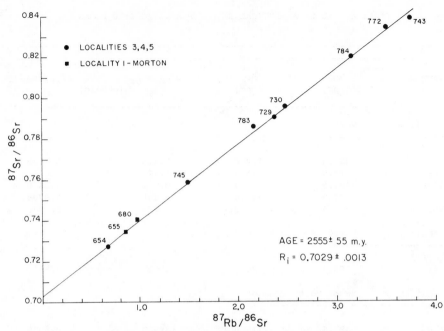

Figure 6. Rb-Sr diagram for samples of adamellite-2.

ed population gave an intermediate age. The degree of discordance in the U-Pb ages is related to the U contents of the fractions (Table 9).

The concordia diagram (Fig. 8) includes the new samples and five previously analyzed samples: 339 and 339U, locality 6; 389D and 389D-2, locality 1; and 337D, locality 2. The new analyses from locality 5 are numbered 1 through 5 (Fig. 8). Samples 339, 339U, 389D-2, and 337D were analyzed by Stern (Goldich and others, 1970); sample 389D was analyzed by Catanzaro (1963). The 10 samples form a linear array along a chord with upper and lower intercepts of 3,490 and 1,730 m.y., respectively. A second interpretation is shown in the line drawn through points 1, 2, and 3 to the origin (Fig. 8). This line suggests that the tonalitic gneiss at locality 5 was formed 3,300 m.y. ago and that the discordance is due to recent Pb loss.

Neither of the interpretations given above are satisfactory, and more detailed consideration of the data indicates other possibilities. Sample 339U was not leached in nitric acid before dissolution, and sample 673.1 is unusual because of its low U concentration. If these samples are not used, the remaining eight samples define a chord (Fig. 9) with upper and lower intercepts of 3,590 and 1,910 m.y., respectively. Similar chords were plotted by Catanzaro (1963), Stern and others (1966), and Goldich and others (1970).

## Pegmatitic Granite Gneiss

Data for three size fractions of a zircon concentrate from a sample of the pegmatitic granite gneiss (sample 781) are given in Tables 8 and 9. Compared with the fractions of the zircon from the tonalitic gneiss (673), the U concentrations are high but show a limited range (809 to 988 ppm). The three fractions give similar ages (Table 8), and the pattern is similar to sample 390L analyzed by Catanzaro (1963). The latter sample was not sized, contained

962 ppm of U, and gave a $^{207}$Pb-$^{206}$Pb age of 3,055 m.y. The fractions from sample 781, as well as the older sample 390L, fall on the chord whose intercepts are 3,590 and 1,910 m.y. (Fig. 9), although they were not included in the regression; the pegmatite zircon, like that from the tonalite gneiss, may have been involved in the event indicated at 1,910 m.y. B.P. The close grouping of the pegmatite zircons, however, suggests that the average $^{207}$Pb-$^{206}$Pb age of 3,043 ± 26 m.y. may be the time of origin, an interpretation that finds some support in the geological relationships discussed earlier. It implies that the zircons from the pegmatite granite gneiss show no effects of the thermal event 1,800 to 1,900 m.y. ago that affected the zircons from the tonalitic gneiss. It should be noted that younger zircons (2,600 m.y. old) from the garnet-biotite gneiss at Granite Falls and from the Sacred Heart Granite analyzed by Stern (Goldich and others, 1970) show a similar discordance pattern which cannot be related to an event 1,800 to 1,900 m.y. ago, and recourse must be made to some mechanism such as the loss of Pb by continuous diffusion (Tilton, 1960) or the dilatancy model (Goldich and Mudrey, 1972).

## Discussion

A two-stage model was proposed (Goldich and others, 1970) to explain the discordance in the zircons from the Precambrian rocks of the Minnesota River Valley. The primary discordance was attributed to the high-grade igneous-metamorphic event 2,600 m.y. ago. The second stage involved relatively recent loss of Pb as a result of the dilatancy effect in crystalline plutonic rocks brought to the surface by uplift and erosion. Water which accumulated in the metamict zircon at depth was partially released as a result of near-surface dilatancy, and some of the radiogenic Pb was removed (Goldich and Mudrey, 1972). Sample 673.1, which was omitted from the regression of the tonalitic gneiss samples in Figure 9, plots on the line postulated in the 1970 inter-

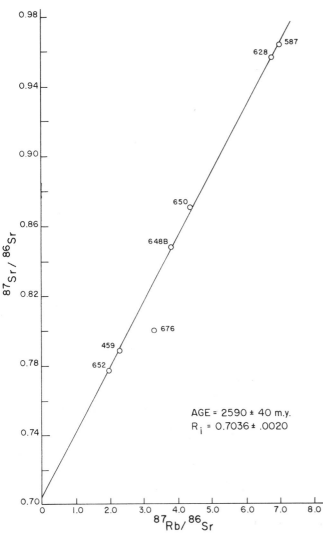

**Figure 7. Rb-Sr diagram for samples of aplitic dike rocks.**

pretation, the chord whose intercepts are 3,590 and 2,600 m.y. in Figure 10.

Michard-Vitrac and others (1977) have published the results of five U-Pb analyses of zircons supplied by G. W. Wetherill. Their results for three single grains of zircon, a sample of 10 zircon grains, and a sample of 1 mg are shown in Figure 10. They concluded that their results are in better agreement with Farhat's (1975) interpretation than with that of Goldich and others (1970). Two of their single-grain results fall close to the chord with 3,590 and 2,600-m.y. intercepts (Fig. 10); none approaches Farhat's chord with 3,230- and 2,530-m.y. intercepts; two samples plot near the chord with 3,590- and 1,190-m.y. intercepts, and the fifth sample falls between the lines. Four of our samples, however, have ages older than any of the zircons analyzed by Michard-Vitrac and co-workers.

Zircon samples 339, 337D, and 389D (Fig. 8) range from 702 to 783 ppm U (Goldich and others, 1970), and the large concentrations raise the question of the possible addition of U in the form of new crystals of zircon and overgrowths during the high-grade metamorphic events 3,050 and 2,600 m.y. ago. This possibility is also suggested by the two chords (Fig. 10) that intersect concordia at approximately 3,600 m.y. ago. Two different zircon populations are indicated, and these may differ not only in the level of U concentration but also in age.

The pegmatitic granite and adamellite-1 gneisses at Morton are closely associated. Zircon from the adamellite has not been dated, but T. W. Stern has analyzed zircon from adamellite which has similar relationships to pegmatitic granite gneiss in the Granite Falls area (Goldich and others, 1980a, this volume). The $^{207}Pb$-$^{206}Pb$ age is 3,050 m.y., but the U-Pb ages, like those from the pegmatite at Morton are discordant, and the original age, therefore, is uncertain. The pegmatitic granite and adamellite-1 may have been emplaced during the 3,050–m.y. B.P. metamorphic event, but they may be older. The U-Pb systems may have been completely reset 3,050 m.y. ago and were not appreciably affected during the 1,800–m.y. B.P. event, or the zircons were mild-

TABLE 7.  APPARENT Rb-Sr AGES FOR THE GRANITIC GNEISSES AND APLITE DIKES

| | Rock Unit | No. Samples | Age (m.y.) | $R_i$ | Explanation |
|---|---|---|---|---|---|
| 1. | Pegmatitic granite | | | | |
| | Sample 674 | 1 | 2,735 | 0.701 | $R_i$ assumed |
| | Sample 390L | 1 | 2,655 | | |
| 2. | Adamellite-1 | 6 | 2,640 ± 115 | 0.7048 ± .0014 | Fig. 5 |
| 3. | Agmatic granodiorite | 3 | 2,670 | 0.7048 | From line 2 |
| 4. | Adamellite-2 | 10 | 2,555 ± 55 | 0.7029 ± .0013 | Fig. 6 |
| 5. | Aplites | 6 | 2,590 ± 40 | 0.7036 ± .0020 | Fig. 7 |
| 6. | Adamellite-2 and aplites | 16 | 2,590 ± 40 | 0.7024 ± .0011 | Not plotted |

## TABLE 8. U-Pb AGES OF ZIRCONS

| | | Age (m.y.) | | |
|---|---|---|---|---|
| | | $\dfrac{^{206}Pb}{^{238}U}$ | $\dfrac{^{207}Pb}{^{235}U}$ | $\dfrac{^{207}Pb}{^{206}Pb}$ |
| Sample | Description | | | |
| **Tonalitic gneiss (locality 5)** | | | | |
| 673.1 | clear      > 80 mesh | 3,150 | 3,245 | 3,305 |
| 673.2 | clear   80-140 | 2,935 | 3,155 | 3,300 |
| 673.3 | clear  140-200 | 2,840 | 3,105 | 3,280 |
| 673.4 | dark   140-200 | 2,690 | 2,970 | 3,170 |
| 673.5 | clear & dark  200-270 | 2,765 | 3,040 | 3,230 |
| **Pegmatitic granite (locality 1)** | | | | |
| 781A | dark      80 mesh | 2,550 | 2,840 | 3,055 |
| 781B | dark   80-140 | 2,565 | 2,835 | 3,035 |
| 781C | dark  140-200 | 2,535 | 2,815 | 3,025 |

*(J. L. Wooden, Analyst)*

## TABLE 9. U-Pb ANALYTICAL DATA

| | Concentration (ppm) * | | Atom Percent[+] | | | |
|---|---|---|---|---|---|---|
| Sample | U | Pb | $^{204}Pb$ | $^{206}Pb$ | $^{207}Pb$ | $^{208}Pb$ |
| 673.1 | 195.4 | 146.9 | 0.0242 | 72.35 | 19.80 | 7.83 |
| 673.2 | 236.0 | 161.6 | 0.0198 | 72.62 | 19.71 | 7.65 |
| 673.3 | 250.0 | 164.2 | 0.0183 | 72.69 | 19.49 | 7.81 |
| 673.4 | 520.3 | 324.0 | 0.1063 | 72.78 | 19.11 | 8.01 |
| 673.5 | 285.5 | 182.3 | 0.0297 | 72.46 | 18.94 | 8.57 |
| 781A | 830.4 | 484.7 | 0.0804 | 72.40 | 17.53 | 9.99 |
| 781B | 809.3 | 477.9 | 0.0925 | 72.29 | 17.42 | 10.19 |
| 781C | 987.8 | 571.6 | 0.0769 | 72.50 | 17.20 | 10.22 |

* *Corrected for blank: Pb = 1.0 ng; U = 1.0 ng.*
+ *Corrected for an assumed 3600 m.y. common lead of atomic composition:*
 *204 : 206 : 207 : 208 = 1.75 : 20.14 : 23.12 : 54.99.*

**Figure 8. Concordia diagram for size fractions (numbered 1 through 5) of zircon from tonalite gneiss, locality 5. Other samples are from Goldich and others (1970) and Catanzaro (1963). The line from the origin intersects concordia at 3,300 m.y. which is the $^{207}$Pb-$^{206}$Pb age, and is a minimum age for the tonalite gneiss.**

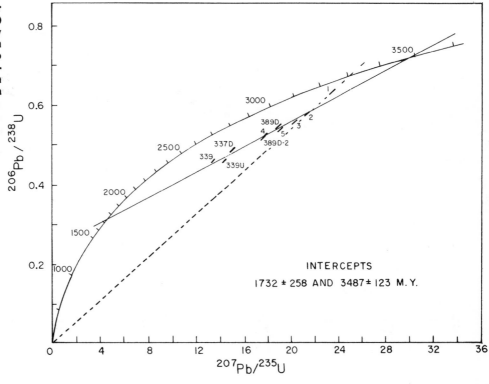

**Figure 9. Concordia diagram for samples of tonalitic gneiss, excluding samples 673-1 and 339U (Fig. 8). Four samples of zircon from the pegmatitic granite gneiss (390L; A, B, and C of sample 781) fall along the chord defined by the eight tonalite samples and give a $^{207}$Pb-$^{206}$Pb age of 3,043 ± 26 m.y.**

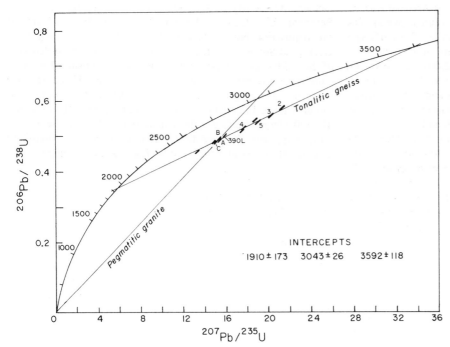

ly affected by the 3,050–m.y. B.P. event, and the age discordance is related to the younger 1,800–m.y. B.P. event. The present data do not permit resolution of this uncertainty.

Introduction of U with the recrystallization of small zircon crystals or growth of a new generation very likely occurred at the time of the intrusion of the pegmatitic granite and adamellite-1 magmas. This would ultimately result in age discordancy that would appear on the chord with 3,560- and 2,600-m.y. intercepts

(Fig. 10). The 2,600- and 1,800-m.y. B.P. events and recent Pb loss further complicate the relationships.

## CONCLUDING REMARKS

The major events and geologic history of the Precambrian rocks in the Morton area of the Minnesota River Valley are sum-

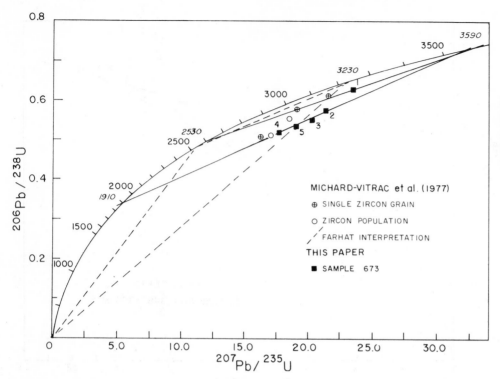

Figure 10. Concordia diagram presenting the results of analyses by Michard-Vitrac and others (1977), Farhat's interpretation (1975), and the analyses of the fractions from tonalite gneiss 673. Fraction 1 and two single crystals fall on a chord postulated by Goldich and others (1970, Fig. 7).

marized in Table 10. The U-Pb age of the oldest Archean gneisses remains unresolved because of the effects of younger events. The $^{207}$Pb-$^{206}$Pb age of 3,300 m.y. is a minimum value. The U-Pb ages are discordant as may be expected in view of metamorphic-igneous events at (1) 3,050 m.y. B.P., (2) 2,600 m.y. B.P., (3) 1,800 m.y. B.P., and (4) recent Pb loss. The Rb-Sr isochron age of approximately 3,500 m.y. for the tonalitic and related gneisses also is not precise and involves some interpretation and evaluation of the effects of losses and gains of both Rb and Sr.

The rare-earth elements Sm and Nd are less affected by metamorphic processes than the more mobile Rb and Sr, and the Sm-Nd system, therefore, is more effective in penetrating the metamorphic veil. Model ages of 3,580 ± 30 m.y. have been reported by McCulloch and Wasserburg (1978) for two samples of residual clay developed on the Morton Gneiss by weathering in Cretaceous time. The samples are from the early study by Goldich (1938) and are channel samples. Preliminary Sm-Nd model ages on samples used in this study range from 3,200 to 3,600 m.y.

The problems encountered in our efforts to date the ancient gneisses of the Minnesota River Valley are similar to those reported by Peterman and others (this volume) in the Watersmeet area of northern Michigan. The antiquity of the tonalitic gneiss is shown in discordant U-Pb ages with a minimum $^{207}$Pb-$^{206}$Pb age of 3,410 m.y. A Sm-Nd model age of 3,600 m.y. is reported by McCulloch and Wasserburg (this volume).

## ACKNOWLEDGMENTS

We are indebted to many people who helped in the field and laboratory. We thank G. N. Hanson for graciously making available facilities at the State University of New York at Stony Brook for the U-Pb analyses of zircons. We thank G. R. Him-melberg, University of Missouri, and C. E. Hedge and Z. E. Peterman, U.S. Geological Survey for constructive criticisms of the manuscript. Numerous discussions with K. R. Ludwig, U.S. Geological Survey, enhanced our appreciation of the systematics of the U-Pb decay systems, but we alone are responsible for shortcomings in the presentation and interpretation. This study was made possible by grants GA-40226 and 40226a#1 from the National Science Foundation, supplementary grants 41-12122 and 41-23302 from the Graduate School Fund of Northern Illinois University, and by logistical support from the Minnesota Geological Survey and the Cold Spring Granite Company.

## APPENDIX 1. LOCATION AND DESCRIPTION OF SAMPLES

### Explanation

Sample localities are shown in Figure 1; quarries in the Morton vicinity are indicated Q-1 and so forth and are located in Figure 3 of Part 1. Key to sample descriptions: mode, modal analysis listed in table of given number in Part 1; chem, chemical analysis listed in table of given number in Part 2; Rb-Sr, determinations of Rb and Sr (in parts per million by weight) and the atomic ratios $^{87}$Rb/$^{86}$Sr and $^{87}$Sr/$^{86}$Sr are listed in table of given number in Part 3 (same for U-Pb); Ba, barium content in parts per million by weight; D, density in grams per cubic centimetre (Ba and D not listed in other tables are included in tables of Part 3).

**Locality 1. Morton Vicinity**—Sec. 31, T. 113 N., R. 34 W., Renville County, lat 44°33′09″N, long 94°59′33″W.

*Tonalitic Gneiss*

*389D.* Very large sample collected by S. S. Goldich and T. W. Stern in 1962; Catanzaro (1963) and Goldich and others (1970); Q-4; chem, 1.

*644.* Gray, medium-grained, thin-banded; Q-3; mode, 2; chem, 1.

TABLE 10.  SUMMARY OF EVENTS IN THE MORTON AREA, MINNESOTA RIVER VALLEY

| Time (m.y.) | Description |
|---|---|
| | Sioux Formation, conglomerate and quartzite, New Ulm (Goldich and others, 1961, 1970). |
| | Uplift and erosion |
| 1,800 ● | • Thermal event accompanied by emplacement of small igneous masses, i.e., gabbro-granophyre and basaltic dikes southeast of Franklin (Goldich and others, 1961; Hanson, 1968). |
| | * Epidote veinlets in Morton Gneiss by hydrothermal activity. |
| | • Uplift and erosion; stabilization of area 2,400-2,500 m.y. ago (K-Ar and Rb-Sr mineral ages, Goldich and others, 1970). |
| Proterozoic | |
| Archean | |
| | • Aplite, pegmatite, and aplite-pegmatite dikes; post-kinematic and traversing structure of the Morton Gneiss (Rb-Sr, 2,590 ± 40 m.y.). |
| | • Adamellite-2, late-kinematic (Rb-Sr, 2,555 ± 55 m.y.). |
| 2,600 ● | • High-grade metamorphism producing final structure of the Morton Gneiss (Rb-Sr rock-mineral, K-Ar mineral ages). Emplacement of Sacred Heart Granite (Pb-Pb, Doe and Delevaux, this volume; U-Pb, Rb-Sr, Goldich and others, 1970) |
| | • * Emplacement of pegmatitic granite and adamellite-1 (U-Pb minimum age, 3,043 ± 26 m.y.). |
| 3,050 ● | • High-grade metamorphism of tonalitic gneiss-amphibolite complex (Rb-Sr secondary isochrons resulting from Rb gain in amphibolites and tonalitic gneisses; U-Pb age of 3,070 m.y. on zircon from amphibolite (Farhat, 1975). |
| | * Emplacement of pegmatitic granite and adamellite-1. |
| | Regional folding forming the tonalitic gneiss-amphibolite complex. Intrusion of basaltic dikes and sills. |
| 3,500 ● | • Intrusion of coarse-grained tonalites in a volcanic pile of andesitic to dacitic flows and/or pyroclastic material with possible intercalations of basaltic flows, surmised from field relationships. (Rb-Sr, 3,475 ± 110 m.y. for tonalitic and related gneisses; U-Pb minimum age, 3,300 m.y.). |

● *Major event;* • *based on age determinations; *position uncertain.*

649B. Dark gray, coarse-grained, contorted structure; some amphibolite contamination; Q-4; mode, 2; chem, 1.

679D. Gray, fine-grained, layered; cut by veins of speckled gneiss; Q-4; mode, 2; Rb-Sr, 2; Ba, 310; D, 2.708.

682. Dark gray, layered; from waste pile, Q-5; Rb-Sr, 2; Ba, 380; D, 2.700.

799. Dark gray, coarse-grained; from a large quarried block, courtesy of Cold Spring Granite Company; chem, 1.

*Granodioritic Gneiss*

633. Dark gray, fine- to medium-grained, thin-banded; from waste pile, Q-6; mode, 3; chem, 3.

633B. Similar to 633; Q-6; Rb-Sr, 2; Ba, 605; D, 2.725.

679. Gray, layered, similar to tonalite 679D; Q-4; mode, 3; chem, 3.

713. Gray, contorted, with conspicuous biotite; Q-4; Rb-Sr, 2; Ba, 625; D, 2.709.

*Granoblastic Granodioritic (Speckled) Gneiss*

593S. Very coarse grained intergrowth of hornblende, quartz, and feldspar; waste pile, Q-4; mode, 4; no other data.

649S. Granoblastic vein in gray, layered tonalitic gneiss; Q-4; chem, 4.

678S. Granoblastic vein; Q-4; mode, 4; chem, 4.

*Amphibolite*—Samples with prefix BN were collected by Nielsen and Weiblen (this volume).

BN-74-11. Q-2; Rb-Sr, 3; Ba, 45.

BN-74-23A. Sample from a large block in Q-3; Rb-Sr, 3; Ba, 70. Same for the following except as noted: *BN-74-23B*—Ba, 50; D, 3.041. *BN-74-23C*—Ba, 40. *BN-74-23D*—Ba, 85; D, 2.986. *BN-74-23E*—Ba, 65. Average chemical composition of BN-74-23A through BN-74-23E, 8.

BN-74-31. Between Q-1 and Q-2; Rb-Sr, 3; Ba, 220; D, 3.033.

BN-74-32. Between Q-1 and Q-2; Rb-Sr, 3; Ba, 180; D, 3.084.

338.I. Biotite-rich amphibolite clast (Goldich and others, 1970); Q-1;

Rb-Sr, 3; Ba, 470; *D*, 2.951.

*594.* Inner part of a large clast in tonalitic gneiss; Q-4; mode, 5; chem, 6 and 7.

   *600G.* Clast in tonalitic gneiss; waste pile, Q-4; mode, 5; chem, 6 and 7.

   *642.* Fine-grained clast; Q-3; mode only, 5.

   *677.* Medium-grained clast; north end of Q-4; Rb-Sr, 3; Ba, 30.

   *801.* Large dikelike mass on waste pile, Q-4; chem, 6 and 7.

### Pegmatitic Granite Gneiss

*390L.* Very large sample collected by S. S. Goldich and T. W. Stern in 1962; Catanzaro (1963), and Goldich and others (1070); estimated mode, 6; chem, 9.

   *674.* Medium- to coarse-grained, foliated with thin discontinuous layers of biotite; Q-3; mode, 6; chem, 9.

   *781.* Coarse-grained; Q-2; large sample (~30 kg) for zircon; U-Pb, 8 and 9; no other data.

### Adamellite-1

*Adamellite-1*—All samples are similar; light-gray to pink; fine- to medium-grained, resembling aplitic texture; leuococcratic masses infolded with the tonalitic gneiss and closely associated with the pegmatitic granite gneiss.

   *600B.* Q-4; mode, 7; chem, 9. Same for the following: *603* from Q-6. *647* from Q-3. *651* from Q-4. *656* from Q-4.

   *720A.* Q-4, Rb-Sr, 4; Ba, 2,380.

### Agmatic Granodiorite

*602.* Dark reddish-brown, fine-grained dikelike mass in Q-3; mode, 7; chem, 10.

   *782.* Similar to 602, from a second block in Q-3; Rb-Sr, 4; Ba, 1,660; *D*, 2.680.

   *646D.* Second sample from block of sample 602 in Q-3; mode only, 7.

   *646L.* Quartz-feldspar vein in granodiorite; Q-3; mode, 7; Rb-Sr, 4; Ba, 1,920.

### Adamellite-2 (Microadamellite Porphyry)

*Adamellite-2 (Microadamellite Porphyry)*—Samples are similar microphanerite porphyry; gray to pink.

   *655.* Q-2; mode, 8; chem, 10.

   *680.* From block on waste pile, Q-4; mode, 8; chem, 10.

   *681.* From block in Q-4; mode, 8; chem, 10; no isotopic data.

   *684.* Q-6; mode only, 8.

### Mixed-rock Samples

*76.* Composite sample; Q-2 (Goldich, 1938); chem, 10.

   *338.* Very large sample; Q-1 (Goldich and others, 1970); chem, 10.

### Epidote from Veins

*Epidote from Veins*—Q-6; Rb, 0.33 ppm; Sr, 4,160 ppm; $^{87}Sr/^{86}Sr$, 0.7142 ± 0.0001.

### Aplite Dike Rocks

*459.* Pink, fine-grained; from waste pile, Q-4; mode, 9; chem, 12.

   *648B.* Gray, outer zone of aplite-pegmatite composite dike in contact with tonalitic gneiss; large block from waste pile, Q-4; mode, 9; chem, 12.

   *648A.* Pink, middle zone of aplite above; mode, 9; chem, 12.

   *650.* From a composite aplite-pegmatite dike; block on waste pile, Q-4; mode, 9; chem, 12.

   *652.* Aplite dike in tonalitic gneiss, Q-4; mode, 9; chem, 12.

**Locality 2. Southeast of Franklin**—NW cor. sec. 22, T. 111 N., R. 32 W., Nicollet County, lat 44°24′37″N, long 94°40′52″W.

*Granodioritic Gneiss*—Three samples are from blocks in small quarry openings from locality KA-51 (Goldich and others, 1961), approximately 13 mi (24 km) southeast of Franklin.

   *337D.* Very large sample (Goldich and others, 1970); Rb-Sr, 2; Ba, 965.

*659.* Dark gray, coarse-grained; mode, 3; chem, 3.

   *659C.* Dark gray, coarse-grained; Rb-Sr 2; Ba, 745.

*Amphibolite*—Two amphibolite samples from large dikelike mass of amphibolite in gneiss; on east side of County Road 51, approximately 300 m south of bridge on Wabasha Creek SE¼, sec. 14, T. 112 N., R. 34 W., Redwood County.

   *731.* Dark gray, medium-grained; chem, 6 and 7.

   *733.* Dark gray, medium-grained; Rb-Sr, 3; Ba, 65; *D*, 3,077.

**Locality 3. North Redwood**—Outcrops on either side of State Highway 101, northwest of North Redwood in SE¼ sec. 19 and SW¼ sec. 20, T. 113 N., R. 35 W., Redwood and Renville Counties, lat 44°34′38″N, long 95°05′10″W.

### Tonalitic Gneiss

*739A.* Dark gray, fine- to medium-grained, foliated; east side of road at bridge; mode, 2; chem, 1.

   *740.* Same as 739A except mode, 2; Rb-Sr, 2; Ba, 210; *D*, 2.714.

   *785.* Dark gray, medium-grained; from small quarry opening on Ellison farm; Rb-Sr, 2; Ba, 590; *D*, 2.703.

*Adamellite-2 Gneiss*—The fine-grained gneissic adamellite-2 samples are similar and range from gray to pink.

   *772.* From abandoned quarry, SW¼ sec. 20; chem, 11.

   *783.* From small quarry opening, Ellison farm, west side of road and south of Minnesota River; chem, 11.

   *784.* Same as 783 except chem, 11.

### Aplite

*741.* Pink, fine-grained; east side of road near bridge; Rb-Sr, 6; Ba, 100.

**Locality 4. Northwest of North Redwood**—NE¼ sec. 11, T. 113 N., R. 36 W., Renville County, lat 44°36′09″N, long 95°08′48″W.

### Tonalitic Gneiss

*657.* Dark reddish-gray, fine-grained, some mineral foliation, not banded; contains amphibolite clasts that are rimmed with pegmatite which also forms veins in the gneiss; mode, 2; chem, 2.

### Granodioritic Gneiss

*636.* Similar to 657; sampled 35 cm from a pegmatite vein; mode, 3; chem, 3.

   *637.* Similar to 636 and 657; sampled 1 m from pegmatite; mode, 3; Rb-Sr, 2; Ba, 815; *D*, 2.708.

### Adamellite-2 Gneiss

*743.* Pink, intrusive in tonalitic and granodioritic gneiss; Rb-Sr, 5; Ba, 800; *D*, 2.609.

   *745.* Pink, medium-grained; near 636; chem, 11.

### Aplite Dike Rocks

*635.* Pink, fine-grained; from composite aplite-pegmatite dike; mode only, 9.

   *676.* Pink, fine-grained; mode, 9; Rb-Sr, 6; Ba, 900.

**Locality 5. Northeast of Delhi**—SE cor. sec. 33, T. 114 N., R. 36 W., Renville County, lat 44°37′46″N, long 95°10′49″W.

### Tonalitic Gneiss

*629B.* Gray, medium- to coarse-grained, roughly banded; in large outcrop west of County Road 15, sampled approximately 7 m east of the southern end of large aplite dike; mode, 2; chem, 2.

*669.* Gray, fine-grained; outcrop 100 m northwest of 629B; mode, 2; chem, 2.

*669C.* Same as 669 except Rb-Sr, 2; Ba, 950.

*673.* Gray, fine- to medium-grained; approximately 8 m west of large aplite dike; mode, 2; chem, 2.

### Granoblastic Granodioritic Gneiss

*667S.* Thin (6 to 7 cm) veins of granoblastic hornblende gneiss in layered gneiss at site of sample 669; mode, 4; chem, 4.

### Amphibolite

*663.* Black, coarse-grained, hornblende-rich, folded; approximately 6 m southwest of sample 673 and 15 m west of aplite dike; mode, 5; chem, 6 and 7.

*665.* Similar to 663; west end of small knob approximately 100 m west of 663; mode, 5; chem, 6 and 7.

*664.* Dark gray, medium-grained raft, tholeiitic; approximately 18 m south of 663 and 22 m west of large aplite dike; mode, 5; chem, 6 and 7.

*672.* Similar to 664; east end of small knob; apprxoimately 18 m east of sample 665; mode, 5; chem, 6 and 7.

*Adamellite-2 Gneiss*—Pink, fine- to medium-grained, but coarser than the aplite; distinct gneissic structure but not layered.

*631.* Outcrop approximately 20 m south of southern end of large aplite dike; weathered; mode only, 8.

*654.* Approximately 45 m west of south end of large aplite dike and 25 m southeast of sample 672; mode, 8; chem, 11.

*729.* Same as 654 except chem, 11.

*730.* Same as 654 except Rb-Sr, 5; Ba, 905; *D*, 2.633.

*Aplite Dike*—Three samples from the large aplite dike show uniformity; pink, fine-grained, some secondary structure in biotite alignment.

*587.* West side near northern end and near contact with gneiss; mode, 9; chem, 12.

*628.* Resampled near 587; mode, 9; Rb-Sr, 6; Ba, 785; *D*, 2.615.

*630.* Middle of dike near southern end; mode only, 9.

**Locality 6. North of Delhi**—Outcrop on west side of County Road 6, approximately 1 km south of bridge on Minnesota River in SE¼ sec. 30, T. 114 N., R. 36 W., Redwood County, lat 44° 38′47″N, long 95° 13′14″W.

*339.* Large sample of gray, banded and contorted tonalite gneiss (Goldich and others, 1970); mode, 2; chem, 2.

*671.* New sample from same outcrop; mode, 2; chem, 2.

*671C.* Same as 671 except Rb-Sr, 2; Ba, 865; *D*, 2. 664.

**Between Localities 5 and 6**—Amphibolite samples collected by Nielsen and Weiblen (this volume) on the east side of the Minnesota River, SW¼ sec. 28, SE¼ sec. 29, NW¼ sec. 32, NE¼ sec 33, T. 114 N., R. 36 W., Renville County.

*BN-74-25.* Rb-Sr, 5; Ba, 60; *D*, 3.040.

*BN-74-26.* Rb-Sr, 5; Ba, 25; *D*, 3.010.

*BN-74-28.* Rb-Sr, 5; Ba, 70; *D*, 3.000.

Average chemical composition of BN-74-25, BN-74-26, and BN-74-28—8, column B.

## REFERENCES CITED

Aldrich, L. T., Davis, G. L., and James, H. L., 1965, Ages of minerals from metamorphic and igneous rocks near Iron Mountain, Michigan: Journal of Petrology, v. 6, p. 445–472.

Brooks, C., 1968, Relationship between feldspar alteration and the precise post-crystallization movement of rubidium and strontium isotopes in a granite: Journal of Geophysical Research, v. 73, p. 4751–4757.

Catanzaro, E. J., 1963, Zircon ages in southwestern Minnesota: Journal of Geophysical Research, v. 68, p. 2045–2048.

Doe, B. R., and Delevaux, M. H., 1980, Lead-isotope investigations in the Minnesota River Valley—Late-tectonic and posttectonic granites: Geological Society of America Special Paper 182 (this volume).

Fairbairn, H. W., Hurley, P. M., and Pinson, W. H., 1961, The relation of discordant Rb-Sr mineral and whole rock ages in an igneous rock to its time crystallization and to the time of $Sr^{87}/Sr^{86}$ metamorphism: Geochimica et Cosmochimica Acta, v. 23, p. 135–144.

Farhat, J. S., 1975, Geochemical and geochronological investigation of the early Archaean of the Minnesota River Valley, and the effect of metamorphism on Rb-Sr whole rock isochrons [Ph.D. thesis]: Los Angeles, University of California, 171 p.

Farhat, J. S., and Wetherill, G. W., 1974, Geochronology of the Minnesota River Valley: Geological Society of America Abstracts with Programs, v. 6, p. 729–730.

———1975, Interpretation of apparent ages in Minnesota: Nature, v. 257, p. 721–722.

Faure, G., 1977, Principles of isotope geology: New York, John Wiley & Sons, 464 p.

Fratta, M., and Shaw, D., 1974, "Residence" contamination of K, Rb, Li, and Tl in diabase dikes: Canadian Journal of Earth Sciences, v. 11, p. 423–429.

Goldich, S. S., 1938, A study in rock-weathering: Journal of Geology, v. 46, p. 17–58.

Goldich, S. S., and Hedge, C. E., 1974, 3,800-Myr granitic gneiss in southwestern Minnesota: Nature, v. 252, p. 467–468.

———1975, Interpretation of apparent ages in Minnesota, Reply: Nature, v. 257, p. 271–272.

Goldich, S. S., and Mudrey, M. G., Jr., 1972, Dilatancy model for discordant U-Pb zircon ages, in Tugarinov, A. I., ed., Recent contributions to geochemistry and analytical chemistry: Moscow, Nauka, p. 415–418.

Goldich, S. S., and others, 1956, $K^{40}/A^{40}$ dating of the Precambrian rocks of Minnesota [abs.]: Geological Society of America Bulletin, v. 67, p. 1698–1699.

———1961, The Precambrian geology and geochronology of Minnesota: Minnesota Geological Survey Bulletin 41, 193 p.

Goldich, S. S., Hedge, C. E., and Stern, T. W., 1970, Age of the Morton and Montevideo gneisses and related rocks, southwestern Minnesota: Geological Society of America Bulletin, v. 81, p. 3671–3695.

Goldich, S. S., and others, 1980a, Archean rocks of the Granite Falls area, southwestern Minnesota: Geological Society of America Special Paper 182 (this volume).

———1980b, Origin of the Morton Gneiss, southwestern Minnesota: Part 1. Lithology: Geological Society of America Special Paper 182 (this volume).

Hanson, G. N., 1968, K-Ar ages for hornblende from granites and gneisses and for basaltic intrusives in Minnesota: Minnesota Geological Survey Report of Investigations 8, 20 p.

Hanson, G. N., and others, 1971, Age of the early Precambrian rocks of the Saganaga Lake–Northern Light Lake area, Ontario-Minnesota: Canadian Journal of Earth Sciences, v. 8, p. 1110–1124.

Hart, S. R., and Davis, G. L., 1969, Zircon U-Pb and whole-rock Rb-Sr ages and early crustal development near Rainy Lake, Ontario: Geological Society of America Bulletin, v. 80, p. 595–616.

Heier, K. S., 1965, Metamorphism and the chemical differentiation of the crust: Geologiska Foereningens Stockholm Foerhandlinger, v. 87, p. 249–256.

Krogh, T. E., 1973, A low-contamination method for hydrothermal decomposition of zircon and extraction of U and Pb for isotopic age determinations: Geochimica et Cosmochimica Acta, v. 37, p. 485–494.

Lambert, I. B., and Heier, K. S., 1968, Geochemical investigations of deep-seated rocks in the Australian shield: Lithos, v. 1, p. 30–53.

Lund, E. H., 1956, Igneous and metamorphic rocks of the Minnesota River Valley: Geological Society of America Bulletin, v. 67, p. 1475–1490.

McCulloch, M. T., and Wasserburg, G. J., 1978, Sm-Nd and Rb-Sr chronology of continental crust formation: Science, v. 200, p. 1003–1011.

———1980, Early Archean Sm-Nd model ages from a tonalitic gneiss, northern Michigan: Geological Society of America Special Paper 182 (this volume).

Michard-Vitrac, A., and others, 1977, U-Pb ages on single zircons from the early Precambrian rocks of West Greenland and the Minnesota River Valley: Earth and Planetary Science Letters, v. 35, p. 449–453.

Moore, L. J., and others, 1973, Trace determinations of rubidium and strontium in silicate glass standard reference materials: Analytical Chemistry, v. 45, p. 2384–2387.

Nielsen, B. V., and Weiblen, P. W., 1980, Mineral and rock compositions of mafic enclaves in the Morton Gneiss: Geological Society of America Special Paper 182 (this volume).

Peterman, Z. E., and others, 1972, Geochronology of the Rainy Lake region, Minnesota-Ontario, in Doe, B. R., and Smith, D. K., eds., Studies in mineralogy and Precambrian geology: Geological Society of America Memoir 135, p. 193–215.

Peterman, Z. E., Zartman, R. E., and Sims, P. K., 1980, Tonalitic gneiss of early Archean age from northern Michigan: Geological Society of America Special Paper 182 (this volume).

Riley, G. H., and Compston, W., 1962, Theoretical and technical aspects of Rb-Sr geochronology: Geochimica et Cosmochimica Acta, v. 26, p. 1255–1281.

Silver, L. T., and Deutsch, S., 1961, Uranium-lead method on zircons, in Kulp, J. L., ed., Geochronology of rocks systems: Annals of the New York Academy of Sciences, v. 91, art 2, p. 279–283.

Steiger, R. H. and Jäger, E., 1977, Subcommission on Geochronology: Convention on the use of decay constants in geo- and cosmochronology: Earth and Planetary Science Letters, v. 37, p. 359–362.

Stern, T. W., Goldich, S. S., and Newell, M. F., 1966, Effects of weathering on the U-Pb ages of zircon from the Morton Gneiss, Minnesota: Earth and Planetary Science Letters, v. 1, p. 369–371.

Tarney, J., Skinner, A. C., and Sheraton, J. W., 1972, A geochemical comparison of major Archaean gneiss units from North-West Scotland and East Greenland: International Geological Congress, 24th, Montreal, Report 1, p. 162–174.

Tilton, G. R., 1960, Volume diffusion as a mechanism for discordant lead ages: Journal of Geophysical Research, v. 65, p. 2933–2945.

Wetherill, G. W., 1956, Discordant uranium-lead ages, I: American Geophysical Union Transactions, v. 37, p. 320–326.

Wooden, J. L., Goldich, S. S., and Suhr, N. H., 1980, Origin of the Morton Gneiss, southwestern Minnesota: Part 2. Geochemistry: Geological Society of America Special Paper 182 (this volume).

York, D., 1966, Least-squares fitting of a straight line: Canadian Journal of Physics, v. 44, p. 1079–1086.

MANUSCRIPT RECEIVED BY THE SOCIETY APRIL 25, 1979
REVISED MANUSCRIPT RECEIVED NOVEMBER 26, 1979
MANUSCRIPT ACCEPTED JANUARY 11, 1980

Geological Society of America
Special Paper 182
1980

# Mineral and rock compositions of mafic enclaves in the Morton Gneiss

BRUCE V. NIELSEN *
PAUL W. WEIBLEN
*Minnesota Geological Survey, 1633 Eustis, St. Paul, Minnesota 55108*

## ABSTRACT

Two major mineralogic types of mafic enclaves (amphibolite and pyroxene-amphibolite) have been characterized in the Morton quartzofeldspathic gneiss at Morton, Minnesota. Some amphibolite contains plagioclase megacrysts with $An_{66}$ cores and $An_{43}$ rims. One indication of compositional equilibration of the amphibolite with host gneiss is the presence of biotite-rich rims on many enclaves. Textural and compositional reaction relationships observed in the mafic enclaves may be interpreted as relict igneous features. Bulk compositional data from three representative enclaves at Morton suggest that the pyroxene amphibolite could be derived from amphibolite by crystal fractionation and that both enclave types could have crystallized from a tholeiitic magma.

## INTRODUCTION

The question of the origin of deformed and metamorphosed mafic rocks in Archean quartzofeldspathic gneisses raises several broad questions concerning Archean petrogenesis. Was the basement for the gneiss terrane a mafic volcanic sequence that was included as remnants in the sedimentary or igneous precursors of the gneiss terrane? Was a metasedimentary, metavolcanic, or a felsic-plutonic terrane intruded by mafic rocks prior to extensive deformation and metamorphism? Or did an early stage of metamorphism of a sedimentary sequence produce leucocratic and melanocratic segregations in the gneisses? Given the stratigraphic, structural, and metamorphic complexity of most Archean gneiss terranes it is not unlikely that mafic rocks of all three of the above origins might be found in a single exposure. The long-range goal of our study of the mafic rocks in the quartzofeldspathic gneisses in the Minnesota River Valley is to obtain the data necessary to determine their different possible origins. We report here mineral and rock compositions and textural data on a limited number of mafic rocks from a restricted group of outcrops in the Morton area. The data presented do not permit drawing broad generalizations concerning all the mafic rocks in the Minnesota River Valley but do allow definition of two distinct textural and compositional types. We suggest that the data may be construed to support the conclusion that they are remnants of a mafic volcanic-intrusive sequence included in quartzofeldspathic gneiss.

## PREVIOUS WORK

Lund (1956) noted the interlayered and fragmented relationships of the amphibolite to the other gneisses. Himmelberg (1965, p. 11) recognized and mapped several 10-m- to 1-km-thick conformable amphibolite lenses in the granitic and hornblende-pyroxene gneisses at Granite Falls. He also noted three unusual occurrences of amphibolite lenses with plagioclase porphyroblasts up to 7.5 cm in length. Grant (1972, p. 184) described the occurrence of amphibolite farther south in the Minnesota River Valley in the vicinity of Delhi, Minnesota. He found that amphibolite could be mapped as conformable "rafts" and smaller lenses that are veined and crosscut by gneissic material. He used the abundance of amphibolite with sharp, angular contacts with the enclosing quartzofeldspathic gneiss to define three amphibolite units within the gneisses at Delhi and Morton. He also found that the amount of amphibolite decreases upward in the section at these localities. On the basis of field occurrence, all three of the authors referred to above assumed igneous rocks to be the precursors for the amphibolites, and they suggested that the amphibolites could be intrusive or interlayered mafic volcanic rocks in the protoliths of quartzofeldspathic, hornblende-pyroxene, and biotite gneisses.

## METHODS

The data presented in this paper are from work by Nielsen (1976) on the amphibolites exposed in the vicinity of Morton, Minnesota. This includes field and petrographic descriptions of

---

* Present address: Lancaster, Minnesota 56735.

specific occurrences of amphibolite along with electron microprobe analyses of certain phases by Nielsen and bulk compositional data provided by S. S. Goldich (1976, written commun.).

The electron microprobe analyses were obtained on the Model 400 Materials Analysis Probe in the Department of Geology and Geophysics at the University of Minnesota. All analyses were made at 20 kV, 0.015 $\mu$A with an approximately 5-$\mu$m-diameter beam. Each analysis is an average of 2 to 3 replicate, 24-s analyses. Data reduction followed that used by Grant and Weiblen (1971). As can be seen from the sums and the stoichiometry, the electron microprobe data are not of the highest quality, but they provide a basis for establishing the major differences in phase chemistry of the amphibolites studied by Nielsen in the Morton area.

## FIELD OCCURRENCE AND GENERAL CHARACTERISTICS OF AMPHIBOLITE IN THE MORTON AREA

Dark gray to black phaneritic amphibolite enclaves in the gneisses at Morton, which were accessible for observation and sampling in 1974 and 1975 (Fig. 1) in the quarries and surface exposures, range up to 20 m in the largest dimension with most less than 1 m. Their form ranges from rounded to angular and equidimensional to lenticular. Several occurrences of conformable lenticular enclaves in an en echelon pattern suggestive of disrupted dikes are presently exposed in the north wall of the active quarry at Morton (loc. 2, Fig. 1). Layering in the gneiss tends to wrap around less than metre-sized, angular enclaves. In many cases foliation and lineations within enclaves do not conform to those in the enclosing gneiss. Many of the angular enclaves occur in clusters of fragments with angular outlines that suggest they could be fitted together to form a single enclave (south wall, quarry 3, Fig. 1). Swirls in the foliation in the gneiss define unique directions of rotation of the original unbroken enclaves. Biotite-rich rims on enclave fragments are best developed on the trailing edges of apparently rotated blocks.

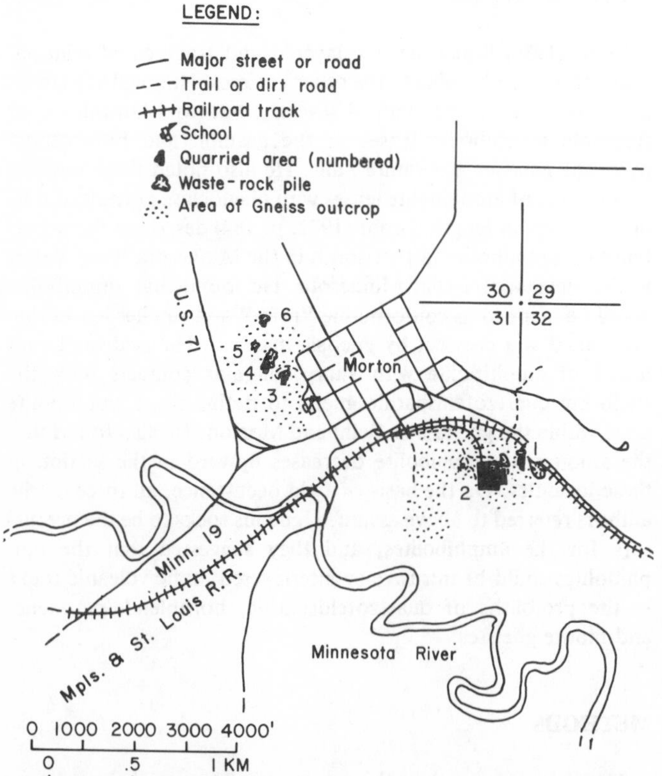

Modal data from 30 enclaves indicate that the mafic enclaves at Morton consist of two compositional types: (1) amphibolite enclaves containing essentially hornblende-plagioclase, with the proportions of hornblende to plagioclase ranging from 1:1 to 2:1 and including as much as 10% biotite (Table 1, cols. 1, 2, 3) and (2) pyroxene amphibolite enclaves containing significant amounts (as much as 30% of clinopyroxene and lesser amphibole (Table 1, cols. 4 through 7).

The amphibolite enclaves found in quarries 1 through 4 (Fig. 1) are all less than 2 m in size, whereas the pyroxene amphibolite enclaves, although less common, are all larger. Only two of the latter were studied by Nielsen (1976). One (enclave BN-1-2) is exposed in the surface exposures between quarries 1 and 2 (Fig. 1) and has a minimum size of 20 x 15 x 5 m, and the other (BN-2) of roughly the same size was exposed in the working floor of quarry 2 (Fig. 1) in 1974.

### Amphibolite Enclaves

The amphibolite enclaves are equigranular, phaneritic rocks with a granulose or weakly lineated texture. A rim of well-foliated biotite may be found along the amphibolite-gneiss boundary. Rounded megacrysts of variably altered plagioclase ranging up to 1 cm in diameter are found in the amphibolites of quarries 3 and 4. Some examples of these megacrysts are found in evenly distributed (several centimetres between megacrysts), subparallel, planar swarms as much as 1 m wide. In the north end of the floor in quarry 3 (Fig. 1), the orientation of the swarms of individual elongate megacrysts in the enclave do not coincide with the lineation in the amphibolite or foliation in the gneiss.

The amphibolite enclaves consist basically of hornblende and plagioclase with variable accessories, the most abundant of which is biotite (Table 1). Two other common accessory phases are ilmenite and sphene. Apatite occurs as fine-grained inclusions in plagioclase and as separate grains. Zircon is occasionally found in biotite and hornblende crystals in most of the enclaves examined. Two phases of sporadic occurrence are pyrite and pyrrhotite. Other minor phases include garnet, quartz, actinolite, clinopyroxene, and chlorite. Sericite and saussurite alteration patches in some of the plagioclase are common to most enclaves.

Hornblende in the amphibolites has an iron-pargasite composition (Table 2a, cols. 1 through 4). The hornblendes are occasionally zoned, being more actinolitic (higher Fe and Si and low Na, K, and Al and light blue-green) at the borders. The bulk of the hornblende is dark-green to green-brown and is subhedral to euhedral but fairly equidimensional in form. Euhedral horn-

**Figure 1. Sketch map of the Morton area (sec. 31, T. 113 N., R. 34 W.), Renville County, southwestern Minnesota, showing the location of quarries and outcrops mentioned in the text.**

TABLE 1. MODAL COMPOSITIONS

| | Amphibole | | | Pyroxene Amphibolite | | | |
|---|---|---|---|---|---|---|---|
| | 1 | 2 | 3 | 4 | 5 | 6 | 7 |
| Plagioclase | 35.7 | 31.6 | 37.4 | 50.4 | 56.4 | 57.4 | 60.0 |
| Dark Amphibole | 53.0 | 66.7 | 61.9 | 11.8 | 10.3 | 9.8 | 5.1 |
| Light Amphibole | tr | tr | – | – | – | 15.5 | tr |
| Biotite | 10.3 | 0.5 | 0.2 | 0.3 | 10.3 | 4.5 | 13.1 |
| Orthopyroxene | – | – | – | 3.8 | – | – | – |
| Clinopyroxene | – | tr | – | 28.9 | 27.7 | 7.8 | 17.5 |
| Sulfide | 0.2 | 0.3 | 0.2 | – | tr | 7.8 | 0.4 |
| Oxide | 0.2 | 0.2 | – | 4.8 | 4.9 | 1.6 | 3.8 |
| Sphene | 0.5 | 0.7 | 0.3 | – | 0.1 | 0.1 | 0.1 |
| Quartz | 0.1 | tr | tr | tr | – | 3.2 | tr |
| Garnet | – | – | – | – | – | tr | – |
| Apatite | tr | tr | tr | tr | tr | – | tr |
| Zircon | tr | tr | – | tr | tr | tr | tr |

Note: Data based on 1,000 to 2,000 point counts on each sample. For this and subsequent tables, numbered column headings are explained in footnotes; the first three items in parentheses form the sample number, and the last item refers to a specific thin section.

1. Outer rim of an angular, 0.5-m-sized amphibolite enclave in the wall of quarry 2 in 1974. Note high biotite content (BN-74-10-1).

2. Amphibolite from center of same enclave as 1 (BN-74-11-3). Note low biotite content. See Figure 2 for data on variation of biotite content between 1 and 2.

3. Amphibolite from about 0.5 m from edge of a 1.5-m-sized rounded amphibolite enclave with plagioclase megacrysts exposed in north end of floor of quarry 3 (Fig. 1; BN-74-23b-B).

4. Pyroxene amphibolite from the center of a 20 x 15 x 5 m-sized enclave (BN-1-2) exposed in a quarry core between quarries 1 and 2 (Fig. 1). Note the significant orthopyroxene content (BN-74-32-37).

5. Pyroxene amphibolite from near edge of enclave BN-1-2. This sample contains no orthopyroxene (BN-74-29-35).

6. Pyroxene amphibolite from near edge of 15 x 10 x 3 m-sized enclave (BN-2) exposed in north end of quarry 2 (Fig. 1) in 1974 (BN-74-4-20).

7. Pyroxene amphibolite from enclave BN-2. Sample taken 1 m below sample 6 on vertical face (BN-74-4-39j).

TABLE 2a. AMPHIBOLE COMPOSITIONS

| | 1 | 2 | 3 | 4 | 5 | 6 | 7 | 8 | 9 |
|---|---|---|---|---|---|---|---|---|---|
| $SiO_2$ | 43.56 | 44.00 | 43.34 | 43.32 | 42.37 | 43.13 | 50.82 | 47.03 | 54.98 |
| $Al_2O_3$ | 8.40 | 9.47 | 9.09 | 12.07 | 12.46 | 10.84 | 2.79 | 5.07 | 3.05 |
| FeO | 19.02 | 19.78 | 18.20 | 21.00 | 17.75 | 20.80 | 18.93 | 19.75 | 16.37 |
| MgO | 8.24 | 8.30 | 8.69 | 6.89 | 9.35 | 6.67 | 11.32 | 9.09 | 13.45 |
| MnO | 0.22 | 0.25 | 0.46 | 0.31 | 0.77 | 0.19 | 0.13 | 0.36 | 0.73 |
| $TiO_2$ | 1.89 | 2.88 | 2.16 | 1.08 | 1.31 | 2.31 | 0.82 | 1.35 | 0.06 |
| CaO | 11.94 | 10.78 | 12.30 | 12.57 | 12.08 | 12.23 | 12.16 | 11.94 | 11.98 |
| $Na_2O$ | 1.26 | 1.31 | 1.15 | 0.86 | 1.28 | 0.88 | 0.14 | 0.94 | 0.41 |
| $K_2O$ | 1.25 | 1.01 | 1.26 | 1.42 | 1.51 | 1.28 | 0.29 | 1.12 | 0.18 |
| Total | 95.78 | 97.78 | 96.65 | 99.52 | 98.88 | 98.33 | 97.40 | 96.65 | 101.21 |
| Atoms per 23 oxygens | | | | | | | | | |
| Si | 6.769 | 6.644 | 6.660 | 6.504 | 6.313 | 6.562 | 7.570 | 7.219 | 7.744 |
| Al IV | 1.231 | 1.356 | 1.340 | 1.496 | 1.687 | 1.438 | 0.430 | 0.781 | 0.250 |
| Al VI* | 0.308 | 0.330 | 0.307 | 0.640 | 0.501 | 0.506 | 0.060 | 0.136 | 0.250 |
| Fe II | 2.472 | 2.304 | 2.340 | 2.545 | 1.900 | 2.647 | 2.226 | 2.535 | 1.884 |
| Fe III | – | 0.194 | – | 0.091 | 0.312 | – | 0.131 | – | 0.044 |
| Mg | 1.902 | 1.868 | 1.990 | 1.542 | 2.076 | 1.512 | 2.513 | 2.079 | 2.823 |
| Mn | 0.029 | 0.032 | 0.060 | 0.039 | 0.097 | 0.024 | 0.016 | 0.047 | 0.087 |
| Ti | 0.221 | 0.327 | 0.250 | 0.122 | 0.149 | 0.264 | 0.092 | 0.166 | 0.808 |
| Ca | 1.988 | 1.744 | 2.025 | 2.022 | 1.929 | 1.994 | 1.941 | 1.964 | 1.808 |
| Na | 0.380 | 0.384 | 0.343 | 0.250 | 0.370 | 0.259 | 0.041 | 0.280 | 0.102 |
| K | 0.248 | 0.195 | 0.247 | 0.272 | 0.287 | 0.248 | 0.055 | 0.219 | 0.032 |
| Total | 15.548 | 15.378 | 15.562 | 15.523 | 15.621 | 15.454 | 15.075 | 15.426 | 15.047 |

Note:

1. Hornblende in amphibolite enclave (BN-74-10-1).

2. Hornblende in amphibolite enclave (BN-74-11-3).

3. Hornblende in amphibolite enclave (BN-74-23-Al).

4. Pargasitic hornblende intergrown with plagioclase (Table 2b, 10, 11, 12) and biotite (Table 2c, col. 3) between plagioclase megacrysts (Table 2b, cols. 4, 5, 6; BN-74-23-Al).

5. Pargasitic hornblende inclusion in plagioclase megacrysts (Table 2b, col. 4; (BN-74-23-Al).

6. Pargasitic hornblende from pyroxene amphibolite enclave BN-1-2 (BN-74-31-36).

7. Actinolitic hornblende from pyroxene amphibolite enclave BN-2 (BN-74-4-18) intergrown with kaersutitic hornblende (col. 8) surrounding relict clinopyroxene (Table 2e, col. 6).

8. Kaersutitic hornblende from pyroxene amphibolite enclave BN-2. See above (BN-74-4-18).

9. Actinolitic hornblende from pyroxene amphibolite enclave BN-2 (BN-74-30-30K).

*Al (VI) (octahedral sites ) = Al (total) + Si (total) - 8. Al (IV) (tetrahedral sites) = Al (total) - Al (VI). Ferric iron estimated from stoichiometry and charge balance (Papike and others, 1974).

TABLE 2b. PLAGIOCLASE COMPOSITIONS

| | 1 | 2 | 3 | 4 | 5 | 6 | 7 | 8 | 9 | 10 | 11 | 12 | 13 |
|---|---|---|---|---|---|---|---|---|---|---|---|---|---|
| $SiO_2$ | 64.28 | 63.24 | 64.05 | 48.00 | 48.25 | 46.00 | 59.50 | 56.00 | 54.00 | 54.00 | 50.75 | 49.50 | 63.74 |
| $Al_2O_3$ | 20.80 | 21.41 | 20.98 | 33.25 | 32.90 | 33.50 | 28.30 | 29.50 | 33.75 | 27.00 | 28.30 | 30.25 | 20.27 |
| CaO | 5.41 | 5.30 | 5.19 | 15.30 | 15.42 | 15.67 | 7.73 | 9.59 | 13.30 | 9.80 | 11.94 | 14.03 | 5.90 |
| $Na_2O$ | 8.78 | 6.55 | 7.11 | 2.35 | 2.22 | 2.18 | 5.57 | 5.05 | 3.70 | 4.50 | 5.10 | 3.70 | 7.75 |
| $K_2O$ | 0.39 | 0.30 | 0.17 | 0.05 | 0.04 | 0.01 | 0.10 | 0.16 | 0.10 | 0.14 | 0.11 | 0.07 | 0.50 |
| Total | 99.66 | 96.80 | 97.50 | 98.95 | 98.83 | 97.36 | 101.20 | 100.30 | 104.85 | 95.44 | 96.20 | 97.55 | 98.16 |
| | | | | | | Atoms per 8 oxygens | | | | | | | |
| Si | 2.860 | 2.866 | 2.883 | 2.214 | 2.227 | 2.164 | 2.607 | 2.497 | 2.329 | 2.531 | 2.398 | 2.315 | 2.875 |
| Al | 1.091 | 1.080 | 1.114 | 1.808 | 1.790 | 1.858 | 1.457 | 1.522 | 1.717 | 1.493 | 1.577 | 1.667 | 1.078 |
| Ca | 0.258 | 0.257 | 0.250 | 0.756 | 0.763 | 0.790 | 0.363 | 0.459 | 0.615 | 0.493 | 0.604 | 0.703 | 0.285 |
| Na | 0.757 | 0.576 | 0.620 | 0.210 | 0.199 | 0.199 | 0.474 | 0.437 | 0.309 | 0.409 | 0.467 | 0.336 | 0.678 |
| K | 0.022 | 0.017 | 0.010 | 0.003 | 0.003 | - | 0.005 | 0.009 | 0.005 | 0.008 | 0.007 | 0.004 | 0.028 |
| Total | 4.988 | 4.796 | 4.877 | 4.991 | 4.982 | 5.011 | 4.906 | 4.924 | 4.975 | 4.934 | 5.053 | 5.025 | 4.944 |
| Or | 2.1 | 2.0 | 1.2 | 0.3 | 0.3 | - | 0.6 | 1.0 | 0.5 | 0.9 | 0.7 | 0.4 | 2.8 |
| Ab | 73.0 | 67.8 | 70.4 | 21.7 | 20.6 | 20.1 | 56.3 | 48.3 | 33.3 | 44.9 | 43.3 | 32.2 | 68.4 |
| An | 24.9 | 30.2 | 28.4 | 78.0 | 79.1 | 79.9 | 43.1 | 50.7 | 66.2 | 54.2 | 56.0 | 67.4 | 28.8 |

Note:
1, 2.  Plagioclase in amphibolite enclave (BN-74-10-1, 2).
3.  Plagioclase in amphibolite enclave (BN-74-11-3).
4.  Plagioclase core of 8-mm-long megacryst in enclave BN-74-23.  See Table 1, col. 3 (BN-74-23-A1).
5.  Plagioclase rim, same grain as analysis 4 (BN-74-23-A1).
6.  Plagioclase core of 2 mm-long megacrysts (BN-74-23-A1).
7-9  Plagioclase rim surrounding cluster of plagioclase megacrysts (cols. 4, 5, 6; 7, outer part of rim; 8, interior of rim; 9, inner edge of rim; BN-74-23-A1).
10-12. Plagioclase rimming individual grains of hornblende within cluster of plagioclase megacrysts (10 next to hornblende; 11, center of rim; 12, outerpart of rim; BN-74-23-A1).
13.  Plagioclase in pyroxene amphibolite enclave BN-2.  Typical composition of cores of grains that are zoned toward cracks and edges down to $An_{20}$.

TABLE 2c. BIOTITE COMPOSITIONS

| | 1 | 2 | 3 | 4 | 5 | 6 |
|---|---|---|---|---|---|---|
| $SiO_2$ | 35.52 | 36.02 | 35.19 | 36.29 | 36.17 | 35.08 |
| $Al_2O_3$ | 13.64 | 13.82 | 15.04 | 12.55 | 14.38 | 15.82 |
| FeO | 24.27 | 23.08 | 22.36 | 21.77 | 22.70 | 21.57 |
| MgO | 10.27 | 10.23 | 10.37 | 11.20 | 8.53 | 9.85 |
| CaO | 0.01 | - | 0.08 | 0.07 | - | 0.05 |
| $K_2O$ | 9.68 | 9.29 | 9.19 | 9.53 | 9.82 | 9.69 |
| $TiO_2$ | 2.81 | 3.20 | 3.07 | 2.72 | 4.51 | 2.86 |
| MnO | 0.31 | 0.27 | 0.30 | 0.23 | 0.09 | 0.27 |
| $Cr_2O_3$ | - | - | 0.04 | - | - | - |
| Total | 96.51 | 95.91 | 95.64 | 94.36 | 96.20 | 95.19 |
| | | | Atoms per 22 oxygens | | | |
| Si | 5.525 | 5.581 | 5.452 | 5.698 | 5.581 | 5.449 |
| Al | 2.501 | 2.524 | 2.747 | 2.323 | 2.616 | 2.897 |
| Fe | 3.157 | 2.991 | 2.897 | 2.859 | 2.929 | 2.802 |
| Mg | 2.381 | 2.362 | 2.394 | 2.621 | 1.961 | 2.280 |
| Ca | 0.002 | - | 0.013 | .012 | - | .008 |
| Na | - | - | - | - | - | - |
| K | 1.921 | 1.836 | 1.816 | 1.909 | 1.933 | 1.920 |
| Ti | 0.329 | 0.373 | 0.358 | 0.321 | 0.523 | 0.334 |
| Mn | 0.041 | 0.035 | 0.039 | 0.031 | 0.012 | 0.036 |
| Cr | - | - | 0.005 | - | - | - |
| Total | 15.857 | 15.702 | 15.721 | 15.774 | 15.555 | 15.726 |

Note:
1.  Biotite in amphibolite enclave (BN-74-10-1).
2.  Biotite in amphibolite enclave (BN-74-11-3).
3.  Biotite intergrown with hornblende (Table 2a, col. 4), both of which are rimmed by plagioclase (Table 2b, BN-74-23-A1).
4.  Biotite that occurs with hornblende and plagioclase outside megacryst clusters in amphibolite enclave (BN-74-23-A1).
5.  Biotite in pyroxene amphibolite BN-1-2  (BN-74-31-36).
6.  Biotite intergrown with hornblende (Table 2a, cols. 7, 8) in pyroxene amphibolite enclave (BN-74-4-18).

| | 1 | 2 | 3 | 4 | 5 | 6 | 7 | 8 | 9 |
|---|---|---|---|---|---|---|---|---|---|
| $SiO_2$ | 0.77 | 0.01 | – | 35.3 | 34.5 | 0.09 | 0.08 | – | – |
| $Al_2O_3$ | 0.18 | 0.05 | – | 0.79 | 1.33 | 0.16 | 0.13 | 0.67 | 0.52 |
| FeO | 44.6 | 46.3 | – | 0.91 | 0.77 | 44.9 | 93.2 | 91.8 | 46.8 |
| MgO | 0.19 | – | – | – | – | – | – | 0.41 | 0.62 |
| CaO | – | – | – | 30.5 | 26.6 | 0.06 | – | – | – |
| $TiO_2$ | 51.1 | 50.70 | – | 32.3 | 33.70 | 51.6 | 0.38 | 1.60 | 49.9 |
| MnO | 3.17 | 4.47 | – | 0.24 | – | 2.01 | 0.02 | – | 0.98 |
| $Cr_2O_3$ | – | – | – | – | – | – | 0.14 | 0.30 | – |
| S | – | – | 53.4 | – | – | – | – | – | – |
| Fe | – | – | 44.4 | – | – | – | – | – | – |
| Total | 100.01 | 101.53 | 97.8 | 100.04 | 96.90 | 98.82 | 93.95 | 94.78 | 98.82 |

| | | | | | Atoms | | | | |
|---|---|---|---|---|---|---|---|---|---|
| O | 6.000 | 6.0000 | – | 20.000 | 20.000 | 6.000 | 32.000 | 32.000 | 6.000 |

| | 1 | 2 | 3 | 4 | 5 | 6 | 7 | 8 | 9 |
|---|---|---|---|---|---|---|---|---|---|
| Si | 0.039 | – | – | 4.577 | 4.563 | 0.005 | 0.032 | – | – |
| Al | 0.012 | 0.003 | – | 0.121 | 0.207 | 0.010 | 0.062 | 0.311 | 0.031 |
| Fe++ | 1.880 | 1.954 | – | 0.099 | 0.085 | 1.919 | 31.530 | 30.000 | 2.009 |
| MG | 0.014 | – | – | – | – | – | – | 0.240 | 0.047 |
| Ca | – | – | – | 4.238 | 3.770 | 0.003 | – | – | – |
| Ti | 1.937 | 1.924 | – | 3.150 | 3.352 | 1.983 | 0.116 | 0.473 | 1.926 |
| Mn | 0.135 | 0.191 | – | 0.026 | – | 0.087 | 0.007 | – | 0.043 |
| Cr | – | – | – | – | – | – | 0.045 | 0.093 | – |
| S | – | – | 2.095 | – | – | – | – | – | – |
| Fe° | – | – | 1.000 | – | – | – | – | – | – |
| Total cations | 4.017 | 4.072 | 3.095 | 12.211 | 11.977 | 4.007 | 31.792 | 31.117 | 4.056 |

Note:
1. Ilmenite in amphibolite enclave (BN-74-10-1).
2. Ilmenite in amphibolite enclave (BN-74-1).
3. Pyrite in amphibolite enclave (BN-74-3).
4. Sphene in amphibolite enclave (BN-74-11-3).
5. Sphene in amphibolite enclave (BN-74-10-1).
6. Ilmenite in pyroxene amphibolite enclave BN-1-2.
7. Magnetite surrounded by ilmenite (col. 6) in pyroxene amphibolite enclave BN-1-2 (BN-74-4).
8. Magnetite in pyroxene amphibolite enclave BN-2 (BN-31-36).
9. Ilmenite in pyroxene amphibolite enclave BN-2 (BN-31-36).

| | 1 | 2 | 3 | 4 | 5 | 6 |
|---|---|---|---|---|---|---|
| $SiO_2$ | 51.47 | 53.36 | 51.10 | 50.61 | 51.85 | 53.10 |
| $Al_2O_3$ | 0.49 | 0.94 | 0.59 | 0.56 | 0.83 | 0.52 |
| FeO | 12.11 | 14.34 | 35.32 | 36.48 | 13.18 | 12.23 |
| MgO | 12.58 | 9.37 | 11.58 | 10.04 | 9.76 | 10.21 |
| MnO | 0.56 | 0.43 | 1.06 | 0.79 | 0.37 | 0.46 |
| $TiO_2$ | – | 0.83 | 0.12 | 0.40 | 0.06 | 0.12 |
| CaO | 21.38 | 20.71 | 1.02 | 0.88 | 22.48 | 21.95 |
| Total | 98.59 | 99.98 | 100.79 | 99.76 | 98.53 | 98.59 |
| | | | Atoms per 6 oxygens | | | |
| Si | 1.975 | 2.204 | 2.014 | 2.026 | 2.004 | 2.034 |
| Al | 0.022 | 0.042 | 0.027 | 0.026 | 0.038 | 0.023 |
| Fe | 0.388 | 0.455 | 1.164 | 1.221 | 0.426 | 0.392 |
| Mg | 0.719 | 0.530 | 0.680 | 0.599 | 0.562 | 0.583 |
| Mn | 0.018 | 0.014 | 0.035 | 0.027 | 0.012 | 0.015 |
| Ti | – | 0.024 | 0.004 | 0.012 | 0.002 | 0.003 |
| Ca | 0.879 | 0.842 | 0.043 | 0.038 | 0.931 | 0.901 |
| Total | 4.001 | 4.111 | 3.967 | 3.949 | 3.975 | 3.951 |
| Wo | 44.74 | 46.09 | 2.28 | 2.03 | 48.51 | 48.03 |
| EN | 36.61 | 29.00 | 36.04 | 32.24 | 29.29 | 31.08 |
| FS | 18.64 | 24.91 | 61.68 | 65.73 | 22.20 | 20.89 |

Note:
1. Relict clinopyroxene in amphibolite enclave (BN-74-10-2).
2. Clinopyroxene in pyroxene amphibolite enclave BN-1-2 (BN-74-32-37).
3, 4. Orthopyroxene associated with clinopyroxene in pyroxene amphibolite enclave BN-1-2 (BN-74-32-47).
5. Clinopyroxene in pyroxene amphibolite enclave BN-2 (BN-74-30e).
6. Clinopyroxene in pyroxene amphibolite enclave BN-1-2 (BN-74-4-18). Pyroxene occurs in a poikilitic relationship with plagioclase but is surrounded by amphibole (Table 2a, cols. 7, 8).

blende shows a much stronger cleavage parting than subhedral types and is less commonly zoned. Banded swarms of very fine grained (less than 5 $\mu$m) euhedral opaques, which appear to be ilmenite, cut across some hornblendes oriented within the (110) plane. Other phases that may be included in hornblende are apatite and, occasionally, small grains of sulfide, zircon, and sphene. Untwinned plagioclase also occurs in embayments in the hornblende.

Reconnaissance electron microscope analyses of plagioclase (excluding megacrysts) range in composition from $An_{20}$ to $An_{40}$ for all amphibolites. A maximum variation of about $An_5$ was found in enclaves BN-74-10, 11 (Table 2b, cols. 1, 2, 3). Plagioclase crystals are subhedral to anhedral and have a greater range in size than hornblende. Some plagioclase is twinned and displays albite, carlsbad, and pericline twinning. Twinned plagioclase grains are generally more euhedral, larger, more calcic, more strongly zoned, and apparently more susceptible to sericite-saussurite alterations than untwinned varieties. Many plagioclase grains are compositionally zoned with the maximum variation again excluding megacrysts, being less than $An_5$. The more anhedral or rounded grains are not zoned or are only weakly zoned.

Biotite has a weight MgO/FeO ratio of approximately 0.5 (Table 2c, cols. 1, 2). As previously pointed out, biotite is concentrated along the outer margins of enclaves (Figs. 2, 3) with the elongate biotite grains favoring an orientation subparallel to the edge.

The opaque minerals found in the amphibolite enclaves are ilmenite with minor magnetite intergrowths (Table 2d, cols. 1, 2) and iron sulfide (col. 3). The opaques are mostly subhedral and anhedral with some sulfide grains being highly rounded; however, very fine grained oxides are euhedral.

Sphene occurs as anhedral grains concentrated near the outer edges of amphibolite enclaves (Table 2d, cols. 4, 5). It is found rimming ilmenite with the rimming becoming progressively less developed toward enclave interiors. Sphene is also found as small grains apparently associated with simplectic intergrowths of amphibole and plagioclase.

Apatite is an ubiquitous mineral that occurs as small euhedral inclusions in plagioclase and hornblende and as larger isolated anhedral crystals in the amphibolite.

Quartz is rare in the interiors of amphibolite enclaves (Table 1, cols. 1, 2, 3) but was found in minor amounts as blebs associated with irregular patches of actinolitic hornblende in some enclaves. Quartz is also associated with the biotite-rich rims of amphibolite enclaves.

Zircon is a rare but widely distributed phase. Qualitative microprobe analysis confirmed a measureable amount of hafnium in zircon with the peak intensity of hafnium about an order of magnitude lower than that for zirconium. Zircon occurs as small, rounded, partially metamict grains in hornblende and biotite that are easily recognizable by the radiation halo commonly formed in the host mineral.

Garnet was found as a minor phase in two enclaves in quarry 2 where it occurs as small- to medium-sized euhedral grains.

Clinopyroxene was found in only *one* amphibolite enclave, as a *single* irregular grain (Table 2e, col. 1) mantled by actinolite in one of eight thin sections made from the enclave. The fact that the bulk of the amphibolite in the enclave is hornblende suggests that it did not form from the same type of pyroxene as the single relict grain.

The plagioclase megacrysts in the amphibolites of quarries 3 and 4 are more complicated than their name suggests. They occur as isolated grains and in clusters. Some clusters of megacrysts

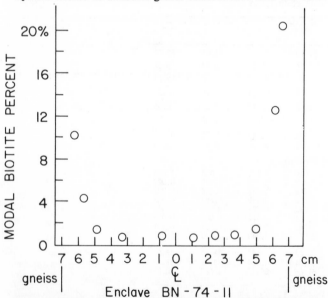

Figure 2. Variations in the relative proportions of modal biotite across a small amphibolite inclusion in the north wall of quarry 2 (Fig. 1). Sample BN-74-11 is from the center of this inclusion; see Table 1, columns 1 and 2, for total mode at edge and center of this inclusion, respectively. The relative abundance of modal biotite at the edges of this inclusion implies that $K_2O$ is concentrated along the margins of this inclusion (see also Fig. 3).

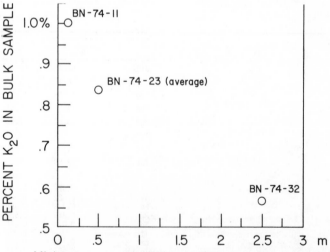

Figure 3. Note that the $K_2O$ content of seven whole-rock samples collected from three amphibolite inclusions appears to be related to their proximity to an apparent gneiss-inclusion contact. The sample distance is measured in two dimensions across the exposed face of the inclusion to the nearest inclusion-gneiss contact. See Table 3, columns 1 through 7, for complete chemical analyses; point BN-74-23 represents the average value for five samples. These data and those summarized in Figure 2 and Tables 1 and 3 indicate that potassium and water were mobile components during a period of chemical equilibrium between the amphibolite and the enclosing quartzofeldspathic gneisses, with water moving out and potassium into the inclusion.

consist of several large plagioclase megacrysts (An $_{80}$; Table 2b, cols. 4, 5, 6) surrounded by a continuous rim of finer-grained plagioclase grains that are zoned from An$_{66}$ to An$_{43}$ from the inside of the rim to the outside in contact with the amphibolite matrix (Table 2b, cols. 7, 8, 9). Some megacrysts within clusters are separated from each other by intergrowths of hornblende (Table 2a, col. 4), biotite (Table 2c, col. 3), and plagioclase (Table 2b, cols. 10, 11, 12). Hornblende also occurs as rectangular inclusions in some megacrysts (Table 2a, col. 5).

## Pyroxene Amphibolite Enclaves

Although the two large, approximately 20-m-long pyroxene-rich enclaves studied are similar in size and gross textural characteristics, their petrographic characteristics are quite different.

**Pyroxene Enclave BN-1-2.** This enclave has a coarse-grained, dark green-gray, to light-colored mottled to foliated appearance in hand specimen. In thin section, however, the dark areas are aggregates of fine- to medium-grained subhedral to euhedral hornblende, generally larger irregular-shaped pyroxene and still larger irregular patches of ilmenite-magnetite and associated biotite. The light areas are aggregates of medium-grained plagioclase, which form elongate patches intergrown with the pyroxene-amphibole aggregates. The irregular ilmenite-magnetite patches are also somewhat elongated parallel to the long direction of the plagioclase patches. The overall texture defines a crude foliation.

Both clinopyroxene and orthopyroxene were found in enclave BN-1-2. Clinopyroxene (Table 2e, col. 2) is by far the most common. It occurs as medium-grained anhedral to subhedral grains. Both homogeneous and grains with fine (10 $\mu$m) exsolution lamellae (with a reciprocal Fe:Ca relationship) were found. The orthopyroxene occurs as medium-grained anhedral to subhedral, relatively homogeneous crystals. It has a distinctly lower Mg/Fe ratio than associated clinopyroxene (Table 2e, cols. 3, 4).

Plagioclase is subhedral to anhedral and coarser grained than the mafic phases. It is commonly twinned and has a compositional range from An$_{34}$ to An$_{40}$. Zoning is generally weaker than the plagioclase found in amphibolite enclaves but may range up to a variation of 5 mol % in a single grain. Inclusions of small euhedral grains of apatite and flecks of sericite are common in the plagioclase.

Biotite occurs as medium-grained flakes and tends to be associated with the large irregular oxide patches. It is least common near the enclave center (Table 1, cols. 4, 5). The biotite is more reddish brown with a higher Fe content (Table 2c, col. 5) and is larger than that found in amphibolite enclaves. Zircons of similar habit to those found in biotites in the amphibolite enclaves are included in enclave BN-1-2 biotites.

Ilmenite and magnetite occur as irregular-shaped single grains and as aggregates. Compositional analyses show ilmenite to be stoichiometrically close to FeTiO$_3$ in composition (Table 2d, col. 6) and the magnetite to have a small percentage of TiO$_2$ (Table 2d, col. 7). Thin veneers of sphene were found around some oxide grains.

**Pyroxene Enclave BN-2.** The texture of this enclave differs from BN-1-2 a few hundred metres away in a number of ways. The most obvious difference is that BN-2 is cut by numerous veins ranging in size from tens of centimetres down to microscopic size

veinlets. Vein mineralogy is largely quartzofeldspathic, but veins of sphene, pyrrhotite, and pyrite are also present.

A series of 11 cores taken along a 1.5-m traverse across BN-2 show a marked variation in mineralogy and texture over distances of only a few centimetres. No orthopyroxene was found in these samples although the clinopyroxene is of similar composition (Table 2e, col. 5) and, in some areas, has a similar texture (including lamellae) to that of BN-1-2. Most of the clinopyroxene, however, is devoid of lamellae and occurs in the form of large (2 cm) amoeboid poikioblasts. The clinopyroxenes in this enclave are all rimmed by zoned actinolitic amphibole with small rounded grains of quartz in the amphibole (Table 2a, col. 7). The amphibole zoning is toward a more kaersutitic composition away from the clinopyroxene (Table 2a, col. 8). Small grains of opaque oxide also occur within the actinolite. Other samples from BN-2 show mats of fibrous tschermakite and fine-grained opaques rimmed by actinolitic hornblende (Table 2a, col. 9), and the textural relations suggest that these mats are pseudomorphs after clinopyroxene.

In addition to the amphibole described above, amphibole and biotite of the same appearance and associations as that found in BN-1-2 are present in BN-2. The amphibole in BN-2, however, has a more kaersutitic composition and is generally coarser grained. In addition, some isolated grains of amphibole are joined with amphibole rims around clinopyroxene to form optically continuous but zoned crystals.

Plagioclase is similar to that in BN-1-2 except that, where it is included in the clinopyroxene, it has a rounded outline and is generally less twinned. The plagioclase is more strongly zoned (An$_{30-20}$) than in BN-1-2 and generally has a lower An content (Table 2b, col. 13). An interesting feature of the zoning is that it does not appear to be related to crystal boundaries but in some cases parallels cracks in crystals, with the An content lowest adjacent to the cracks. This suggests that at least some zoning may not be a primary igneous feature.

The opaque minerals in BN-2 have the same irregular habit as those in BN-1-2, with magnetite and ilmenite in close association. The magnetite is more titaniferous than that in enclave BN-1-2 (Table 2d, col. 8) and is commonly surrounded by biotite, but the ilmenite is significantly lower in manganese (Table 2d, col. 9) than that in BN-1-2. There are minor occurrences of pyrite, pyrrhotite, and sphene in enclave BN-2.

## Bulk Compositions of Amphibolite in the Morton Area

Bulk chemical analyses (Table 3, cols. 1 through 8) indicate that both the amphibolite and the pyroxene-amphibolite enclaves resemble tholeiitic basalts and that the range of compositions are similar to those found for the amphibolites of the old gneiss terrane in west Greenland (Gill and Bridgewater, 1976). In detail, the amphibolite enclaves (Table 3, cols. 1 through 7) resemble a tholeiitic magma composition proposed by Rivalenti (1975, his magma Md-2) as a parent magma for one group of amphibolites in the Fiskenaesset area (Table 3, col. 9). The amphibolite enclave compositions also mimic rather closely certain Vermilion greenstone compositions (Table 3, col. 10). The pyroxene amphibolite bulk composition is notably higher in TiO$_2$ (Table 3, col. 8) and has a higher Fe/Mg ratio than the amphibolites. Mass balance calculations indicate that the pyroxene amphibolite

TABLE 3. BULK COMPOSITION OF MAFIC ENCLAVES IN THE MORTON GNEISS

| | 1 | 2 | 3 | 4 | 5 | 6 | 7 | 8 | 9 | 10 |
|---|---|---|---|---|---|---|---|---|---|---|
| $SiO_2$ | 48.0 | 48.0 | 48.3 | 48.1 | 47.8 | 48.04 | 48.7 | 49.3 | 48.19 | 50.14 |
| $Al_2O_3$ | 14.3 | 14.5 | 14.6 | 14.5 | 14.2 | 14.42 | 13.7 | 13.4 | 14.83 | 13.54 |
| $TiO_2$ | 1.09 | 1.03 | 1.08 | 1.00 | 1.04 | 1.05 | 1.35 | 2.54 | 1.42 | 1.59 |
| $Fe_2O_3$ | 3.45 | 2.96 | 3.12 | 3.40 | 3.25 | 3.24 | 3.52 | 4.04 | 2.61 | 2.77 |
| FeO | 10.07 | 9.72 | 9.76 | 9.72 | 10.27 | 9.91 | 10.24 | 12.75 | 11.10 | 11.81 |
| MnO | 0.25 | 0.22 | 0.23 | 0.25 | 0.26 | 0.24 | 0.23 | 0.23 | 0.23 | 0.30 |
| MgO | 7.19 | 7.26 | 7.19 | 7.21 | 6.96 | 7.16 | 6.07 | 4.42 | 5.59 | 5.07 |
| CaO | 10.55 | 10.72 | 10.83 | 10.13 | 10.10 | 10.47 | 9.85 | 8.55 | 10.45 | 8.25 |
| $Na_2O$ | 2.92 | 2.90 | 2.99 | 2.98 | 2.98 | 2.95 | 3.36 | 3.50 | 2.73 | 2.95 |
| $K_2O$ | 0.85 | 0.72 | 0.63 | 1.06 | 0.93 | 0.84 | 1.13 | 0.57 | 0.61 | 0.25 |
| $P_2O_5$ | 0.10 | 0.09 | 0.11 | 0.10 | 0.12 | 0.10 | 0.15 | 0.23 | 0.18 | 0.24 |
| $Cr_2O_3$ | 0.04 | 0.04 | 0.03 | 0.03 | 0.02 | 0.03 | 0.01 | 0.01 | - | - |
| SrO | 0.02 | 0.01 | 0.02 | 0.02 | 0.01 | 0.02 | 0.01 | 0.03 | - | - |
| BaO | 0.01 | 0.01 | 0.01 | 0.01 | 0.01 | 0.01 | 0.01 | 0.00 | - | - |
| $H_2O+$ | 1.33 | 1.20 | 1.28 | 1.36 | 1.38 | 1.31 | 1.09 | 0.29 | 0.81 | 2.95 |
| $H_2O-$ | 0.04 | 0.05 | 0.05 | 0.05 | 0.05 | 0.05 | 0.06 | 0.04 | - | - |
| $CO_2$ | 0.05 | 0.05 | 0.02 | 0.15 | 0.10 | 0.07 | 0.03 | 0.04 | - | 0.06 |
| Total | 100.22 | 99.43 | 100.20 | 100.02 | 99.43 | 99.86 | 99.45 | 99.90 | 96.75 | 99.92 |

CIPW MINERAL NORM

| | 1 | 2 | 3 | 4 | 5 | 6 | 7 | 8 | 9 | 10 |
|---|---|---|---|---|---|---|---|---|---|---|
| Quartz | - | - | - | - | - | - | - | 0.03 | - | 3.21 |
| Nepheline | - | - | - | - | - | - | 0.03 | - | - | - |
| Orthoclase | 5.08 | 4.33 | 3.76 | 6.35 | 3.81 | 5.01 | 6.79 | .03 | 3.68 | 1.52 |
| Albite | 24.39 | 24.38 | 25.34 | 23.75 | 24.58 | 16.78 | 27.99 | 29.26 | 23.59 | 25.76 |
| Anorthite | 23.99 | 25.19 | 24.95 | 24.34 | 24.70 | 28.32 | 19.47 | 19.50 | 26.96 | 23.70 |
| Diopside | 23.29 | 23.31 | 23.56 | 21.35 | 21.25 | 19.33 | 24.15 | 18.16 | 20.60 | 14.15 |
| Hypersthene | 0.89 | 1.69 | 0.99 | 2.77 | 6.50 | 19.40 | - | 18.30 | 11.09 | 23.83 |
| Olivine | 14.80 | 14.35 | 14.39 | 13.89 | 11.77 | 2.52 | 13.06 | - | 7.04 | - |
| Magnetite | 5.06 | 4.37 | 4.57 | 5.00 | 4.82 | 4.74 | 5.19 | 5.88 | 3.86 | 4.14 |
| Ilmenite | 2.09 | 1.99 | 2.07 | 1.93 | 2.02 | 2.01 | 2.61 | 4.84 | 2.75 | 3.12 |
| Chromite | 0.06 | 0.06 | 0.04 | 0.04 | 0.03 | 0.04 | 0.01 | 0.01 | - | - |
| Apatite | 0.24 | 0.21 | 0.26 | 0.24 | 0.29 | 0.23 | 0.35 | 0.54 | 0.43 | 0.58 |
| Calcite | 0.12 | 0.12 | 0.05 | 0.35 | 0.23 | 1.61 | 0.07 | 0.09 | - | - |

Note:
1-5.     Amphibolite enclave BN-74-23 (Table 1, col. 3). Note the higher $K_2O$ and $H_2O$ content of samples 1, 4, and 5 which are all within less than 0.5 m from the edge of the enclave. Samples 2 and 3 are both about 1 m from the edge of the enclave.
6.     Average of cols. 1 through 5.
7.     Amphibolite from center of a 0.5-m enclave (Table 1, cols, 1, 2; BN-74-11).
8.     Pyroxene amphibolite from quarry core (Table 1, col. 4; BN-74-32).
9.     Average composition of center and contact of a 15-m dike near Fiskenaesset, west Greenland (Rivalenti, 1975).
10.     Variolitic, pillowed greenstone, Vermilion district (Sims, 1972).

major-element compositions can be derived from the amphibolite compositions by crystal fractionation (Nielsen, 1976, p. 85).

## Conjectures on Possible Origins and Parental Material for the Mafic Enclaves

The rather complete textural and compositional equilibration of the amphibolite enclaves obscures their origin; however, the textural and compositional data discussed above can be used to define three possible reaction relationships that may be useful in deciphering the origin of the mafic enclaves.

1. **Biotite-rich Rims on the Amphibolite Enclaves.** These may be construed to indicate the type of metamorphic reaction that would be expected to occur across a hydrated basalt/granite or amphibolite/quartzofeldspathic gneiss interface; that is, water and potassium were mobile components during the equilibration of the amphibolite and the enclosing quartzofeldspathic gneisses, with water moving out of and potassium into the amphibolite.

2. **Rare Clinopyroxene Mantled by Actinolite in Hornblende Amphibolite Enclaves.** This occurrence may be explained by assuming the hornblende in this enclave to be a product of a metamorphic reaction involving water, groundmass pyroxenes, and plagioclase and the actinolite mantle on clinopyroxene a result of partial hydration of a phenocryst or xenocryst.

3. **Clusters of Plagioclase Megacrysts.** These clusters resemble occurrences of plagioclase in glomeroporphyritic rocks (for example, Logan intrusions, Weiblen and others, 1972) and "leopard rocks" (Heimlich and Manzer, 1973). Nielsen (1976) suggested that several of the unique characteristics of the megacrysts could be attributed to an igneous origin: (a) the An content (Table 2b, cols. 4, 5, 6) would be normal for phenocrysts; (b) the high Mg/Fe ratios of hornblende (Table 2a, col. 5) that occur as inclu-

sions in the megacrysts could reflect the high temperature composition of melt trapped in plagioclase phenocrysts (Lofgren, 1974), whereas the lower Mg/Fe ratio hornblende (Table 2a, col. 4) intergrown with biotite (Table 2c, col. 3) found between megacrysts could reflect the composition of late-stage melt; (c) the zoned plagioclase rims (Table 2b, cols. 7, 8, 9) surrounding the megacrysts could be produced during the partial equilibration of phenocrysts with surrounding melt in an original igneous rock or partial equilibration of megacryst clusters with host amphibolite during metamorphism.

In summary, all of the textural and compositional data are consistent with the view that the precursors of both the amphibolite and pyroxene amphibolite enclaves were igneous rocks. They do not provide, however, conclusive evidence on this question. The data are not inconsistent with *one* relatively straightforward working hypothesis for the origin of the two enclave types: the enclaves could have been thin tholeiitic flows that were intruded by dikes from the same magma. During an episode of low-grade metamorphism, the flows would have been more susceptible to hydration than the dikes; this would have given rise to the presently observed difference in degree of hydration of the amphibolite compared to the pyroxene amphibolite enclaves. With incorporation of the enclaves in the protolith of the quartzofeldspathic gneisses, any process of fragmentation would have been more effective for layered flows compared to massive dikes giving rise to the presently observed size and shape distribution of the two enclave types.

## ACKNOWLEDGMENTS

We are grateful to S. S. Goldich for the rock analyses and for his persistent study of the Minnesota River Valley exposures throughout the course of his distinguished career. His work has illuminated the significance of what might have been dismissed as a group of isolated inconsequential basement gneisses.

## REFERENCES CITED

Gill, R.C.O. and Bridgewater, D., 1976, The Ameralik dykes of West Greenland, the earliest known basaltic rocks intruding stable continental crust: Earth and Planetary Science Letters, v. 29, p. 276–282.

Grant, J. A., 1972, Minnesota River Valley, southwestern Minnesota, *in* Sims, P. K., and Morey, G. B., eds., Geology of Minnesota: A centennial volume: Minnesota Geological Survey, p. 177–196.

Grant, J. A., and Weiblen, P. W., 1971, Retrograde zoning in garnet near the second sillimanite isograd: American Journal of Science, v. 270, p. 281–296.

Heimlich, R. A., and Manzer, G. K., 1973, Flow differentiation within leopard rock dikes, Bighorn Mountains, Wyoming: Earth and Planetary Science Letters, v. 17, p. 350–356.

Himmelberg, G. R., 1965, Precambrian geology of the Granite Falls-Montevideo area, Minnesota [Ph.D. thesis]: Minneapolis, University of Minnesota, 101 p.

Lofgren, G., 1974, An experimental study of plagioclase crystal morphology: Isothermal crystallization: American Journal of Science, v. 274, p. 243–273.

Lund, E. H., 1956, Igneous and metamorphic rocks of the Minnesota River Valley: Geological Society of America Bulletin, v. 67, p. 1475–1490.

Nielsen, B. V., 1976, The mafic enclaves of the Morton Gneiss, Morton, Minnesota [Ms. thesis]: Minneapolis, University of Minnesota, 121 p.

Papike, T. J., Lameton, K. L., and Baldwin, K., 1974, Amphibole and pyroxenes: Characterization of other than quadrilateral components and estimates of ferric iron from microprobe data: Geological Society of America Abstracts with Programs, v. 6, p. 1053–1054.

Rivalenti, G., 1975, Chemistry and differentiation of mafic dikes in an area near Fiskenaesset, West Greenland: Canadian Journal of Earth Sciences, v. 12, p. 721–730.

Sims, P. K., 1972, Metavolcanic and associated synvolcanic rocks in Vermilion district *in* Sims, P. K., and Morey, G. D., eds., Geology of Minnesota: A centennial volume: Minnesota Geological Survey, p. 63–75.

Weiblen, P. W., Mathez, E. A., and Morey, G. B., 1972, Logan intrusions *in* Sims, P. K., and Morey, G. B., eds., Geology of Minnesota: A centennial volume: Minnesota Geological Survey, p. 374–406.

MANUSCRIPT RECEIVED BY THE SOCIETY APRIL 25, 1979
REVISED MANUSCRIPT RECEIVED NOVEMBER 26, 1979
MANUSCRIPT ACCEPTED JANUARY 11, 1980

Geological Society of America
Special Paper 182
1980

# Lead-isotope investigations in the Minnesota River Valley—Late-tectonic and posttectonic granites

**B. R. DOE**
**M. H. DELEVAUX**
*U.S. Geological Survey, Box 25046, MS 963, Federal Center, Denver, Colorado 80225*

## ABSTRACT

In the Minnesota River Valley, an epizonal, anorogenic granite that is often referred to as the granite of section 28 (lat 44°49.73′N, long 95°33.90′W) has been found to have a Pb-Pb age of $1.84 \pm 0.05$ b.y. on the basis of data obtained on HF leached and unleached feldspars and HCl leached and unleached whole rocks. The Th-Pb age of the feldspar–whole-rock pair is 1.9 b.y., which is in satisfactory agreement with the Pb-Pb age; but the U-Pb age is greater than the Pb-Pb age and probably indicates that uranium has been leached from the whole rock within the past several hundred million years, perhaps as a consequence of dilatancy resulting from uplift and erosion. Another granite, the mesozonal, late-tectonic Precambrian Sacred Heart Granite in the Minnesota River Valley (lat 44°41.3′N, long 95°21.5′W) is found to have a Pb-Pb age of $2.605 \pm 0.006$ b.y. on the basis of data obtained on HF leached and unleached feldspars and HCl leached and unleached whole rocks. Both the Th-Pb and U-Pb isochron ages are much older than the Pb-Pb age. An older age was expected for the U-Pb system as it had been previously found in the epizonal granite, but to also find it for the Th-Pb system was surprising.

As is predicted for these kinds of granites in this type of tectonic environment on the basis of Mesozoic and Cenozoic analogues, the initial leads in the granites indicate that they were derived from source material having values of $^{238}U/^{204}Pb$ <9 normalized to the present day. This feature is common in Mesozoic and Cenozoic igneous rocks penetrating Precambrian terranes but is rarely observed in pre-Mesozoic igneous rocks. The Sacred Heart Granite is the oldest igneous rock known to show this effect and is the first representative of a mesozonal granite. The uranium depletion event appears to have been a granulite-facies metamorphism, but the age of that metamorphism cannot be determined from the available data. The model-lead-age information, however, suggests that it occurred before 2.78 b.y. ago. The source materials for both granites also underwent an earlier stage of extensive but unknown duration during which $^{238}U/^{204}Pb$ >9. In Phanerozoic rocks, such values are characteristic of ensialic tectonic environments. Similar development of ensialic environments was apparently occurring also in perancient times.

## INTRODUCTION

As recently as 1964 the continental crust was commonly assumed to have a "normal" value of $^{238}U/^{204}Pb$ > 9 on the basis of analyses of surface rocks, ores, and detrital feldspar. Such values would result in radiogenic leads or values of $^{206}Pb/^{204}Pb$ greater than expected in so-called single-stage lead-isotope evolutions where the value of $^{238}U/^{204}Pb$ appears to have been near to 9 (normalized to the present-day) since the Earth formed. Surprisingly, however, epizonal Cenozoic and Mesozoic igneous rocks penetrating Precambrian crust were then commonly found to have nonradiogenic leads or values of $^{206}Pb/^{204}Pb$ less than expected in so-called single-stage evolutions; this isotopic nature indicates that these igneous rocks were derived from source material deficient in uranium relative to lead (Doe, 1964, 1967; Hamilton, 1966); that is, $^{238}U/^{204}Pb$ <9. This characteristic was deduced to have been the result either of uranium loss relative to thorium and lead during high-rank metamorphism (granulite facies or above) (Doe, 1968; Doe and others, 1968) or of a more mafic character (andesitic or quartz dioritic) for the lower crust than for the upper crust (Doe, 1968; Zartman and Wasserburg, 1969). These two possibilities were found to be distinguishable by Doe and Zartman (1979), who learned from examination of lead-isotope data that both andesites in mature orogenes (where the lower crust did not appear to have been metamorphosed above upper amphibolite facies in the Precambrian) and oceanic basalts were derived from source material generally with values of Th/U ≤ 4 (case 1).

They also confirmed the findings of Heier and colleagues (Heier and Thoreson, 1971) that high-pressure "granulites" would generally have Th/U ≥ 4 (case 2). They further found that igneous rocks derived from such a source material would have lead-isotope ratios reflecting the large Th/U values of the granulitic source material. Mesozoic and Cenozoic basaltic and silicic igneous rocks of the San Juan Mountains of southwestern Colorado primarily seem to be an example of case 1, although magmas involved in caldera formation and a basaltic rock containing plagioclase phenocrysts appear to have been substantially modified in the upper crust (Lipman and others, 1978). Plutons of the Boulder batholith and Scottish highlands appear to be an example of case 2 (Doe and others, 1968; Hamilton, 1966). Some other granites—such as at Tick Canyon in the San Gabriel Mountains and parts of the Sierra Nevada batholith, both in California—appear to reflect derivation from "normal" mature orogene materials (case 3), that is, materials typified by the upper continental crust (Doe, 1968; Doe and Delevaux, 1973). For a long time, the possibility remained that calc-alkaline silicic magma was derived from the lithospheric mantle. Recent lead-isotope investigations of increased analytical sensitivity are showing that silicic igneous rocks are principally derived from different sources than are basalts. (See Leeman, 1975, for the Snake River Plain area, Idaho; and Lipman and others, 1977, for the San Juan volcanic area, Colorado). The confusion as to source arose because the isotopic signature of the mafic magmas apparently was derived mainly in a continental mantle keel that formed at the time of the oldest crustal rocks in the area. This keel contains similar, but now known to be analytically separable, isotopic values as the associated continental crust.

The lead-isotope picture has been less clear for mesozonal and catazonal igneous rocks; there are no good examples of these kinds of plutons that have nonradiogenic initial leads. Mesozonal and catazonal igneous rocks of Mesozoic and Cenozoic age are rather rare, and restricting their occurrences to rejuvenated cratons narrows the possibility of finding them even more. As the lead-isotope evolution pattern was followed back in time, the record became rapidly obscured for granitic material that was derived from sources depleted in uranium relative to lead either as a result of Precambrian high-rank metamorphism or as a result of the source material being of intermediate composition (Doe, 1967). Zartman and Wasserburg (1969) managed to find two examples of billion-year-old granites having lead isotopic compositions indicative of derivation from source material that was highly metamorphosed at a much older time (the granites at Mellon, Wisconsin, and in the Port Cartier–Mount Reed area, Quebec). A third candidate for a supposed billion-yr-old granite at Gold Butte, Nevada, has subsequently been found to be about 1.46 b.y. old (M. A. Lanphere, 1967, written commun.), an age compatible with the observed lead isotopic ratios. The surviving examples remain the only ones for pre-Mesozoic igneous rocks known to date. Some examples are known for ores, however, and more granite examples are to be expected. The more common case among older granites is represented by the Paleozoic mesothermal plutons that have intruded the Precambrian Baltimore Gneiss in Maryland and that have somewhat radiogenic initial leads—as measured in feldspars, a phase poor in uranium and thorium but rich in lead—rather than nonradiogenic leads (Doe and others, 1965). The question posed is whether the more deeply emplaced igneous rocks in rejuvenated cratons resulted in different lead-isotopic compositions than are found in the more shallow-seated igneous rocks because different source materials were involved, because the degrees of partial melting were different, or because the conditions of magma formation in the pre-Mesozoic were different from those in the Mesozoic and Cenozoic. The Minnesota

River Valley appears to be an excellent area to explore the quandary further.

The setting of the granite in the NW¼ sec. 28, T. 116 N., R. 39 W., about 1.6 km north-northwest of Granite Falls (lat 44°49.73′N, long 95°33.90′W), hereafter simply designated as granite of section 28, suggested to us that the lead in the gran also should be nonradiogenic (indicative of a $^{238}U/^{204}Pb$ <9) has been found for Mesozoic and Cenozoic analogues. Its fine grained groundmass is suggestive of an epizonal intrusion, and i is intruded in an area containing gneissic rocks known to be greater than 3 b.y. old (Goldich and others, 1970). Its approximate age of 1,850 m.y. means that it has intruded rocks that were metamorphosed at least a billion or more years previously, some of which were subjected to metamorphism in the hornblende-granulite facies (Himmelberg and Phinney, 1967). The darkest gray rock was collected from material blasted in making a telephone-post hole. The gray color seems to be an indicator of freshness.

The Sacred Heart Granite of Lund (1956) in the Minnesota River Valley (lat 44°41.3′N, long 95°21.5′W) is more problematical in its classification, but it appears to be a mesozonal, late-tectonic batholith intruded into the same general terrane as the epizonal, anorogenic granite pluton of section 28. The Sacred Heart Granite, however, is substantially older (late Archean, Goldich and others, 1970). The sample collected was from a small quarry on Wurtcher's Farm. The rock collected was the darkest gray material of a free block that probably was within 3 m of the surface. In spite of the older age, this sample provides a test of just how well plutonism in the Precambrian is mirrored in epizonal processes operative in the Mesozoic and Cenozoic.

## ANALYTICAL PROCEDURES

Samples are cut into cubes with a diamond saw used only for cutting rocks. They are then cleaned twice ultrasonically in distilled water (characteristically they have little turbidity after the second time), and about 20 g of sample is pulverized in a steel mortar. Feldspars are separated through the use of heavy liquids. Powders of whole-rock samples are split into 1- to 3-g aliquots in microsplitters.

Whole rocks (about 2 g) are decomposed using hot HF plus a few drops each of $HClO_4$ and $HNO_3$ in Teflonware inside Teflon tanks purged with clean nitrogen gas. The sample is dried, 100 ml of $H_2O$ is added along with some concentrated $HNO_3$ and perchloric acid, the mixture is heated until all flourides are in solution, and it is then taken to dryness again. A third drying is performed after about 40 ml of $H_2O$ and some $HNO_3$ have been added. The sample is taken back into the solution and a hydroxide precipitate is made. The precipitate is dissolved in 1.5-$N$ HBr and passed through a bromide resin column. The uranium and thorium pass through the bromide column and the lead is retained on the column. The lead is stripped from the column with either 6-$N$ HCl or 1-$N$ $HNO_3$ and dried. It is picked up in dilute $HNO_3$-$HClO_4$ and electrodeposited at 1.8 volts onto a platinum anode following procedures established by the National Bureau of Standards (NBS). The contamination level is 20 to 30 ng of lead. The solution containing uranium and thorium is dried and converted to the nitrate by drying again with $HNO_3$ and perchloric acid. It is redissolved in 40 to 50 ml of 6-$N$ HNO$_3$. The solution is passed through a resin column of Dowex-1 resin, extracted with more HNO$_3$, and stripped from the column with H$_2$O. The procedure is repeated on a second smaller column. Some uranium is intentionally lost in the procedure to enhance the intensity of the thorium signal.

On leached whole rocks, a split of about 5 g is leached with hot 1:1 HCl for about 2 hr, and the purification procedures previously described are performed. K-feldspars were first leached with hot 1:1 HC1 for at least 1 h, and then with 1:1 HNO$_3$ for a similar time. When the leach is analyzed, the two acid fractions are combined and put through the purification procedure. The HF leaching is done using a slight variant of the method of Ludwig and Silver (1977). The feldspar is leached with hot 5% HF for 25 min. The procedure is repeated. The HF leach analysis is of the first fraction, whereas the residue is the remainder after the second leach.

The purified lead is dissolved from the anode in 20% HNO$_3$ with a split drop of peroxide and dried. Phosphoric acid is then added to the beaker and the aqueous phase is evaporated. A drop of silica-gel solution is added to one spot on the bottom of the beaker, transferred to a single rhenium filament, and dried. The lead is then analyzed by thermal emission at 1,200°C in a mass spectrometer having a 12-in. radius of curvature. A uranium-thorium mixture is picked up in 4% HNO$_3$ (an acidity at which the acid does not migrate away from the hot spot) and loaded onto two rhenium side filaments. The NBS directions for uranium mass analysis were followed, except that uranium and thorium were analyzed together: the thorium signal grows as the uranium signal dies away.

The concentrations are determined by isotope dilution using "spikes" of $^{206}$Pb, $^{235}$U, and $^{230}$Th obtained from the Oak Ridge National Laboratory.

A systems monitoring of precision is made through repeated analyses of the basalt reference sample BCR-1 (Table 1). Some monitoring of accuracy of isotope ratio measurements is also possible on samples on which concentration work is done, because an uncontaminated $^{208}$Pb/$^{204}$Pb value is obtained. On the crystalline rocks, uranium, thorium, and lead are not homogeneously distributed as they are in BCR-1, and differences in $^{208}$Pb/$^{204}$Pb of as much as 5% are noted in gram-size samples coming out of either side of the microsplitter. This difficulty has been observed by others (such as Oversby, 1976) and deserves attention that we were not able to devote to it.

## DATA FOR EPIZONAL GRANITE OF SECTION 28

The lead-isotope data for the epizonal granite are given in Table 2 and shown in Figures 1 and 2. A York regression at 1$\sigma$ uncertainties, using correlated errors derived from the lead-isotope data for $^{207}$Pb/$^{204}$Pb versus $^{206}$Pb/$^{204}$Pb, is 0.1124, equivalent to a Pb-Pb age of 1.84 ±0.05 b.y. using the conventional decay constants adopted by the IUGG Subcommission on Geochronology in 1976 (Steiger and Jäger, 1977). Using the current decay constants, but with reduced uncertainties, the age is in remarkably good agreement with the Rb-Sr age of Goldich and others (1970) of 1.83 ± 0.16 b.y., and is in good agreement with the $^{207}$Pb/$^{206}$Pb age on a zircon of 1.80 b.y. The Th-Pb age of

about 1.9 b.y. (Fig. 4) is also in good agreement with the Pb-Pb age, but the U-Pb age (Fig. 3) is too old and appears to reflect recent uranium loss. Such recent uranium loss is common in near-surface crystalline rocks and may be due to loss of uranium as a result of a dilatancy effect resulting from uplift and erosion.

Using the model of Stacey and Kramers (1975), the secondary

TABLE 1. DATA FOR THE U-Th-Pb SYSTEM ON THE BASALT REFERENCE SAMPLE AND SELECTED FELDSPARS

| Phase | Concentrations | | | Ratios (atomic) | | | |
|---|---|---|---|---|---|---|---|
| | U | Th | Pb | $\frac{^{206}Pb}{^{204}Pb}$ | $\frac{^{207}Pb}{^{204}Pb}$ | $\frac{^{208}Pb}{^{204}Pb}$ | $\frac{^{208}Pb*}{^{204}Pb}$ |
| BCR-1 Reference sample | | | | | | | |
| Whole rock | 1.72 | 5.93 | 13.67 | 18.802 | 15.622 | 38.696 | 38.55 |
| | 1.72 | 5.93 | 13.74 | 18.819 | 15.623 | 38.688 | 38.62 |
| | 1.71 | 5.92 | 13.55 | 18.815 | 15.629 | 38.767 | 38.65 |
| | 1.73 | 5.99 | 13.72 | 18.817 | 15.630 | 38.721 | 38.72 |
| B-32 Guilford granite, Maryland | | | | | | | |
| K-feldspar[†] | | | | 18.229 | 15.570 | 37.872 | |
| B-77 Notch Cliff pegmatite, Maryland | | | | | | | |
| K-feldspar[†] | | | | 18.053 | 15.588 | 38.326 | |
| KA-354 Vermilion granite, Minnesota | | | | | | | |
| K-feldspar[†] | | | | 13.573 | 14.615 | 33.400 | |

*Note:* Feldspar analyses from Doe and others (1965).
*From concentration analysis.
[†]Residue from HF leach.

TABLE 2. DATA FOR THE U-Th-Pb SYSTEM FOR THE EPIZONAL GRANITE OF SECTION 28 (SAMPLE KA604-73) AND THE SACRED HEART GRANITE (SAMPLE 73MRV-1)

| Phase | Concentrations | | | Ratios (atomic) | | | |
|---|---|---|---|---|---|---|---|
| | U | Th | Pb | $\frac{^{206}Pb}{^{204}Pb}$ | $\frac{^{207}Pb}{^{204}Pb}$ | $\frac{^{208}Pb}{^{204}Pb}$ | $\frac{^{208}Pb*}{^{204}Pb}$ |
| Epizonal granite | | | | | | | |
| K-feldspar | 0.10 | 0.22 | 52.9 | 15.079 | 15.280 | 35.214 | 35.35 |
| | | | | *15.04* | *15.28* | *35.19* | |
| | | | | 15.056[†] | 15.301[†] | 35.228[†] | -- |
| | | | | 15.612[§] | 15.355[§] | 35.813[§] | -- |
| Whole rock | 1.64 | 12.5 | 20.9 | 17.287 | 15.554 | 39.190 | 39.00 |
| | | | | 15.821[#] | 15.372[#] | 35.884[#] | -- |
| | | | | 22.430[**] | 16.117[**] | 50.920[**] | -- |
| Sacred Heart granite | | | | | | | |
| K-feldspar | 0.11 | 0.10 | 47.3 | 14.307 | 15.118 | 34.338 | 34.30 |
| | | | | *14.24* | *15.11* | *34.32* | |
| | | | | 14.109[†] | 15.073[†] | 34.064[†] | -- |
| | | | | 15.175[§] | 15.253[§] | 35.407[§] | -- |
| Whole rock | 3.46 | 25.5 | 31.2 | 19.641 | 16.043 | 45.140 | 45.04 |
| | 3.28[††] | 21.9[††] | 30.3[††] | -- | -- | -- | 44.61 |
| | | | | 14.902[#] | 15.208[#] | 34.523[#] | -- |
| | | | | 34.544[**] | 18.646[**] | 78.877[**] | -- |

*Note:* Numbers in italics are the feldspar lead corrected for 2.6 b.y. of uranogenic and thorogenic lead produced in situ.
*From concentration analysis.
[†]Residue from HF leach.
[§]HF leach fraction.
[#]Residue from HCl leach.
[**]HCl leach fraction.
[††]Second fraction.

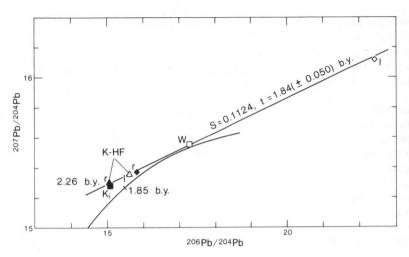

Figure 1. $^{207}Pb/^{204}Pb$ versus $^{206}Pb/^{204}Pb$ evolution diagram for the epizonal granite. The curved line is the two-stage lead-isotope evolution curve of Stacey and Kramers (1975). Open square (w) is the whole-rock sample, the open diamond (l) is the HCl-soluble lead from the whole rock, and the solid diamond (r) is the lead in the residue from that leach. Open triangle (l) is the lead in the leachate from the HF leach of the K-feldspar (K), and solid triangle (r) is the lead in the residue from that leach. Solid square (K$_i$) is the initial lead at 1.85 b.y. ago, calculated from observed values of uranium, thorium, and lead concentrations and from the lead isotopic composition of feldspar not leached by HF. A tick mark is shown on the two-stage evolution curve to indicate where the initial lead for the granite should plot if the evolution model was obeyed. S is the slope of the least-squares line through the data.

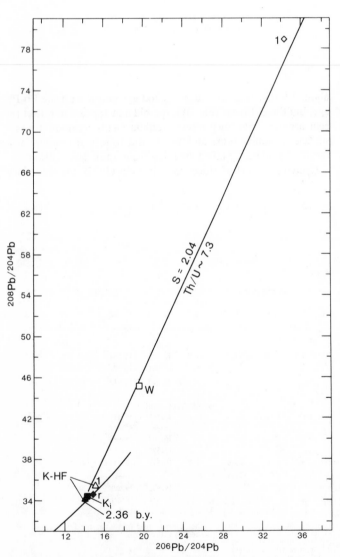

Figure 2. $^{208}Pb/^{204}Pb$ versus $^{206}Pb/^{204}Pb$ evolution for the epizonal granite. Solid curved line is the two-stage evolution diagram of Stacey and Kramers (1975). Symbols are same as in Figure 1.

isochron model lead age of the HF leached feldspar is 2.25 b.y.—about 0.4 b.y. older than the age of the rock. The old model lead age thus bears out the prediction from plumbotectonics models that the rock was derived from source material that had a $^{238}U/^{204}Pb$ <9 over an extended period of time prior to the time of formation of the epizonal granite. The position of the data above the orogene evolution curve shows that the source material at one time was also evolving under conditions of $^{238}U/^{204}Pb$ >9. In order to get some idea how much the values of $^{238}U/^{204}Pb$ might depart from 9, we might reasonably (1) assume that the source material for the magma of the epizonal granite was formed 3.7 b.y. ago, the age determined by Goldich and Hedge (1974) to be the age of the Montevideo Gneiss of Goldich (1970); (2) further assume that no events affected the $^{238}U/^{204}Pb$ and $^{232}Th/^{238}U$ of the source material between the time of its formation and the time of formation of the epizonal granite; and (3) finally, assume that the precursor of the source material had the properties attributed to material older than 3.7 b.y. by Stacey and Kramers (1975) ($^{238}U/^{204}Pb$ = 9.74, $^{232}Th/^{238}U$ = 3.78), resulting in values of $^{206}Pb/^{204}Pb$ = 11.152, $^{207}Pb/^{204}Pb$ = 12.998, $^{208}Pb/^{204}Pb$ = 31.23 at 3.7 b.y. The value of $^{238}U/^{204}Pb$ for the source material is then calculated to be 10.29 and the value of $^{232}Th/^{238}U$ to be 4.40.

A solution to the problem under these conditions can *only* be obtained, however, if the epizonal granite is assumed to be 2.25 b.y. old, whereas, in fact, it must be nearer to 1.84 b.y. in age. To give the right age for a $^{238}U/^{204}Pb$ of about 10.29, the observed $^{206}Pb/^{204}Pb$ should be about 16.1 rather than the retarded values of 15.1 actually observed in the K-feldspar. There must have been an event—granulite-facies metamorphism?—that reduced the value of $^{238}U/^{204}Pb$ (and increased $^{232}Th/^{238}U$?) in the source material before 2.25 b.y. ago. Just when this event occurred cannot be resolved from the existing data, but the value of $^{238}U/^{204}Pb$~10.3 is probably a minimum for the high-value stage, because the following stage of low $^{238}U/^{204}Pb$ would have the effect of decreasing the calculated, as compared to the real, value.

The possible alternative explanation that the low $^{238}U/^{204}Pb$ and high $^{232}Th/^{238}U$ were the result of selective partial melting, at about 1.85 b.y. ago, of phases in the source material having low $^{238}U/^{204}Pb$—such as K-feldspars and plagioclase—is not pro-

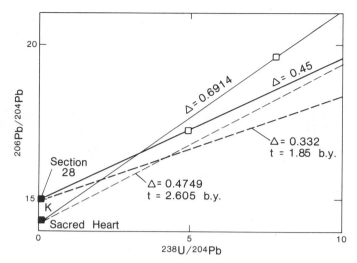

Figure 3. $^{206}$Pb/$^{238}$U isochron plot for whole rock (open squares) and K-feldspar (solid squares) of the granite of section 28 and the Sacred Heart Granite of Lund (1956). Dashed line is the proper slope for the age of the rock. The steeper observed slope is equivalent to an "age" of 2.4 ± 0.08 b.y. for the granite of section 28 and 3.3 ± 0.08 b.y. for the Sacred Heart Granite, probably signifying relatively recent uranium loss from the whole rock. The quoted uncertainty is the extreme for the analytical uncertainties of the two samples for each rock.

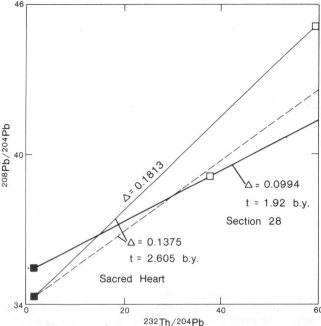

Figure 4. $^{208}$Pb/$^{232}$Th isochron plot for the granite of section 28 and the Sacred Heart Granite of Lund (1956). Symbols are the same as for Figure 3. Dashed line is the proper slope for the age of the rock. The steeper observed slope is equivalent to an "age" of 3.37 ± 0.08 b.y. for the Sacred Heart Granite and probably signifies relatively recent loss of thorium from the whole rock. The proper slope for the age of the granite of section 28 essentially coincides with the observed slope.

bable. Uranium, in particular, largely resides in phases that are easily mobilized. All the information that we have is that partial melting tends selectively to enrich the melt in radiogenic lead over the whole rock rather than to deplete it (see, for example, Lipman and others, 1978). Additionally, phases such as feldspar seem to have values of Th/U <4 rather than >4 (Doe and others, 1965). Whatever the process, the anorogenic, epizonal plutons of the Minnesota River Valley are the oldest known to portray retarded lead isotope evolution.

## DATA FOR SACRED HEART GRANITE

The lead isotope data for the Sacred Heart Granite of Lund (1956) are given in Table 2 and shown on Figures 5 and 6. A York regression using the lead-isotope data for $^{207}$Pb/$^{204}$Pb versus $^{206}$Pb/$^{204}$Pb is 0.1748 ± 0.0007, equivalent to a Pb-Pb age of 2.605 ± 0.006 b.y. using the recently adopted decay constants. The age is in agreement with the Rb-Sr age of about 2.7 b.y. (converted to the new decay constants) given by Goldich and others (1970). There is no evidence of any effect on the Pb-Pb age having been caused by the roughly 1.8-b.y.-old metamorphism known to have occurred northwest of the Sacred Heart Granite. The HCl leach lead should have been sensitive to such a metamorphism, but no such effect is seen. This lack of effect is consistent with other data in that only slight discrepancies have been noted between the age of the rock and K-Ar ages of biotite (given as about 2,300 m.y. by Goldich and others, 1970), so the event must have been very weak in this area. C. W. Naeser (1974, written commun.) gives a fission-track age of 460 ± 90 (2σ) m.y. on apatite containing 25 ppm uranium from an amphibolite inclusion in the Sacred Heart Granite. The age represents, in effect, the time since this rock cooled below 100 °C. This age event—presumably uplift—also

appears to have had no effect on the rock leads of the Sacred Heart Granite, although such an event might account for the fairly young K-Ar ages on biotite.

Unlike the epizonal granite, both the Th-Pb and U-Pb "ages" are much too old—about 3.7 to 3.8 b.y. (Figs. 3, 4). The old Th-Pb "age" is a surprise—although the old U-Pb "age" is less so—because the sample of granite came from the grayish-looking (presumably freshest) part of a quarry and because Rosholt and Bartel (1969) showed that the Th-Pb system has some resistance to alteration. That the two "ages" are about the same is even a greater surprise, because thorium is normally much less soluble than uranium. The similarity in the old ages appears to be a rare coincidence, and perhaps some unusual mineralogical control in the sample of the Sacred Heart Granite accounts for the unusual behavior of the thorium.

When the model of Stacey and Kramers (1975) is used, the secondary isochron model lead age of the feldspar lead is 2.78 b.y., which is some 170 m.y. older than the determined age of the rock. This anomaly is roughly half that found for the 1.8-b.y.-old epizonal granite. The data for the Sacred Heart Granite plot above the curve for mature orogene evolution of Stacey and Kramers (1975); this indicates that at some time before 2.6 b.y. ago the value of $^{238}$U/$^{204}$Pb must also have been greater than about 9. If we accept that the Sacred Heart Granite formed at 2.605 b.y. ago and the epizonal granite at 1.84 b.y. ago and if we assume that both come from the same source material and have lead isotopic compositions representative of that source material, then the value of $^{238}$U/$^{204}$Pb for the source material is 5.65, which

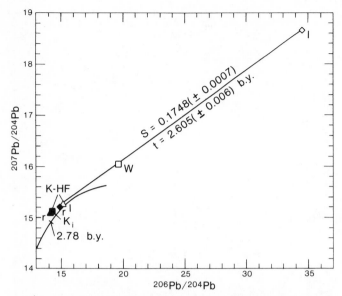

**Figure 5.** $^{207}$Pb/$^{204}$Pb versus $^{206}$Pb/$^{204}$Pb evolution diagram for the Sacred Heart Granite of Lund (1956). Curved line is the two-stage lead-isotope evolution curve of Stacey and Kramers (1975). Open square (W) is the whole-rock sample, open diamond (l) is the HCl soluble lead from the whole rock, and solid diamond (r) is the lead in the residue from that leach. Open triangle (l) is the leachate in the HF leach of the K-feldspar (K), and solid triangle (r) is the lead in the residue from that leach. Solid square (K$_i$) is the initial lead calculated back to 2.60 b.y. ago from observed values of uranium, thorium, and lead concentrations and lead isotopic composition of feldspar not leached by HF. A tick mark is shown on the two-stage evolution curve to indicate where the initial lead for the granite should plot if the evolution model was obeyed. S is the slope of the least-square line through the data.

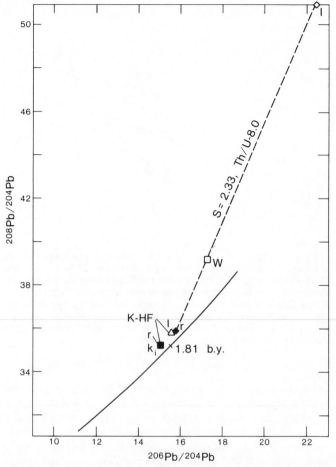

**Figure 6.** $^{208}$Pb/$^{204}$Pb versus $^{206}$Pb/$^{204}$Pb evolution diagram for the Sacred Heart Granite of Lund (1956). The solid curved line is the two-stage evolution diagram of Stacey and Kramers (1975). Symbols are same as in Figure 5.

is also the answer one would get in assuming that the epizonal granite was derived from the Sacred Heart Granite. This value is in contrast to the overall value of $^{238}$U/$^{204}$Pb of about 12.55 needed to produce the observed values of $^{206}$Pb/$^{204}$Pb *and* $^{207}$Pb/$^{204}$Pb for the initial lead in the Sacred Heart Granite, using the values of Stacey and Kramers for 3.7 b.y. ago and assuming that the isochron model lead age of 2.78 b.y. is correct. The assumption is necessary because there is no other age at which the observed lead could be produced in one step. Such an assumption further assumes that the value of $^{238}$U/$^{204}$Pb was 0 between 2.78 and 2.60 b.y. ago, which is unrealistic; however, no better estimate can be made unless the age of the uranium-reduction event (granulite-facies metamorphism?) can be estimated.

There is increasing evidence of an important event around 3 b.y. in the region (given as 3.12 b.y. by Goldich and others, 1975, using Th-Pb systematics on the Montevideo Granite Gneiss that pervades to the northwest). If 3.12 b.y. is taken to represent the age of the uranium loss from the source rocks of the Sacred Heart Granite to a value for $^{238}$U/$^{204}$Pb $\sim$ 5.65, then the value for the stage between 3.7 and 3.12 b.y., which has high $^{238}$U/$^{204}$Pb, is about 14.5. Although this calculation furnishes an approximation of what may have happened to $^{238}$U/$^{204}$Pb in this region, it should be viewed as just that—an example. Individual igneous bodies of similar age and location in rejuvenated plutons commonly have different initial leads. The epizonal granite also is not only separated from the Sacred Heart Granite by about 25 km and a

tectonic hiatus, but the two granites are nearly 800 m.y. different in age. The calculations also presuppose that the exact values of initial lead have been determined for the plutons, an assumption that is by no means certain because of the lack of agreement between the HF-leached K-feldspar and the total-feldspar values after correction for radiogenic lead produced in situ.

The $^{208}$Pb/$^{204}$Pb model age of the HF leached feldspar using the Stacey-Kramers model is 2.36 b.y., which is 240 m.y. younger than the accepted age of 2.60 b.y. Because the conditions assumed in the Stacey-Kramers model did not apply well to uranogenic lead, it comes as no great surprise that it does not apply well to thorogenic lead either. There is, in fact, some suggestion that the value of $^{232}$Th/$^{204}$Pb may be somewhat enhanced in granulite-facies metamorphism or that the stage of high $^{238}$U/$^{204}$Pb may have been accompanied by a high $^{232}$Th/$^{204}$Pb (Doe and Zartman, 1979).

## DISCUSSION

The source material for the 2.60-b.y.-old Sacred Heart Granite and 1.85-b.y.-old granite of section 28, both of which were intruded into a terrane of perancient rocks greater than 3 b.y. in age, have been determined to have undergone an early stage or stages having a value of $^{238}U/^{204}Pb >9$ followed by a stage or stages with $^{238}U/^{204}Pb <9$ and Th/U >4. The first stage or stages are characteristic of ensialic environments with resultant $^{207}Pb/^{204}Pb$ values greater than for average orogene evolution, whereas the second stage or stages seem to be characteristic of lower continental crust formed as a result of cratonization by high-rank metamorphism with consequent loss of uranium. By analogy with the Phanerozoic plumbotectonics model synthesized by Doe and Zartman (1979), it is clear that even by late Archean time the basement rocks in the Minnesota River Valley had become a craton with many of the complexities found in rejuvenated cratons during the Mesozoic and Cenozoic. Quite different isotopic characteristics have been found in Archean plutonic rocks, about 2.7 b.y. in age, just a few hundred kilometres to the north (for example, the Algoman granite, Doe and others, 1965; Icarus pluton, Arth and Hanson, 1975; and the Giants Range Granite, Ludwig and Silver, 1977), where no extensive pre-history is known before the igneous activity and where the tectonic environment has a more ensimatic character (Fig. 7). For a given value of $^{206}Pb/^{204}Pb$, these granitic rocks comparatively have much lower values of $^{207}Pb/^{204}Pb$ and $^{208}Pb/^{204}Pb$ than do those plutonic rocks in the Minnesota River Valley, and the isotopic data tend to plot below the evolution curves for average orogenes. The plutonic rocks in northern Minnesota can best be interpreted in analogy with the Phanerozoic plumbotectonics model as having been derived from mafic or intermediate source materials soon after the source materials were formed. Such source materials typically have both lower values of $^{207}Pb/^{204}Pb$ than do the average orogene and reasonably normal values of $^{208}Pb/^{204}Pb$ such as those observed in the granites.

Such an interpretation for the northern Minnesota plutons is compatible with the rare-earth studies of Arth and Hanson (1975). Likewise, Stacey and others (1976), in investigating massive sulfide ore deposits that are found in Precambrian greenstone and ophiolite-like belts and that appear to be either syngenetic or very early diagenetic in origin, found that lead in these deposits could be explained in terms only somewhat simpler than those used to explain lead in Phanerozoic deposits. The main differences are (1) that there tends to be a greater component of mantle lead in the Precambrian massive sulfide ores—assuming extrapolation of present-day oceanic-mantle lead to the Precambrian is valid—and (2) that mixing of continental and oceanic mantle lead into orogenes most closely resembles end-member mixing of oceanic mantle and a nearby older continental component of only one age. The simple end-member mixing may occur perhaps because the age make-up of the Precambrian continents near the orogenes also tended to be simpler than the make-up of continents by Phanerozoic time. To a first-order approximation, we seem to see development of oceanic mantle, upper continental crust, lower continental crust, and orogenes as early as 2.5 to 3.0 b.y. ago, which resulted in uranium, thorium, and lead distributions much like those observed in the Cenozoic record. The lead isotopes then apparently reflect the tectonic environment of igneous activity (and no doubt ore formation) as far back as late Archean time as well as they reflect tectonic environments today.

Using modern lead-isotope analyses from Table 1, the Paleozoic igneous rocks in the Proterozoic Y and Z Baltimore

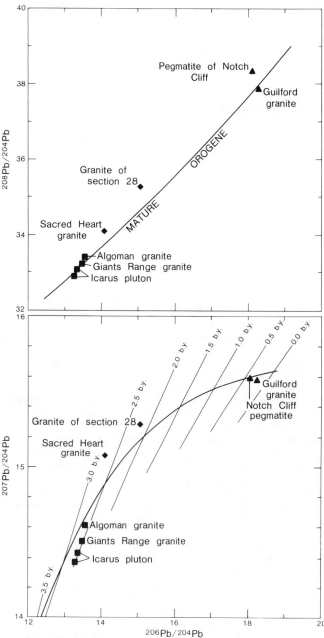

Figure 7. Lead-isotope evolution plots including data on feldspars from late-tectonic Paleozoic igneous rocks from the Precambrian Baltimore Gneiss terrane (solid triangle; Table 1), from Sacred Heart and section 28 Precambrian granites of the Minnesota River Valley (solid diamond; Table 2), and from the Icarus pluton, Giants Range and Algoman granites (Table 1) of northern Minnesota (solid square). All data plotted from Tables 1 and 2 are residues from HF leach experiments. Data for the Icarus pluton are from Arth and Hanson (1975); data for the Giants Range granite are from Ludwig and Silver (1977).

Gneiss still tend to have radiogenic rather than nonradiogenic leads, and some of them still plot above the evolution curve for $^{208}Pb/^{204}Pb$ versus $^{206}Pb/^{204}Pb$ (Fig. 7). This configuration for $^{208}Pb/^{204}Pb$ and $^{206}Pb/^{204}Pb$ suggests that the granites were derived from mature orogenic material that probably underwent granulite facies metamorphism in Proterozoic Y or Z time, resulting in increased values of Th/U. The interval between the high-rank metamorphism and the formation of the Paleozoic granites was probably not great enough to permit the low value of U/Pb in the source material resulting from the metamorphism to be reflected in the $^{206}Pb/^{204}Pb$ values of the Paleozoic granites.

## ACKNOWLEDGMENTS

The sample localities were recommended by S. S. Goldich. Classification of the plutons and tectonic environments in Minnesota was discussed with G. B. Morey, Paul K. Sims, and S. S. Goldich.

## REFERENCES CITED

Arth, J. G., and Hanson, G. N., 1975, Geochemistry and origin of the early Precambrian crust of northeastern Minnesota: Geochimica et Cosmochimica Acta, v. 39, p. 325–362.

Doe, B. R., 1964, Provincial aspects of lead isotopes in granitic rocks in the United States: American Geophysical Union Transactions, v. 45, p. 108.

——1967, The bearing of lead isotopes on the source of granitic magma: Journal of Petrology, v. 8, p. 51–83.

——1968, Lead and strontium isotope studies of Cenozoic volcanic rocks in the Rocky Mountain region—A summary: Colorado School of Mines Quarterly, v. 63, p. 149–174.

Doe, B. R., and Delevaux, M. H., 1973, Variations in lead-isotopic compositions in Mesozoic granitic rocks of California: Geological Society of America Bulletin, v. 84, p. 3513–3526.

Doe, B. R., and Zartman, R. E., 1979, Plumbotectonics, The Phanerozoic, in Barnes, H., ed., Geochemistry of hydrothermal ore deposits (second edition): New York, Chichester, Brisbane, Toronto, John Wiley and Sons, p. 22–71.

Doe, B. R., Tilton, G. R., and Hopson, C. A., 1965, Lead isotopes in feldspars from selected granitic rocks associated with regional metamorphism: Journal of Geophysical Research, v. 70, p. 1947–1968.

Doe, B. R., and others, 1968, Lead and strontium isotope studies of the Boulder batholith, southwestern Montana: Economic Geology, v. 63, p. 884–906.

Goldich, S. S., and Hedge, C. E., 1974, 3800 m.y. granitic gneiss in southwestern Minnesota: Nature, v. 252, p. 467–468.

Goldich, S. S., Doe, B. R., and Delevaux, M. H., 1975, Possible further evidence for 3.8 b.y.-old rocks in the Minnesota River Valley of southwestern Minnesota: U. S. Geological Survey Open-File Report 75-65, 11 p.

Goldich, S. S., Hedge, C. E., and Stern, T. W., 1970, Age of the Morton and Montevideo Gneisses and related rocks, southwestern Minnesota: Geological Society of America Bulletin, v. 81, p. 3671–3696.

Hamilton, E. I., 1966, The isotopic composition of lead in igneous rocks; 1, The origin of some Tertiary granites: Earth and Planetary Science Letters, v. 1, p. 30–37.

Heier, K. S., and Thoresen, K., 1971, Geochemistry of high grade metamorphic rocks, Lofoten-Vesteralen, North Norway: Geochimica et Cosmochimica Acta, v. 35, p. 89–100.

Himmelberg, G. R., and Phinney, W. R., 1967, Granulite-facies metamorphism, Granite Falls—Montevideo area, Minnesota: Journal of Petrology, v. 8, p. 325–348.

Leeman, W. P., 1975, Radiogenic tracers applied to basalt genesis in the Snake River Plain—Yellowstone National Park region—Evidence for a 2.7-b.y.-old upper mantle keel: Geological Society of America Abstracts with Programs, v. 7, p. 1165.

Lipman, P. W., and others, 1978, Petrologic evolution of the San Juan volcanic field, southwestern Colorado: Pb and Sr isotopic evidence: Geological Society of America Bulletin, v. 89, p. 59–82.

Ludwig, K. R., and Silver, L. T., 1977, Lead-isotope inhomogeneity in Precambrian igneous K-feldspars: Geochimica et Cosmochimica Acta, v. 41, p. 1457–1473.

Lund, E. H., 1956, Igneous and metamorphic rocks of the Minnesota River Valley: Geological Society of America Bulletin, v. 67, p. 1475–1490.

Oversby, V. M., 1976, Isotopic ages and geochemistry of Archean acid igneous rocks from the Pilbara, Western Australia: Geochimica et Cosmochimica Acta, v. 40, p. 817–830.

Rosholt, J. N., and Bartel, A. J., 1969, Uranium, thorium, and lead systematics in Granite Mountains, Wyoming: Earth and Planetary Science Letters, v. 7, p. 141–147.

Stacey, J. S., and Kramers, J. D., 1975, Approximation of terrestrial lead isotope evolution by a two-stage model: Earth and Planetary Science Letters, v. 26, p. 207–221.

Stacey, J. S., and others, 1976, Plumbotectonics IIA, Precambrian massive sulfide deposits: U.S. Geological Survey Open-File Report 76-476, 26 p.

Steiger, R. H., and Jäger, E., 1977, Subcommission on geochronology: Convention on the use of decay constants in geo- and cosmochronology: Earth and Planetary Science Letters, v. 36, p. 359–362.

Zartman, R. E., and Wasserburg, G. J., 1969, The isotopic composition of lead in potassium feldspars from some 1.0-b.y. North American igneous rocks: Geochimica et Cosmochimica Acta, v. 33, p. 901–942.

MANUSCRIPT RECEIVED BY THE SOCIETY APRIL 25, 1979
REVISED MANUSCRIPT RECEIVED NOVEMBER 26, 1979
MANUSCRIPT ACCEPTED JANUARY 11, 1980

Geological Society of America
Special Paper 182
1980

# Boundary between Archean greenstone and gneiss terranes in northern Wisconsin and Michigan

**P. K. SIMS**

*U.S. Geological Survey, Box 25046, MS 912, Federal Center, Denver, Colorado 80225*

## ABSTRACT

New geologic and geochronologic data clearly establish that the two Archean basement-rock terranes recognized previously in Minnesota also occur in northern Wisconsin and Michigan, on the east side of the midcontinent gravity high. Greenstone-granite complexes (about 2,700 m.y. old) that are remarkably similar to those in the greenstone terrane of northern Minnesota are present on the south side of the Gogebic Range in northern Wisconsin and adjacent Michigan and in the northern complex of the Marquette district. Migmatitic gneisses and amphibolite, which are similar to the high-grade rocks in the gneiss terrane in southern Minnesota, are sporadically exposed south of the greenstone terrane and appear to compose the basement in most of northern and central Wisconsin and northern Michigan. The gneisses in Wisconsin and Michigan have been dated at two widely separated localities; they have radiometric ages of more than 3,000 m.y. and, for one rock type, an age of approximately 3,500 m.y. The boundary between the greenstone and gneiss terranes is covered by younger supracrustal rocks, but is interpreted to trend approximately eastward across northern Wisconsin to the vicinity of Marquette, Michigan, and beyond. Subsequent to its welding, probably 2,600 to 2,700 m.y. ago, abundant mafic dikes of Precambrian X and Y ages were emplaced in a wide zone parallel to the boundary, which indicates that it was predominantly a zone of extensional tectonics throughout much of post-Archean–pre-Phanerozoic time.

## INTRODUCTION

Two contrasting Archean basement rocks have been recognized at the surface and in the subsurface in Minnesota, in the western part of the Lake Superior region (Sims and Morey, 1973; Morey and Sims, 1976). Greenstone-granite complexes (2,750 to 2,700 m.y. old), which are typical of large parts of the Canadian Shield, make up the basement in the northern part, and migmatitic gneisses and amphibolite (in part 3,500 m.y. or more old) form the basement in the southern part. The boundary between the greenstone and gneiss terranes trends east-northeastward across

Minnesota at approximately its midpoint. We proposed that the same two Archean terranes probably also form the basement in northern Wisconsin and Michigan (Morey and Sims, 1976; Sims, 1976a) on the east side of the midcontinent gravity high (Craddock and others, 1963). New geologic and geochronologic data from the eastern part confirm this contention and extend our knowledge concerning the geographic extent of the respective basement terranes, the location of their mutual boundary, and the ages and tectonic history of their component parts.

The principal purposes of this paper are to (1) review the distinguishing characteristics of the two Archean terranes, (2) describe the geology and geochronology of rocks that make up the two terranes in northern Wisconsin and Michigan, and (3) discuss some aspects of the early crustal evolution of this part of the North American craton. The geology and radiometric ages of the rocks in the boundary zone in this part of the region will be discussed in a later report.

The Lake Superior region is on the southern margin of the Canadian Shield at the approximate juncture of exposed Precambrian rocks and overlapping platform Phanerozoic rocks, which thicken southward and westward. Nearly ubiquitous Quaternary glacial deposits cover the bedrock in the region, and, except for small local areas, outcrops are generally small and widely separated.

Rocks belonging to the three subdivisions of the Precambrian (James, 1958; Goldich, 1968; Sims and Morey, 1972) are exposed in the Lake Superior region (Sims, 1976a, Fig. 2). Archean or lower Precambrian (Precambrian W) rocks, which are more than 2,500 m.y. old, constitute the bedrock in northern Minnesota and in adjacent Ontario; they also crop out discontinuously as inliers through younger rocks in central and southern Minnesota and in northern Wisconsin and Michigan. Middle Precambrian (Precambrian X) supracrustal rocks compose the surface bedrock in much of eastern Minnesota and northern Wisconsin and Michigan. Except for the large Wolf Creek batholith in central Wisconsin (Van Schmus and others, 1975a) and scattered platform deposits of quartzite, upper Precambrian (Precambrian Y and Z) rocks are confined to a narrow, elongate belt that transects the older rocks in and adjacent to Lake Superior (see Fig. 1). The rocks in this belt consist mainly of thick successions of flood basalts, cogenetic gabbroic intrusive rocks, and associated red-bed sedimentary rocks. They constitute the midcontinent gravity high, which extends from Lake Superior southwestward into Kansas, and are generally interpreted as having been deposited in an intracontinental rift system of major proportions (King and Zietz, 1971). Summaries of the geology and pre-Phanerozoic tectonic evolution of the region have been given by Sims (1976a) and Van Schmus (1976).

## DISTINGUISHING CHARACTERISTICS OF THE TWO ARCHEAN TERRANES

The distinguishing characteristics of the two Archean (Precambrian W) terranes, as defined in Minnesota, are outlined in Table 1. In Minnesota, there is no definite evidence for Archean greenstone having been deposited on an older sialic basement, and, accordingly, the greenstone-granite complexes, on the one hand, and the migmatitic gneisses, on the other, appear to compose geographically separate and distinct Archean crustal blocks.

The boundary between the two terranes has not been delineated precisely in Minesota because of an extensive cover of younger Precambrian rocks and unconsolidated Pleistocene glacial deposits. In our original paper on this subject (Morey and Sims, 1976, Fig. 8), it was placed just north of Mille Lacs Lake, immediately north of the outcrop of the McGrath Gneiss. An alternative interpretation, suggested later (Sims, 1976a, Fig. 5), placed the boundary about 30 km north of Mille Lacs Lake, and this position is also shown in Figure 1 of this paper. The uncertainty in the location of the boundary has resulted from the sparse exposures in the critical areas and a lack of radiometric age data on the strongly deformed and sheared granitic rocks in the zone between the two proposed positions for the buried boundary. The cataclastic rocks in this zone are associated with metagraywacke (G. B. Morey, 1977, oral commun.); they could be sheared granite belonging to the 2,700-m.y.-old greenstone-granite complexes or granitic gneiss having affinities with the 3,500-m.y.-old gneisses.

The primary attributes of the gneisses in Minnesota were somewhat modified during the middle Precambrian by tectonothermal events generally assigned to the Penokean orogeny (Goldich and others, 1961; Cannon, 1973). In east-central Minnesota, the basement gneisses were reactivated, at least locally, to form one or more gneiss domes (see Morey and Sims, 1976); the reactivation produced open isotopic systems, which resulted in discordant ages being given by different radiometric techniques, a problem now known to be rather general in the gneiss terrane as a whole. In the Minnesota River Valley, on the other hand, which is the type area for the gneiss terrane in Minnesota (Morey and Sims, 1976), evidence for a Penokean tectonic overprint is lacking (see Grant, 1972). However, evidence for a low-grade thermal event, or events, is widespread and is indicated by numerous K-Ar and Rb-Sr biotite ages in the range 1,850 to 1,700 m.y. or younger and, probably, by the scatter in points on Rb-Sr whole-rock isochron diagrams (see Goldich and others, 1970, Figs. 9, 11). Several small plutons ranging in composition from gabbro to granite were emplaced into the gneiss at approximately this time (about 1,850 m.y. ago), which indicates a potential source of widespread heating. Another possible explanation for the low ap-

TABLE 1. CHARACTERISTICS OF ARCHEAN (PRECAMBRIAN W) TERRANES IN MINNESOTA

| | Greenstone terrane* | Gneiss terrane† |
|---|---|---|
| Rock assemblages | Mafic lavas and subvolcanic rocks (low-K tholeiite); felsic volcaniclastics; and volcanogenic graywacke, shale, and iron-formation | Granitic gneiss, hornblende-pyroxene gneiss, biotite gneiss, and cordierite-garnet biotite gneiss; most rocks are migmatitic and have a potassic neosome. |
| | Syntectonic granite-tonalite; younger anorogenic syenodioritic rocks | Late- or posttectonic granite-quartz monzonite (∿2,650 m.y. old) |
| Metamorphism | Greenschist facies; amphibolite facies adjacent to plutons | Regional granulite and amphibolite facies |
| Structural style | Tight, steeply plunging folds; commonly multiple deformations | Moderately open, dominantly gently plunging folds |
| Age | 2,750 to 2,700 m.y. old | >3,000 m.y. old; in part at least 3,500 to 3,800 m.y. |
| Tectonic environment | Ensimatic, possibly ocean ridge and island-arc or continental borderland environments | Unknown, but not protocrust |

*Note:* Data modified after Morey and Sims (1976).

*Type area: Vermilion district (Sims, 1976b).

†Type area: Minnesota River Valley (Grant, 1972; Goldich and others, 1970; Goldich and Hedge, 1974).

parent mineral ages suggested by Goldich and others (1970) is epeirogeny, with separate blocks within the gneiss terrane having been stabilized at different geologic times. They suggested that one part of the area exposed in the Minnesota River Valley in southwestern Minnesota was stabilized about 2,600 m.y. ago, whereas another part was stabilized much later, 1,700 to 1,850 m.y. ago.

The rocks composing the two terranes in northern Wisconsin and Michigan, as delineated in Figure 1, have the same general characteristics as those in the respective terranes in Minnesota. The greenstone-granite complexes in the northern part of this area are remarkably similar in age, rock assemblages, metamorphic grade, and structural style to those in northern Minnesota; in the same way, the gneisses that compose the gneiss terrane are quite similar to those in southwestern and central Minnesota, although they differ in detail. In particular, some differences exist in structural style of the gneisses: steep dips and plunges seem to be more prevalent in both Wisconsin and Michigan than in Minnesota. Because of the relatively poor exposures, this could, in part at least, simply reflect a sampling bias. Also, as suggested by Van Schmus and Anderson (1977), the isotopic systems in the gneisses in Wisconsin, as well as in Michigan, appear to be more highly disturbed than those in the rocks in southwestern Minnesota, apparently because of a stronger tectonothermal overprint in this part of the region during the middle Precambrian.

Intensive geochronologic studies still in progress have definitely shown that some of the gneisses in Michigan and Wisconsin are approximately as old as those in the Minnesota River Valley. Samples of tonalite gneiss from a gneiss dome in the western part of northern Michigan near Watersmeet have a radiometric age greater than 3,400 m.y., as discussed in a following page. In central Wisconsin, where earlier attempts to date the gneisses were inconclusive (Van Schmus and others, 1975b), zircon from

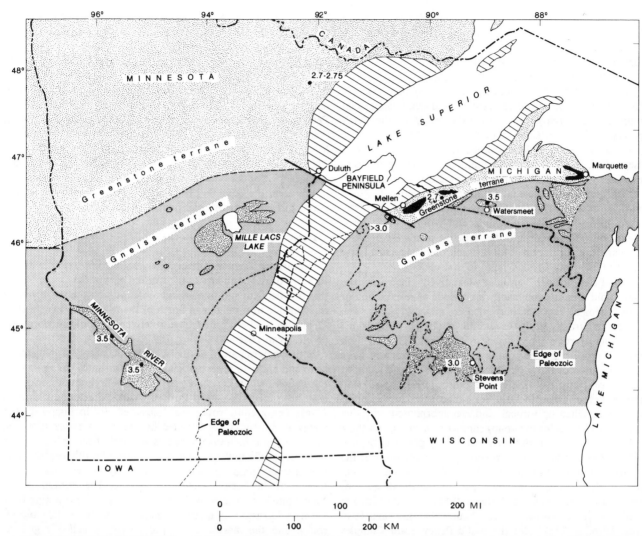

**Figure 1. Map showing distribution of Archean (Precambrian W) basement rocks at surface and in subsurface in Lake Superior region. Modified from Sims (1976a, Fig. 5). Except for midcontinent gravity high (line pattern), younger rocks are omitted. Heavy solid lines = possible Keweenawan transform faults reactivated from older continental fault systems; dense random pattern = areas of exposed Precambrian W gneisses; black = areas of exposed Precambrian W greenstone (Michigan and Wisconsin only); dark stipple = minimum inferred extent of gneiss terrane at surface and in subsurface; light stipple = minimum inferred extent of greenstone terrane at surface and in subsurface. Bayfield Peninsula is underlain by Precambrian Y or Z sedimentary rocks. Radiometric ages in billions of years.**

migmatitic gneisses in the Pittsville area and in Linwood Township define a concordia-intercept age of approximately 2,800 m.y. (Van Schmus and Anderson, 1977). Rb-Sr systematics observed for whole-rock samples from the Pittsville area, although clearly reflecting the effects of a middle Precambrian event, can be interpreted as indicating a primary age in the range of 3,000 to 3,200 m.y. for the gneisses (Van Schmus and Anderson, 1977). These older Archean ages are critical to determining the southern extent of the gneiss terrane in Wisconsin; they indicate that the basement gneisses definitely extend at least 210 km southward from the boundary with the greenstone terrane, as defined in Figure 1, and outcrops of similar-appearing gneissic rocks are known to extend at least 15 km farther south. By comparison, the southernmost exposed Archean gneisses in southwestern Minnesota are about 200 km south of the boundary (see Fig. 1).

Migmatitic gneisses and amphibolite that have not yet been dated are exposed at many other localities in central and northern Wisconsin and adjacent Michigan. As noted previously, radiometric dating is extremely difficult because of disturbed isotopic systems. However, it is improbable that any of the migmatitic gneisses of the type that occur in southwestern Minnesota and the Pittsville, Wisconsin, area are younger than Archean, because, so far as known, the Penokean tectonothermal episode was not a gneiss-forming event. Instead, even at the highest metamorphic (sillimanite) grade, bedding and other primary structures were not destroyed in the middle Precambrian (Marquette Range Supergroup) rocks nor were these rocks migmatized. A major problem that remains is to determine the extent to which middle Precambrian plutonic rocks were deformed in middle Precambrian time, resulting in a secondary structure that could resemble that in the older gneisses. In north-central

Wisconsin, at least, it has been demonstrated (Van Schmus and others, 1975b; Van Schmus and Anderson, 1977) that middle Precambrian granitic rocks have intruded the Archean gneisses, and some of them have a secondary foliation; distinguishing these from the older gneisses is essential to unraveling crustal development in this part of the region. A related problem, not yet satisfactorily resolved, is the nature, extent, and significance of an approximately 2,100-m.y.-old event recorded from the Felch trough area, Dickinson County, Michigan (see Fig. 2). In this area, P. O. Banks and Van Schmus (see Van Schmus and Woolsey, 1975, p. 1728) have distinguished an episode of high-grade metamorphism that occurred about 2,000 to 2,100 m.y. ago, which "resulted in the establishment of metamorphic Rb-Sr whole-rock isochrons with significantly elevated 'initial' $^{87}Sr/^{86}Sr$ ratios (ca. 0.730-0.735)." Van Schmus and Woolsey (1975) proposed extending this event into the Republic area west of Marquette to account for a rotated Rb-Sr isochron on Archean basement rocks (Compeau Creek Gneiss) in that region.

In the Felch trough area, I tentatively interpret the 2,100-m.y.-old event as dominantly an episode of reactivation of Archean basement rocks accompanied by metamorphism. The porphyritic red granite of James and others (1961), which is one of the principal rock types dated in this area, possibly is a remobilized product of the granite gneiss of the Norway Lake area. Mobilization of the gneiss, accompanied by introduction of abundant potassium (see James and others, 1961, p. 29) and cataclastic deformation ($\sim$2,000 m.y. ago), produced gneiss domes that punched up through and metamorphosed overlying Precambrian W (Dickinson Group) supracrustal rocks; but there is no direct evidence that this event necessarily preceded deposition of the Marquette Range Supergroup. Interpretation of the metamorphic history of the rocks in the Felch trough area is complicated by their being in the inner part of the Peavy node of metamorphism (see James and others, 1961, Pl. 6), which has been interpreted as having formed about 1,900 m.y. ago (Van Schmus, 1976, p. 615) when the mafic Peavy Pond Complex (Bayley, 1959) was emplaced into rocks of the Marquette Range Supergroup.

## GEOLOGY IN VICINITY OF BOUNDARY

In northern Wisconsin and Michigan, the boundary between the two basement terranes is covered by younger supracrustal rocks, but its position is known moderately accurately. In the vicinity of Marenisco, Michigan, it lies between the elliptical body of exposed Archean gneiss at Watersmeet (Wgn unit, Fig. 2) and the adjacent northeast-trending belt of mixed metavolcanic rocks, metagraywacke, and granitic rocks of late Archean age southeast of Marenisco (Wr unit, Fig. 2). Northeastward from there, the boundary curves gently eastward beneath overlapping upper Precambrian (Keweenawan) rocks and then extends eastward beneath the middle Precambrian succession (Xsv unit, Fig. 2) in the Marquette trough. Its approximate position is shown on Figure 1. About 50 km southwest of Marenisco in the Mellen, Wisconsin, area, the boundary between the two terranes previously was inferred to be truncated and displaced northwestward a distance of about 160 km by a major right-lateral fault system (Mineral Lake fault), as shown on Figure 1 (Sims, 1976a, p.

1105). This interpretation was based on the presence of granitic gneisses south of the fault thought to be part of the gneiss terrane, and this designation is given in Figures 2 and 4. Recently, geologic reconnaissance has shown that the gneisses are folded on northwest-trending axes, subparallel to the fault, and U-Pb zircon data indicate that they are 2,700 m.y. old rather than 3,000 m.y. old, as suggested by preliminary Rb-Sr whole-rock data (Z. E. Peterman, 1979, written commun.). The granitic gneisses, therefore, are deformed facies of the granitic rocks north of the Mineral Lake fault and compose a part of the greenstone terrane. This revised interpretation does not require a large horizontal displacement of the boundary by the Mineral Lake fault; instead, in this area, the boundary bends to a northwestward trend, subparallel to the Mineral Lake fault.

## Description of Archean Rocks

**Greenstone Terrane.** Greenstone-granite complexes are exposed north of the boundary in the western part of the Upper Peninsula of Michigan and in adjacent Wisconsin on the south side of the Gogebic iron range and north of the Marquette trough (Fig. 2), in what has been called the northern complex of the Marquette district (Van Hise and Bayley, 1895).

In the western part of northern Michigan, steeply dipping mafic lavas and mafic to felsic pyroclastic rocks assigned to the Ramsay Formation (Wv in Fig. 2) are coextensive with granitic rocks that have been formally named Puritan Quartz Monzonite (Schmidt, 1976). The granitic rocks are designated Wg in Figure 2. The metavolcanic rocks are intruded by the quartz monzonite, and their metamorphic grade increases generally from chlorite grade to garnet grade toward the quartz monzonite; this suggests that their metamorphism is related both spatially and genetically to intrusion of the Puritan Quartz Monzonite. The Puritan Quartz Monzonite is associated with abundant migmatite resulting from intimate intrusion of quartz monzonite into the older volcanic rocks, and the mass could appropriately be called a granite-migmatite complex. The greenstone and quartz monzonite are unconformably overlain by the middle Precambrian sedimentary rocks of the Gogebic Range, including the economically significant Ironwood Iron-Formation. In Wisconsin, adjacent to the Michigan State line, similar greenstone and quartz monzonite are exposed. On the shores of Gile Flowage about 6 km west of Ironwood, Michigan, the volcanic rocks are mafic-intermediate pyroclastic deposits and subordinate mafic, pillow lavas.

Intensely deformed granite, mafic metavolcanic rocks, and metagraywacke (Wr unit, Fig. 2), now believed correlative with the Puritan Quartz Monzonite and associated greenstone in the Gogebic Range (described above) compose a 7-km-wide, northeast-trending belt southeast of Marenisco (Fig. 3). Previously, these rocks were mapped by Fritts (1969) as strata near Cup Lake, strata near Banner Lake, and granite near Thayer and were assigned a middle Precambrian age. They differ from typical Archean greenstone-granite complexes in the Lake Superior region in having been deformed during both the late Archean and Penokean orogenies and in having a pervasive amphibolite-grade metamorphism. An Archean age for them is indicated by a firm U-Pb zircon age of 2,700 m.y. (Z. E. Peterman, 1978, written commun.) on a cataclastic tonalite and by Rb-Sr whole-rock age data on associated granitic rocks, which were called granite near

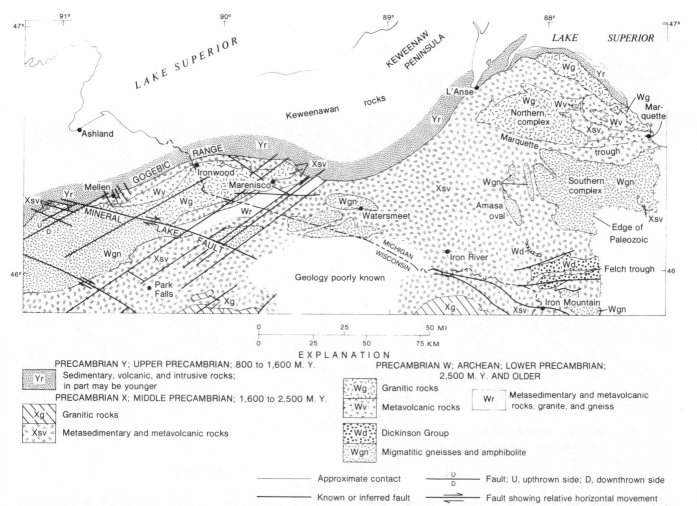

**Figure 2. Generalized geologic map of part of northern Wisconsin and Michigan. Modified from Sims (1976a, Fig. 2). Wr unit is approximately correlative with combined Wr and Wg units.**

Thayer by Fritts (1969). Because of their proximity to the boundary and their apparently unique tectonic fabric, these rocks are being studied in detail by field and radiometric methods.

In the Mellen, Wisconsin, area near the western end of the Gogebic Range (see Fig. 4), bedded mafic-intermediate tuff (now interlayered biotite schist and hornblende schist) and subordinate mafic pillow lavas (amphibolite) are exposed locally south of the city, particularly along and near the Bad River. Quartz monzonite comparable in texture to the Puritan Quartz Monzonite from the type locality also is exposed north of the Mineral Lake fault along the Bad River, about 5 km southwest of Mellen. Also, as noted previously, the granitic gneisses south of the Mineral Lake fault in this area are deformed and metamorphosed facies of the quartz monzonite. These gneisses are similar mineralogically and chemically to those exposed southeast of Marenisco.

The granitic rocks along the south side of the Gogebic Range constitute a batholith that is about 100 km long and as much as 20 km wide. Analysis of seven samples of the quartz monzonite from western northern Michigan gave a Rb-Sr whole-rock age of 2,710 ± 104 m.y. and an initial $^{87}Sr/^{86}Sr$ ratio of 0.7015 ± 0.0017 (Sims and others, 1977). Preliminary Rb-Sr data by S. Chaudhuri of Kansas State University on granitic rocks collected along U.S.

Route 51 south of Ironwood suggest a similar, but perhaps slightly younger age (W. R. Van Schmus, 1977, written commun.). The associated metavolcanic rocks include tholeiitic lavas. Except for the sparsity of sedimentary rocks of graywacke affinity, the volcanic succession resembles that in the type area of the greenstone terrane in northern Minnesota; the Puritan Quartz Monzonite and associated rocks resemble in gross aspects the Vermilion granite-migmatite massif in northern Minnesota, as described by Southwick (1972).

In the northern complex near Marquette, greenstone-granite complexes are widely exposed over an area of about 2,500 km² (Fig. 2); they are covered locally—in the Dead River, Clark Creek, and Baraga basins—by strata of the middle Precambrian Marquette Range Supergroup. Except for recent geologic mapping of a small part of the area along the north margin of the Marquette trough by members of the U.S. Geological Survey, modern studies of the lower Precambrian rocks in the northern complex are lacking.

Steeply dipping metavolcanic rocks, assigned to the Mona Schist and Kitchi Schist (Van Hise and Bayley, 1895) and designated Wv in Figure 2, are coextensive with younger granitic rocks in the eastern part of the northern complex. The Mona

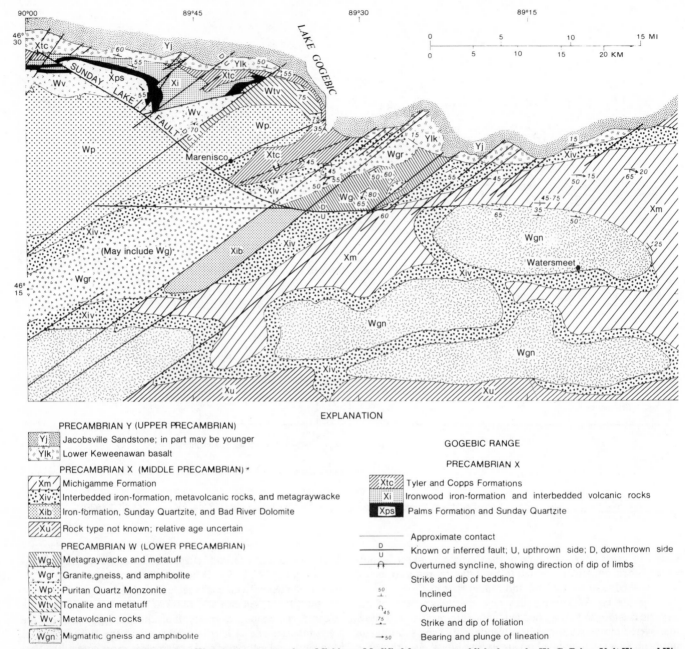

**Figure 3. Geologic map of Marenisco-Watersmeet area, northern Michigan. Modified from an unpublished map by W. C. Prinz. Unit Wgr and Wg are approximately correlative with units Wv, Wtv, and Wp.**

Schist is composed of pillowed and massive basalt and layered amphibolite; the amphibolite is interpreted as metamorphosed mafic tuff (Gair and Thaden, 1968, p. 12). The Kitchi Schist is mainly a pyroclastic rock of latitic to dacitic composition that is approximately the same age as the lower part of the Mona Schist (Puffett, 1974). The available data indicate that the regional metamorphic grade is greenschist facies but that amphibolite-facies metamorphic assemblages were formed adjacent to the intrusive granitic rocks.

Except for the eastern part, the granitic rocks of the northern complex have neither been formally named nor studied in detail.

In the eastern area, the principal rock type has been formally designated the Compeau Creek Gneiss (Gair and Thaden, 1968). It is dominantly foliated tonalite and granodiorite but locally it is uniform massive granitoid rock and includes some amphibolite and dark varieties of tonalite and granodiorite. Gair and Thaden (1968, p. 25–26) have proposed that the more conspicuously layered variety of the Compeau Creek Gneiss probably was formed by intrusion of a siliceous magma into the Mona Schist, with incorporation of some of the volcanic rocks into the magmatic body. The Compeau Creek Gneiss greatly resembles the Precambrian W Vermilion granite of northern Minnesota. Both granitic

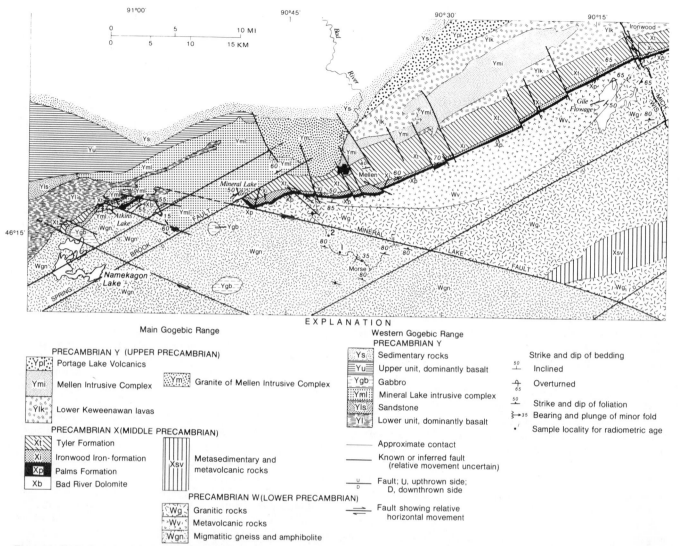

**Figure 4. Geologic map of Mellen area, northern Wisconsin. Geology of western part of Gogebic Range modified from Aldrich (1929), Hubbard (1975), and Olmsted (1967). Geology of Archean (Precambrian W) rocks based on reconnaissance by P. K. Sims and M. G. Mudrey, Jr. (1976, written commun.).**

rock bodies intruded dominantly subaqueous volcanic rocks having tholeiitic affinities, with the development of local migmatite, and have steep thermal metamorphic gradients. The Compeau Creek Gneiss in the northern complex of the Marquette district has not been dated radiometrically, but similar rocks from the southern complex, which are also called Compeau Creek Gneiss by Gair and Thaden and which intrude the Precambrian W Bell Creek Gneiss (Cannon and Simmons, 1973), have been dated by the whole-rock Rb-Sr method as being approximately 2,600 to 2,800 m.y. old (Van Schmus and Woolsey, 1975).

Judged from a brief reconnaissance, the granitic rocks in the western part of the northern complex are largely foliated granodiorite and tonalite cut by abundant pink leucogranite and pegmatite. They resemble the intermediate granitic rocks in the western part of the Giants Range batholith in northern Minnesota (Sims and Viswanathan, 1972).

**Gneiss Terrane.** Migmatitic gneisses and amphibolite are exposed south of the boundary in western northern Michigan and in the

southern complex and adjacent areas south of the Marquette trough. In these areas, the gneisses make up the cores of gneiss domes or fault-bounded antiformal blocks surrounded by superjacent bedded rocks of the Precambrian X Marquette Range Supergroup.

In northern Michigan, gneiss is exposed sparsely in the northern part of the Watersmeet gneiss dome, the northernmost of four gneiss domes in the Marenisco-Watersmeet area (Fig. 3). The gneiss is overlain unconformably by amphibolite and biotite schist, and those rock types are intruded by granite, which in part is gneissoid. The gneiss is a medium-gray, irregular layered tonalite containing feldspar megacrysts. Locally, it is a streaked gneiss with abundant clots of fresh biotite. Radiometric studies of the tonalite gneiss indicate that it is lower Precambrian (Archean) but was reworked in the middle Precambrian (Sims and Peterman, 1976; Peterman and others, this volume). An old Archean age of about 3,400 m.y. for the gneiss is indicated by $^{207}Pb/^{206}Pb$ ages on three size fractions of zircon; a primary age of about

3,500 m.y. is suggested by a concordia plot of the zircon data. The intrusive granite has a U-Pb zircon age of about 2,600 m.y. The whole-rock Rb-Sr systems in both the gneiss and the granite are highly disturbed, but data for samples from two localities define secondary isochrons of about 1,750 m.y. in age having elevated $^{87}Sr/^{86}Sr$ ratios (Peterman and others, this volume).

In the southern complex south of the Marquette trough, the dominant gneiss is the Precambrian W Bell Creek Gneiss, which has been described by Cannon and Simmons (1973). A similar, possibly equivalent gneiss designated formally as the Margeson Creek Gneiss (Gair and Wier, 1956), forms the core of the Amasa oval, about 15 km west of the southern complex (Fig. 2). In the southern complex, the dominant phase of the Bell Creek is a very coarse grained megacrystic granitic rock of granite-granodiorite composition that contains concordant layers, some 300 m or more thick, of mafic gneiss. Lenses of highly metamorphosed iron formation occur locally in the mafic gneiss; thin layers of biotitic or hornblendic quartzite are rare. A less common facies of the gneiss, which crops out along the northern margin of the southern complex, is a fine- to medium-grained megacrystic granite gneiss. Migmatite is rare, but a 300-m-wide layer at least 5 km long was mapped in the southern complex (Cannon and Simmons, 1973). The Compeau Creek Gneiss, which was named and described by Gair and Thaden (1968) from its type section in the northern complex north of the Marquette trough, occurs along much of the north margin of the southern complex; it intrudes the Bell Creek Gneiss at several localities. The gneisses dip steeply and, according to Taylor (1967), were deformed at least twice in early Precambrian time. Radiometric data on the Bell Creek Gneiss indicate that it is at least 2,600 m.y. old and that it has undergone Rb and Sr redistribution, probably about 2,100 m.y. ago (Van Schmus and Woolsey, 1975). One or more later thermal events affected the gneiss; Rb-Sr biotite (metamorphic) ages are consistent at 1,665 ± 40 m.y.

South of the southern complex in the Felch trough area, a megacrystic gneiss informally called granite gneiss of the Norway Lake area (James and others, 1961) is basement to the supracrustal Dickinson Group (see Fig. 2). Attempts to date the gneiss have been unsuccessful (W. R. Van Schmus, 1979, oral commun.), but it is older than 2,700 m.y., the age of the granite that intrudes the Dickinson Group.

## Structure

The nature and structural attitude of the juncture between the two Archean terranes is not known, but the tectonism of the middle Precambrian supracrustal rocks within and adjacent to the boundary zone provides some insight into its character.

In the Marenisco-Watersmeet area, rocks belonging to the two basement terranes are exposed within 5 km of each other, but the boundary between the two is covered by middle Precambrian rocks of the Marquette Range Supergroup (Xiv unit, Fig. 3). The geology of the rocks in and adjacent to the boundary zone is complex, as a result of intense deformation during the middle Precambrian (Penokean) orogeny. Immediately south of the boundary in the gneiss terrane, the Archean gneisses form domes mantled by the supracrustal middle Precambrian strata. In this area, as well as in areas to the east (W. F. Cannon 1977, oral commun.), the gneiss domes were uplifted subsequent to the develop-

ment of an east-trending, southward-dipping regional schistosity in the metasedimentary rocks. The trend of this schistosity is about parallel to the juncture between the two basement rocks. During reactivation of the basement gneisses, the gneisses were cataclastically deformed and partly recrystallized, with the generation of minor anatectic magma (Sims and Peterman, 1976). The supracrustal strata adjacent to the domes were metamorphosed to garnet grade approximately concurrently with the doming.

On the north side of the boundary in the greenstone-granite terrane, both the Archean rocks and the supracrustal middle Precambrian rocks are strongly deformed in a 10-km-wide northeast-trending belt south of Marenisco (Fig. 3). The tectonothermal event took place approximately concurrently with development of the gneiss domes to the southeast. The bedded rocks within this narrow belt are folded on northeast-trending axes and are overturned to the northwest; the folds have southeast-dipping axial surfaces and an axial plane penetrative cleavage. This deformation was superposed on an older Archean deformation, which can be particularly well documented in the Archean metagraywacke unit (Wg, Fig. 3). The granitic rocks, which are mineralogically and chemically similar to the Puritan Quartz Monzonite north of Marenisco (Fig. 3), were partly deformed cataclastically and locally are now cataclastic gneisses. The Penokean dynamothermal metamorphism disturbed the Rb-Sr whole-rock and mineral systems, with the consequent development of rotated isochrons (Z. E. Peterman, 1978, written commun.).

The field and radiometric data indicate that the boundary between the two basement terranes in the Marenisco-Watersmeet area was the locus of intense deformation and metamorphism during the Penokean orogeny. Differential movements in this zone, which apparently initiated deposition of the middle Precambrian sequence in the Lake Superior region (Sims, 1976a), continued after deposition and culminated about 1,800 m.y. ago. Probably the Penokean deformational patterns—fold vergence and penetrative cleavage—resulted from tectonic transport northward and northwestward of the gneiss crustal segment against the more stable greenstone-granite crustal segment. These tectonic features suggest that the boundary zone, at least in this area, is a steep, southward-dipping fault zone.

In the western part of the Marquette trough, Klasner and Cannon (1974) modeled the geology across the boundary from gravity profiles, without taking into account possible differences in the compositions of the basement rocks on either side of the juncture. This is probably unnecessary, however, for in this area the granitic rocks on the north side of the boundary apparently are quite similar in density to the granitic gneisses (mainly Bell Creek Gneiss) on the south side. As shown in Figure 3 of their paper, Klasner and Cannon interpreted the middle Precambrian supracrustal rocks as occupying fault-bounded depressions in the basement rocks, which are subparallel to the boundary. According to their data, the boundary here must be a steeply inclined surface, probably a fault, or more likely a broad zone of faults.

## Mafic and Ultramafic Dikes and Plutons

Dike swarms and small plutons of two and possibly three different Precambrian ages are abundant in the vicinity of the boundary between the two Archean terranes and appear to decrease in

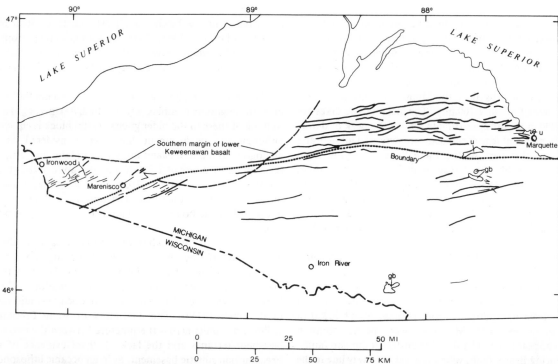

**Figure 5. Mafic dikes and plutons in vicinity of boundary between Archean gneiss terrane and greenstone terrane, northern Michigan. Heavy lines = Precambrian Y (Keweenawan) diabase dikes; thin lines = Precambrian X and Precambrian W metadiabase dikes; gb = gabbroic pluton; u = ultramafic pluton. Southern margin of lower Keweenawan basalt east of long. 89° is inferred from aeromagnetic data.**

abundance away from it. Many of the dikes that have been delineated on detailed geologic maps in northern Michigan are shown in Figure 5; because of a general lack of exposures in adjacent Wisconsin, dikes are not shown for this area.

Diabase dikes of upper Precambrian (Keweenawan) age are abundant. In the eastern part of the region, these are negatively polarized, and many of those shown in Figure 5 are inferred from negative aeromagnetic anomalies in published aeromagnetic maps (Case and Gair, 1965; Zietz and Kirby, 1971). In the western part of northern Michigan, on the other hand, dikes of Keweenawan age produce strong positive anomalies (Schmidt, 1976). Most dikes of this age are virtually unmetamorphosed; they cut rocks as young as lower Keweenawan. In the vicinity of Ironwood and areas farther west, Keweenawan dikes are weakly metamorphosed, apparently by the thermal event related to emplacement of the Precambrian Y Mellen Intrusive Complex (see Fig. 4). The diabase dikes are described (Gair and Thaden, 1968; Puffett, 1974; Schmidt, 1976) as containing dominant plagioclase ($An_{48-70}$) and clinopyroxene, mainly augite but partly pigeonite. Small amounts of olivine or its alteration products and magnetite are present. A chemical analysis of one sample of diabase (Puffett, 1974, Table 11) indicates a basaltic composition.

Altered diabasic dikes of Precambrian X age cut Archean rocks and some Precambrian X rocks. In the Marquette area, they cut rocks as young as the Clarksburg Volcanics Member of the Michigamme Formation (Marquette Range Supergroup) but not younger rocks (Gair and Thaden, 1968); in the western part of northern Michigan, they cut the Ironwood Iron-Formation but not the younger Michigamme Formation and its correlatives (W. C. Prinz, 1976, oral commun.). The dikes of this age are metamorphosed to different degrees, up to sillimanite grade

(James, 1955), but the original diabasic texture generally is preserved. The least metamorphosed rocks from the Ironwood area consist mainly of plagioclase ($An_{58-70}$) and clinopyroxene and contain several percent of magnetite and altered olivine (Schmidt, 1976). Sparse chemical data indicate that the metadiabase is basaltic in composition.

In the Ironwood, Michigan, area, Schmidt (1976, p. 30) has distinguished a group of metadiabasic rocks that have been metamorphosed twice, which he infers are Precambrian W in age. If they indeed are Precambrian W in age, this group may be widespread in the region. They are included with metadiabase of Precambrian X age in Figure 5. The dikes are composed mainly of hornblende and plagioclase, but, because of their multiple metamorphism under differing metamorphic temperatures, they also contain a variety of secondary minerals. Rhythmic layering was recognized in one of the larger dikes south of Ironwood (Schmidt, 1976). The chemical composition of this group has not been determined, but according to the mineralogy the dikes probably are basalt.

Ultramafic plutons are known from two localities near the boundary in the Marquette area. At Deer Lake about 25 km west of Marquette, a northeast-trending serpentinized peridotite body about 8 km long and 1 km wide is associated with and intrudes the Precambrian W Kitchi Schist (Case and Gair, 1965). The peridotite is presumed to be Precambrian W in age, but the age designation is equivocal. Gold has been mined from the body at the Ropes gold mine (Broderick, 1945). A second body of serpentinized peridotite on Presque Isle near Marquette is about 1 km² in area. It is overlain by the Precambrian Y Jacobsville Sandstone. Its age is not known, but Gair and Thaden (1968) infer from its weak metamorphism that it probably is Precambrian X.

Radiometric age data on the mafic dikes are lacking, and ages can only be inferred by indirect means, as discussed above. The diabase dikes in the region north of the Marquette trough have been determined from their paleomagnetism (Pesonen and Halls, 1975) to have pole positions similar to the reversely magnetized Precambrian Y Logan Sills along the Minnesota-Ontario border, suggesting an age of about 1,100 m.y. The metadiabase dikes are older than 1,900 m.y., the approximate time of the peak of deformation and metamorphism (Penokean orogeny of Goldich and others, 1961) in this part of the Lake Superior region (Van Schmus, 1976). Possibly the Precambrian X metadiabase dikes are approximately correlative with the 2,000- to 2,200-m.y.-old tholeiitic dikes in northern Minnesota (Hanson and Malhotra, 1971; Sims and Mudrey, 1972) and the 2,150-m.y.-old Nipissing diabase of Canadian usage in the Lake Huron area, Canada (Van Schmus, 1965).

The pronounced preferred orientation of the dikes, regardless of age, is subparallel to the trend of the juncture between the Archean greenstone and gneiss terranes and is evident in Figure 5. It clearly reflects a protracted interval of extensional deformation continuing at least intermittently from perhaps about 2,200 to about 1,100 m.y. ago. In the Marenisco area, the Keweenawan diabase dikes appear to be older than the northeast dip-slip faults, such as the Spring Brook fault, which formed relatively late in the Keweenawan volcanic-sedimentary episode.

## DISCUSSION

The boundary between the two Archean basement terranes is a fundamental crustal structure of regional significance, which provides a unifying thread for interpreting the tectonic evolution of the southern part of the Canadian Shield (Sims, 1976a).

Data from northern Wisconsin and Michigan support the suggestion, made earlier (Morey and Sims, 1976), that the two basement terranes were welded together by voluminous granite emplaced into the crust at or near the close of the Archean. In Wisconsin and western northern Michigan (see Fig. 2), a large granitic pluton composed principally of Puritan Quartz Monzonite (Schmidt, 1976) occurs adjacent to and subparallel to the boundary; it was emplaced about 2,700 m.y. ago (Sims and others, 1977), apparently as part of the extensive plutonic activity that took place throughout the greenstone terrane at that time (Sims, 1976b). The batholith is aligned subparallel to the boundary in a manner similar to that of the late Archean Giants Range batholith and its buried westward extension in northern Minnesota. The granitic rocks in the northern complex near Marquette also seem to have their long dimensions subparallel to the juncture (see Fig. 2). This parallelism in trend of the batholithic rocks and the boundary strongly suggests that the greenstone and gneiss terranes were in juxtaposition at the time the granitic rocks were emplaced. Granitic rocks 2,600 to 2,700 m.y. old occur sporadically within the gneiss terrane. As noted earlier, 2,600-m.y.-old granite cuts older Archean gneiss in the Watersmeet gneiss dome. It is associated with and apparently coeval with associated biotite schist and amphibolite. The granite is nearly identical to the Puritan Quartz Monzonite in the batholith north of Marenisco and to the smaller bodies of granite southeast of Marenisco. Granite and associated supracrustal

rocks occur in the Felch trough area north of Iron Mountain (James and others, 1961). Late Archean granitic plutons are present in the Minnesota River Valley, the largest—the Sacred Heart batholith—has been dated at 2,650 m.y. (Goldich and others, 1970).

The tectonic regime at the time of accumulation of the greenstone-granite complexes (2,750 to 2,700 m.y. ago) and of the welding of these to the older gneiss crustal block is equivocal; for, although the environment has been considered as possibly analogous to modern island-arc settings (Jahn and others, 1974), the apparent scarcity of 2,700-m.y.-old granite in the gneisses and the absence of some other petrotectonic assemblages indicative of paleoconvergent junctures (Dickinson, 1971) negate the likelihood of the boundary having been a convergent plate juncture.

Subsequent to the welding of the two terranes at the approximate end of the Archean, the boundary apparently was dominantly a zone of extensional tectonics. The middle Precambrian basin, in which the great iron formations of the Lake Superior region were formed, was developed over and subparallel to it (Sims, 1976a). The continuity of some Precambrian X sedimentary units across the juncture between the two respective basement terranes and the lack of direct evidence of a middle Precambrian simatic basement, as in an oceanic lithosphere plate, support an intracratonal environment of deposition for these rocks (Sims, 1976c). Probably the boundary was the locus of substantial necking of the crust at this time, but evidence for separation at the juncture is lacking. However, abundant mafic dikes, probably from a subcrustal source, were emplaced in steeply dipping tensional fractures roughly parallel to the boundary in both segments of the basement. Dike emplacement may have extended over a considerable time span in the middle Precambrian but mainly occurred prior to deposition of the Michigamme Formation, which is stratigraphically in the upper part of the Marquette Range Supergroup.

The Penokean orogeny (1,800 to 1,900 m.y. ago), which deformed the middle Precambrian supracrustal rocks, was localized directly or indirectly by the basement terranes (Sims, 1976a). The supracrustal rocks lying on the older Archean gneiss basement were intensely deformed and metamorphosed during this orogeny, whereas the supracrustal rocks overlying the greenstone basement terrane were little deformed and metamorphosed. The tectonism within the gneiss terrane resulted from reactivation of the Archean gneisses, with the development of gneiss domes and depressions, recrystallization and partial melting, and extensive cataclasis (Sims and Peterman, 1976); the basement was deformed together with the supracrustal rocks. The driving force for the tectonism apparently was thermal expansion and contraction of the basement gneisses, for the tectonism was virtually confined to this segment of the crust. During this tectonic episode, the greenstone crustal segment remained stable, except in the boundary zone between the two basement blocks where the gneiss block impinged upon the greenstone block, as in the Marenisco area of western northern Michigan.

During Keweenawan time (Precambrian Y), mafic dikes again were emplaced in abundance within a wide zone adjacent to the boundary (Fig. 5). Thus, although the axis of the midcontinent rift system in this part of the region was northeast and many of the attendant faults similarly are oriented northeast, the boundary remained a zone of dominantly extensional deformation during Keweenawan time and continued to have a pronounced influence on the tectonics of the Lake Superior region.

## ACKNOWLEDGMENTS

The pioneering geochronologic studies of S. S. Goldich and associates and his later work with students and other colleagues provided the time framework needed to unravel the complex geology of the Precambrian rocks in Minnesota and adjacent areas within the Lake Superior region. More particularly, his recognition of very old Archean rocks in the Minnesota River Valley was the key to delineating the two contrasting Archean terranes now generally recognized as composing the basement in the Lake Superior region.

Demarcation of the two basement terranes in northern Michigan and Wisconsin was aided greatly by recently completed radiometric age data obtained by Z. E. Peterman and R. E. Zartman of the U.S. Geological Survey and W. R. Van Schmus of the University of Kansas. Field relationships in critical areas were discussed with Peterman, W. C. Prinz, and W. F. Cannon of the U.S. Geological Survey and M. G. Mudrey, Jr., of the Wisconsin Geological and Natural History Survey.

## REFERENCES CITED

Aldrich, H. R., 1929, Geology of the Gogebic iron range of Wisconsin: Wisconsin Geological and Natural History Survey Bulletin 71, 279 p.

Bayley, R. W., 1959, Geology of the Lake Mary quadrangle, Iron County, Michigan: U. S. Geological Survey Bulletin 1077, 112 p.

Broderick, T. M., 1945, Geology of the Ropes gold mine, Marquette County, Michigan: Economic Geology, v. 40, p. 115-128.

Cannon, W. F., 1973, The Penokean orogeny in northern Michigan, in Young, G. M., ed., Huronian stratigraphy and sedimentation: Geological Association of Canada Special Paper 12, p. 251-271.

Cannon, W. F., and Simmons, G. C., 1973, Geology of part of the southern complex, Marquette district, Michigan: U.S. Geological Survey Journal of Research, v. 1, p. 165-172.

Case, J. E., and Gair, J. E., 1965, Aeromagnetic map of parts of Marquette, Dickinson, Baraga, Alger, and Schoolcraft Counties, Michigan, and its geologic interpretation: U. S. Geological Survey Geophysical Investigations Map GP-467.

Craddock, Campbell, Thiel, E. C., and Gross, Barton, 1963, A gravity investigation of the Precambrian of southeastern Minnesota and western Wisconsin: Journal of Geophysical Research, v. 68, p. 6015-6032.

Dickinson, W. R., 1971, Plate tectonics in geological history: Science, v. 174, p. 107-113.

Fritts, C. E., 1969, Bedrock geologic map of the Marenisco-Watersmeet area, Gogebic and Ontonagon Counties, Michigan: U. S. Geological Survey Miscellaneous Geologic Investigations Map I-576, scale 1:48,000.

Gair, J. E., and Thaden, R. E., 1968, Geology of the Marquette and Sands quadrangles, Marquette County, Michigan: U.S. Geological Survey Professional Paper 397, 77 p.

Gair, J. E., and Wier, K. L., 1956, Geology of the Kiernan quadrangle, Iron County, Michigan: U.S. Geological Survey Bulletin 1044, 88 p.

Goldich, S. S., 1968, Geochronology in the Lake Superior region: Canadian Journal of Earth Sciences, v. 5, p. 715-724.

Goldich, S. S., and Hedge, C. E., 1974, 3,800-Myr granitic gneiss in southwestern Minnesota: Nature, v. 252, no. 5483, p. 467-468.

Goldich, S. S., Hedge, C. E., and Stern, T. W., 1970, Age of the Morton and Montevideo gneisses and related rocks, southwestern Minnesota: Geological Society of America Bulletin, v. 81, p. 3671-3696.

Goldich, S. S., and others, 1961, The Precambrian geology and geochronology of Minnesota: Minnesota Geological Survey Bulletin 41, 193 p.

Grant, J. A., 1972, Minnesota River Valley, southwestern Minnesota, in Sims, P. K., and Morey, G. B., eds., Geology of Minnesota—A centennial volume: Minnesota Geological Survey, p. 177-196.

Hanson, G. N., and Malhotra, R., 1971, K-Ar ages of mafic dikes and evidence for low-grade metamorphism in northeastern Minnesota: Geological Society of America Bulletin, v. 82, p. 1107-1114.

Hubbard, H. A., 1975, Lower Keweenawan volcanic rocks of Michigan and Wisconsin: U.S. Geological Survey Journal of Research, v. 3, p. 529-541.

Jahn, Bor-ming, Shih, Chi-Yu, and Murthy, V. R., 1974, Trace element geochemistry of Archean volcanic rocks: Geochimica et Cosmochimica Acta, v. 38, p. 611-627.

James, H. L., 1955, Zones of regional metamorphism in the Precambrian of northern Michigan: Geological Society of America Bulletin, v. 66, p. 1455-1488.

———1958, Stratigraphy of pre-Keweenawan rocks in parts of northern Michigan: U.S. Geological Survey Professional Paper 314-C, p. 27-44.

James, H. L., and others, 1961, Geology of central Dickinson County, Michigan: U.S. Geological Survey Professional Paper 310, 176 p.

King, E. R., and Zietz, Isidore, 1971, Aeromagnetic study of the midcontinent gravity high of the central United States: Geological Society of America Bulletin, v. 82, p. 2187-2207.

Klasner, J. S., and Cannon, W. F., 1974, Geologic interpretation of gravity profiles in the western Marquette district, northern Michigan: Geological Society of America Bulletin, v. 85, p. 213-218.

Morey, G. B., and Sims, P. K., 1976, Boundary between two Precambrian W terranes in Minnesota and its geologic significance: Geological Society of America Bulletin, v. 87, p. 141-152.

Olmsted, J. F., 1967, Geologic map of part of the Marengo quadrangle: Wisconsin Geological and Natural History Survey Open-File Map, scale 1:24,000

Pesonen, L. J., and Halls, H. C., 1975, Paleomagnetism of Keweenawan diabase dikes from Baraga and Marquette Counties, Michigan: Annual Institute on Lake Superior Geology, 21st, Northern Michigan University, Marquette, Michigan, p. 34.

Peterman, Z. E., Zartman, R. E., and Sims, P. K., 1976, Early Archean tonalite gneiss from northern Michigan, U.S.A.: Geological Society of America Special Paper 182 (this volume).

Puffett, W. P., 1974, Geology of the Negaunee quadrangle, Marquette County, Michigan: U.S. Geological Survey Professional Paper 788, 53 p.

Schmidt, R. G., 1976, Geology of Precambrian rocks, Ironwood-Ramsay area, Michigan: U.S. Geological Survey Bulletin 1407, 40 p.

Sims, P. K., 1976a, Precambrian tectonics and mineral deposits, Lake Superior region: Economic Geology, v. 71, p. 1092-1118.

———1976b, Early Precambrian tectonic-igneous evolution in Vermilion district, northeastern Minnesota: Geological Society of America Bulletin, v. 87, p. 379-389.

———1976c, The Penokean fold belt in Lake Superior region, an intracratonic Precambrian mobile belt: Geological Society of America Abstracts with Programs, v. 8, p. 1109.

Sims, P. K., and Morey, G. B., 1972, Résumé of geology of Minnesota, in Sims, P. K., and Morey, G. B., eds., Geology of Minnesota—A centennial volume: Minnesota Geological Survey, p. 3-17.

———1973, A geologic model for the development of early Precambrian crust in Minnesota: Geological Society of America Abstracts with Programs, v. 5, p. 812.

Sims, P. K., and Mudrey, M. G., Jr., 1972, Diabase dikes in northern Minnesota, *in* Sims, P. K., and Morey, G. B., eds., Geology of Minnesota—A centennial volume: Minnesota Geological Survey, p. 256–259.

Sims, P. K., and Peterman, Z. E., 1976, Geology and Rb-Sr ages of reactivated Precambrian gneisses and granite in the Marenisco-Watersmeet area, northern Michigan: U.S. Geological Survey Journal of Research, v. 4, p. 405–414.

Sims, P. K., and Viswanathan, Subramanian, 1972, Giants Range batholith, *in* Sims, P. K., and Morey, G. B., eds., Geology of Minnesota—A centennial volume: Minnesota Geological Survey, p. 120–139.

Sims, P. K., Peterman, Z. E., and Prinz, W. C., 1977, Geology and Rb-Sr age of Precambrian W Puritan Quartz Monzonite, northern Michigan: U.S. Geological Survey Journal of Research, v. 5, p. 185–192.

Southwick, D. L., 1972, Vermilion granite-migmatite massif, *in* Sims, P. K., and Morey, G. B., eds., Geology of Minnesota—A centennial volume: Minnesota Geological Survey, p. 108–119.

Taylor, W.E.G., 1967, The geology of the lower Precambrian rocks of the Champion area of upper Michigan: Northwestern University Report 13 (National Aeronautics and Space Administration Research Grant NGR 14-007-27), 34 p.

Van Hise, C. R., and Bayley, W. S., 1895, Preliminary report on the Marquette iron-bearing district of Michigan: U.S. Geological Survey 15th Annual Report, p. 485–650.

Van Schmus, W. R., 1965, The geochronology of the Blind River–Bruce mines area, Ontario, Canada: Journal of Geology, v. 73, p. 755–780.

——1976, Early and middle Proterozoic history of the Great Lakes area, North America: Royal Society of London Philosophical Transactions, ser. A., v. 280, p. 605–628.

Van Schmus, W. R., and Anderson, J. L., 1977, Gneiss and migmatite of Archean age in the Precambrian basement of central Wisconsin: Geology, v. 5, p. 45–48.

Van Schmus, W. R., and Woolsey, L. L., 1975, Rb-Sr geochronology of the Republic area, Marquette County, Michigan: Canadian Journal of Earth Sciences, v. 12, p. 1723–1733.

Van Schmus, W. R., Medaris, L. G., and Banks, P. O., 1975a, Geology and age of the Wolf River batholith, Wisconsin: Geological Society of America Bulletin, v. 86, p. 907–914.

Van Schmus, W. R., Thurman, E. M., and Peterman, Z. E., 1975b, Geology and Rb-Sr chronology of middle Precambrian rocks in eastern and central Wisconsin: Geological Society of America Bulletin, v. 86, p. 1255–1265.

Zietz, Isidore, and Kirby, J. R., 1971, Aeromagnetic map of the western part of the northern peninsula, Michigan, and part of northern Wisconsin: U.S. Geological Survey Geophysical Investigations Map GP-750, scale 1:250,000.

MANUSCRIPT RECEIVED BY THE SOCIETY APRIL 25, 1979
REVISED MANUSCRIPT RECEIVED NOVEMBER 26, 1979
MANUSCRIPT ACCEPTED JANUARY 11, 1980

Geological Society of America
Special Paper 182
1980

# Tonalitic gneiss of early Archean age from northern Michigan

Z. E. PETERMAN
R. E. ZARTMAN
P. K. SIMS
*U.S. Geological Survey, Box 25046, MS 963, Federal Center, Denver, Colorado 80225*

## ABSTRACT

The Marenisco-Watersmeet area in the western part of the northern peninsula of Michigan contains a greenstone and granite terrane (Puritan Quartz Monzonite) of late Archean age on the north and a gneiss terrane (gneiss at Watersmeet) on the south. A granite and gneiss belt (collectively called the granite near Thayer) crops out between these contrasting terranes. Lower Proterozoic metasedimentary and metavolcanic rocks of the Marquette Range Supergroup are extensive.

Radiometric dating of the tonalitic phase of the gneiss at Watersmeet establishes an early Archean age and a complex subsequent history. U-Th-Pb systematics provide a firm minimum age of 3,410 m.y. with the possibility of a much greater age—3,500 to 3,800 m.y. Cataclasis and recrystallization during the early Proterozoic Penokean orogeny are recorded on a regional scale by whole-rock and mineral Rb-Sr ages of 1,750 m.y. Intense cataclasis of granodioritic gneiss in the Watersmeet dome locally produced metamorphic zircon with concordant ages of 1,755 m.y.

Zircons from a tonalitic phase of the granite near Thayer are dated at 2,750 m.y. Zircons from leucogranite dikes, which are abundant in the tonalitic phase of the gneiss at Watersmeet, are slightly younger at 2,600 m.y. These intrusive rocks are approximately contemporaneous with the development of the greenstone-granite terrane of late Archean age.

## INTRODUCTION

The existence of ancient (3,400 to 3,800 m.y. old) gneisses in southwestern Minnesota, Labrador, and western Greenland is well established (Goldich and others, 1970; Moorbath and others, 1972; Baadsgaard, 1973, 1976; Goldich and Hedge, 1974; Barton, 1975; Hurst and others, 1975). This paper concerns a fourth occurrence of tonalitic gneiss, at least 3,400 m.y. old, in the Marenisco-Watersmeet area in the upper peninsula of Michigan.

In addition to these very old rocks, an increasing number of

gneissic rocks with ages greater than 2,800 m.y., but generally less than 3,100 m.y., are being found in the Canadian Shield when radiometric methods capable of penetrating the effects of the Kenoran orogeny have been employed (Fig. 1). In the western Superior province, the ages of several of these rocks originally were interpreted to indicate progressively younger greenstone belts and associated granitic intrusions to the south (Krogh and Davis, 1971). However, most of the dated rocks can be interpreted as possible pre-greenstone basement, largely of tonalitic composition (Baragar and McGlynn, 1976).

Zircon U-Pb and whole-rock Rb-Sr ages of samples from

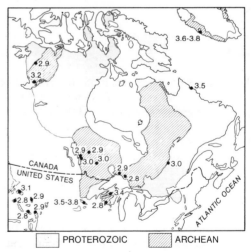

Figure 1. Localities (dots) in the Canadian Shield and in the Wyoming age province where ages of 2.8 b.y. or greater have been determined. Compiled from Baadsgaard (1973, 1976), Barton (1975), Catanzaro and Kulp (1964), Frith and Doig (1975), Frith and others (1977), Goldich and Hedge (1974), Goldich and others (1970), Heimlich and Banks (1968), Hurst and others (1975), Johnson and Hills (1976), Krogh and Davis (1971), Krogh and others (1974, 1976), Moorbath and others (1972), Nikic and others (1975, 1980), Nunes and Tilton (1971), Peterman and Hildreth (1978), Reed and Zartman (1973), Stueber and Heimlich (1977), Van Schmus and Anderson (1977), Wanless and others (1974).

several localities in the Superior and Slave provinces have emphasized the regional importance and widespread extent of rocks formed late in the Archean—2,600 to 2,760 m.y. ago—both in greenstone and gneiss belts (Steiger and Wasserburg, 1969; Hart and Davis, 1969; Green and Baadsgaard, 1971; Hanson and others, 1971; Peterman and others, 1972; Krogh and Davis, 1971; Krogh and others, 1974, 1976; Jahn and Murthy, 1975). Granitic rocks of late Archean or Algoman age, as well as older granitic and metamorphic rocks, are also abundant in the Wyoming age province (Catanzaro and Kulp, 1964; Heimlich and Banks, 1968; Naylor and others, 1970; Reed and Zartman, 1973; Peterman and Hildreth, 1978; Stueber and Heimlich, 1977).

The age, extent, and composition of the gneissic terranes and their relation to adjoining and intervening greenstone belts are crucial factors in understanding the geologic evolution of the Superior province and adjacent provinces. The nature of the contacts between these two contrasting terranes commonly has been obscured by tectonism, metamorphism, and plutonism. Models that are being considered (Goodwin and West, 1974) include ensialic development of the greenstone sequences on older crust, marginal accretion of these sequences with or without subduction, and ensimatic formation in rift environments.

In the Lake Superior region, the southernmost extension of the Superior province, an east-northeast–trending boundary between gneisses of early Archean age and a greenstone-granite terrane of late Archean age has been delineated on the basis of geochronology, regional geology, and gravity and aeromagnetic data (Morey and Sims, 1976; Sims, 1976). This boundary zone apparently localized the large early Proterozoic (Precambrian X) basin in which the economically important iron formations of the Lake Superior region were deposited (Sims, 1976). During the early Proterozoic Penokean orogeny, Archean basement gneisses were reactivated and partly mobilized to penetrate sedimentary and volcanic sequences that were originally deposited on the gneisses.

## GEOLOGY

The Marenisco-Watersmeet area in the western part of the upper peninsula of Michigan lies astride the boundary between rocks of the greenstone-granite assemblage and the older gneisses (Fig. 2). In the western part of the area, lower Proterozoic metasedimentary rocks lie unconformably on granite and metavolcanic rocks of late Archean age. The granitic batholith composed of the Puritan Quartz Monzonite is a large intrusion extending several kilometres west of the area. It has been dated (Sims and others, 1977) using the whole-rock Rb-Sr method at 2,650 ± 140 m.y.

In the eastern part of the area (Fig. 2), lower Proterozoic metavolcanic and metasedimentary rocks surround and overlie a roughly elliptical body of Archean gneiss informally named the gneiss at Watersmeet (Sims and Peterman, 1976). Several rock types are present but a common unit, especially at localities 1 and 2 (Fig. 2), is a gray tonalitic biotite augen gneiss (M-45L, Table 1). As shown later, this is the only phase in the dome that has been demonstrated conclusively to be of early Archean age. This rock is characterized by composite augen of quartz and plagioclase within a finer grained groundmass of quartz, plagioclase, biotite,

and traces of microcline. Apatite, zircon, sphene, altered allanite, and opaque minerals are common accessories. Migmatitic tonalitic gneisses occur at locality 3, but the common varietal mineral is hornblende rather than biotite. Slightly discordant leucogranite dikes are common but subordinate to the gneisses. Small bodies of discordant pegmatite are present locally. All of the rock types in the dome show evidence of cataclasis and recrystallization under amphibolite-grade conditions as described by Sims and Peterman (1976). Cataclasis was especially intense at locality 4 (Fig.2) where a strongly foliated biotite granodiorite gneiss (M-48, Table 1) occurs with sheared leucogranite and infolded biotite schist (metagraywacke). Similar metagraywacke is exposed near locality 2 where it appears to rest unconformably on the tonalitic augen gneiss and both units are cut by the leucogranite.

The nature of the precursor of the tonalite gneiss is obscured by shearing and recrystallization. However, its composition is that of an igneous rock (Table 1); gross compositional banding that would be expected from sedimentary layering is lacking. We conclude that the original rock was of igneous origin, but whether it was plutonic or volcanic is not known.

Small granitic and gneiss bodies crop out between the gneiss at Watersmeet and the Puritan Quartz Monzonite (Fig. 2). These were named the granite near Thayer by Fritts (1969) and were interpreted as having intruded lower Proterozoic metasedimentary rocks. Radiometric dating and field relations indicate, however, that these bodies are Archean and intrude metavolcanic rocks and metagraywacke. We now interpret the granite and associated cataclastic gneiss as being a strongly deformed equivalent of the Puritan Quartz Monzonite.

In the northern part of the area (Fig. 2), Archean and lower Proterozoic rocks are covered by little-disturbed Keweenawan (middle Proterozoic) volcanic and sedimentary rocks that are approximately 1,100 m.y. old.

## GEOCHRONOLOGY

Procedures for the Rb and Sr analyses are similar to those described by Peterman and others (1967) except that a $^{84}Sr$ spike was used. The precision of analyses (1 $o$) is estimated to be ±1.0% of the $^{87}Rb/^{86}Sr$ ratio and ±0.03% of the $^{87}Sr/^{86}Sr$ ratio. The mean $^{87}Sr/^{86}Sr$ ratio for the Eimer and Amend strontium is 0.7080. The U-Th-Pb method is slightly modified from the procedure of Krogh (1973). Uncertainties (1$o$) are approximately ±0.7% for the parent-daughter ratios and ±0.05% for the $^{207}Pb/^{206}Pb$ ratios. The isotope and decay constants used in all of the age calculations are those recommended by the IUGS Subcommission on Geochronology (Steiger and Jäger, 1977).

### Rb-Sr Results

A preliminary Rb-Sr study of samples from localities 1 and 2 (Fig. 3) by Sims and Peterman (1976) suggested an Archean age for rocks in the Watersmeet dome and provided evidence for a severe event 1,700 to 1,800 m.y. ago. Additional data have substantiated these conclusions (Table 2, Fig. 3), and some further discussion is warranted. Tonalitic augen gneiss, leucogranite, and pegmatite are the main rock types at locality 1. Samples

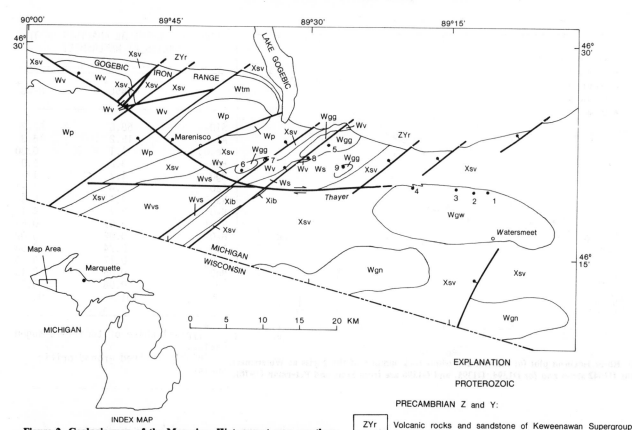

**Figure 2. Geologic map of the Marenisco-Watersmeet area, northern Michigan. Samples taken from each numbered locality are described in Tables 2 and 3.**

EXPLANATION

PROTEROZOIC

PRECAMBRIAN Z and Y:

| ZYr | Volcanic rocks and sandstone of Keweenawan Supergroup |

PRECAMBRIAN X:

| Xsv | Metasedimentary and metavolcanic rocks |

| Xib | Iron-formation, quartzite, and dolomite--age relative to Xv and Xs uncertain |

ARCHEAN

| Wp | Puritan Quartz Monzonite |

| Wgg | Granite near Thayer--cataclastic gneiss and associated granite |

| Wtm | Tonalite and metavolcanic rocks |

| Wvs | Dominantly amphibolite and metagraywacke with local bodies of unit Wgg--Amphibolite is equivalent to unit Wv and metagraywacke is equivalent to unit Ws. |

| Ws | Metagraywacke and tuff |

| Wv | Metavolcanic rocks; amphibolite southeast of Marenisco |

| Wgw | Gneiss at Watersmeet |

| Wgn | Gneiss |

———— Contact

Fault--Bar and ball on downthrown side

Fault, showing relative horizontal displacement

•₃ Sample locality

D1042 and M-45L are of the tonalitic augen gneiss, and D1042j, D1042h, D1042b, and M-45H are of the leucogranite. Hence, the isochron is defined mainly by the leucogranite, and these data, including the microcline and plagioclase analyses, are consistent with a Rb-Sr isochron age of 1,720 m.y. and an initial $^{87}Sr/^{86}Sr$ ratio of 0.773. As will be shown later, the leucogranite is probably late Archean in age, whereas the tonalitic augen gneiss is approximately 1 b.y. older. The Rb-Sr systems of these contrasting rock types were brought into near concordancy by the cataclasis and recrystallization at about 1,720 m.y. ago. However, the apparent rotation of the isochron may not have been as great as first appears because of the characteristically different Rb/Sr ratios of the tonalite and leucogranite. Although the tonalite D1042 falls close to the secondary isochron, M-45L plots significantly below the line suggesting that cataclasis and recrystallization did not result in complete isotopic homogenization of the two rock systems.

Samples from locality 2 (Fig. 3) are apparently less potassic variants of the leucogranite, but the Rb-Sr data fail to define a line. However, sample D1396R has clearly been an open system during the dynamic metamorphism as indicated by its position on Figure 3.

Rocks at locality 4 are more thoroughly sheared and recrystallized than samples from other localities. A strongly foliated biotite gneiss of granodioritic composition (Table 1)

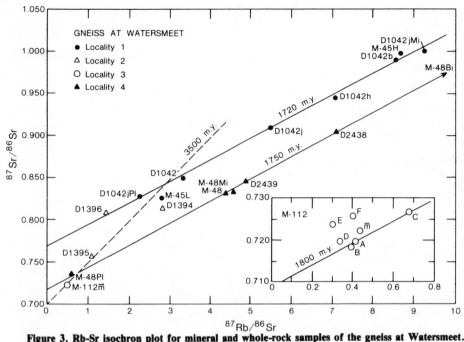

Figure 3. Rb-Sr isochron plot for mineral and whole-rock samples of the gneiss at Watersmeet. Data for the D1042 series and for D1394, D1395, and D1396 are from Sims and Peterman (1976).

TABLE 1.  CHEMICAL ANALYSES OF THE GNEISS AT WATERSMEET

|                    | M-45L[*] | M-48[†] |
|--------------------|----------|---------|
| $SiO_2$            | 68.6     | 62.2    |
| $Al_2O_3$          | 15.2     | 16.3    |
| $Fe_2O_3$          | 1.1      | 0.30    |
| $FeO$              | 3.2      | 6.0     |
| $MgO$              | 1.7      | 1.3     |
| $CaO$              | 2.4      | 1.9     |
| $Na_2O$            | 4.1      | 5.3     |
| $K_2O$             | 2.7      | 4.1     |
| $H_2O+$            | 0.87     | 0.65    |
| $H_2O-$            | 0.02     | 0.02    |
| $TiO_2$            | 0.74     | 0.58    |
| $P_2O_5$           | 0.17     | 0.10    |
| $MnO$              | 0.05     | 0.15    |
| $CO_2$             | 0.08     | 0.52    |
| SUM                | 101      | 99      |

[*]Typical phase of tonalitic augen gneiss.
[†]Highly sheared granodioritic gneiss.

superficially resembles the tonalitic augen gneiss at locality 1. However, we have no independent evidence that would indicate that these are necessarily the same unit. The granodioritic gneiss is cut by highly sheared leucogranite and contains "inclusions" of biotite schist that probably represent infolded metagraywacke. Samples from this locality define a fairly precise Rb-Sr isochron of 1,760 ± 30 m.y. with an intitial Sr-isotope ratio of 0.7215 ± 0.0008. The isochron is defined by biotite, microcline, and plagioclase from M-48 as well as whole-rock samples of M-48, D2439, and D2438 (Table 2). As discussed later, zircon from M-48 gives concordant U-Pb ages that agree with the Rb-Sr isochron age. This concordancy of systems attests to the intensity of shearing and recrystallization at this locality.

As concluded earlier, whole-rock samples of the gneiss at Watersmeet were open systems with significant migration of Rb and Sr as a consequence of intense shearing and recrystallization. The scale of isotopic homogenization was local and, in the less-sheared rocks, may have been on the order of centimetres as suggested by data for sample M-112 (Table 2, Fig. 3) at locality 3. This rock is a migmatitic gneiss with thin and discontinuous plagioclase- and quartz-rich layers in a mafic hornblende-biotite gneiss of tonalitic composition. Among the samples analyzed, this one showed the least shearing, and we thought the Rb-Sr systems on a small scale might have retained some record of the pre-1,800 m.y. history. Six 4-cm-long adjacent samples were taken normal to the banding. Samples A, B, C, E, and F are hornblende-biotite gneiss, and sample D is a 4-cm-wide leucocratic band (Table 2). The point M-112m̄ (Fig. 3) is the calculated whole-rock point. Assuming that the bands are coeval and that no post-1,800 m.y. disturbance of these systems has occurred, the data show that Sr isotope homogenization was not attained across the 25-cm width of the rock. The disposition of points A, B, and C (Fig. 3) is con-

sistent with equilibration at approximately 1,800 m.y. ago, but the Sr in these layers was not in equilibrium with the Sr in layer D, the leucocratic band. Points E and F are inconclusive alone because of the limited range in Rb/Sr ratios.

The behavior of the Rb-Sr systems in the granite near Thayer (Fig. 4) is similar to systematics for the gneiss at Watersmeet. A 2,750-m.y. isochron based on the U-Pb zircon ages is shown for reference. Whole-rock samples from locality 5 are aligned along this isochron. However, whole-rock and mineral data from localities 6 and 7 define a precise secondary isochron corresponding to an age of 1,750 ± 35 m.y. and an initial Sr-isotope ratio of 0.7101 ± 0.0005. The isochron is strongly influenced by the biotite point, but it is changed very little (1,780 ± 50 m.y.) if the biotite point is excluded. Three of the granitic samples from locality 5, although with a limited range in Rb/Sr ratios, suggest a secondary isochron approximately 1,700 to 1,800 m.y. old. Biotite from M-71 gives a slightly lower age of 1,500 m.y. when calculated with the corresponding whole-rock point. The biotite is chloritized, which may explain the low Rb content (Table 2) and lower age.

## U-Th-Pb Results

U-Th-Pb analyses were completed on zircon from both the tonalitic gneiss and the leucogranite in the gneiss at Watersmeet and from three samples of the granite near Thayer (Table 3, Fig. 5). The results establish an early Archean age for the tonalite gneiss and illustrate the complex effects of subsequent geologic events on these systems. The patterns of discordance are not easily explained by two-event episodic models or by lead loss related to continuous diffusion.

Zircons separated from three samples of tonalitic gneiss (M-83,

TABLE 2. Rb-Sr DATA FOR WHOLE ROCKS AND MINERALS

| Sample no. | Rock type | Rb, ppm | Sr, ppm | $^{87}Rb/^{86}Sr$ | $^{87}Sr/^{86}Sr$ |
|---|---|---|---|---|---|
| | Gneiss at Watersmeet | | | | |
| [1] M-45H(R) | Leucogranite | 136 | 46.6 | 8.65 | 0.9986 |
| [1] M-45L(R) | Tonalite gneiss | 184 | 194 | 2.78 | 0.8256 |
| [4] M-48(R) | Granodiorite gneiss | 96.3 | 64.8 | 4.36 | 0.8303 |
| [4] M-48(Pl) | Do. | 18.4 | 91.2 | 0.585 | 0.7365 |
| [4] M-48(Mi) | Do. | 89.1 | 56.9 | 4.59 | 0.8305 |
| [4] M-48(Bi) | Do. | 356 | 6.43 | 270 | 7.723 |
| [4] D2439(R) | Do. | 87.2 | 52.3 | 4.89 | 0.8466 |
| [4] D2438(R) | Bio. gneiss inclusion | 314 | 130 | 7.10 | 0.9044 |
| [3] M-112A(S) | Tonalite gneiss | 30.6 | 213 | 0.416 | 0.7199 |
| [3] M-112B(S) | Do. | 29.4 | 214 | 0.398 | 0.7185 |
| [3] M-112C(S) | Do. | 46.4 | 200 | 0.675 | 0.7268 |
| [3] M-112D(S) | Do. | 23.5 | 201 | 0.339 | 0.7199 |
| [3] M-112E(S) | Do. | 31.7 | 268 | 0.302 | 0.7239 |
| [3] M-112F(S) | Do. | 33.9 | 239 | 0.411 | 0.7257 |
| [3] M-112(R-calc.) | Do. | 33.0 | 218 | 0.439 | 0.7224 |
| | Granite near Thayer | | | | |
| [5] M-71(R) | Granite | 101 | 94.3 | 3.15 | 0.8312 |
| [5] M-71(Bi) | Do. | 144 | 38.7 | 11.05 | 1.0012 |
| [6] M-68(R) | Granite gneiss | 102 | 136 | 2.19 | 0.7659 |
| [6] M-68(Bi) | Do. | 474 | 26.4 | 59.4 | 2.1655 |
| [6] M-150(R) | Leucotonalite | 14.8 | 275 | 0.156 | 0.7128 |
| [7] M-147-1A(R) | Tonalite | 51.4 | 304 | 0.491 | 0.7226 |
| [7] M-147-1B(R) | Do. | 28.6 | 239 | 0.346 | 0.7197 |
| [8] M-85(R) | Garnetiferous granite | 93.3 | 61.7 | 4.45 | 0.8867 |
| [9] M-124(R) | Tonalite | 45.9 | 646 | 0.206 | 0.7092 |

*Note:* Prefix number in bracket is locality number (Fig. 2). Suffix letter on sample number identifies material analyzed: R, whole rock; Pl, plagioclase; Mi, microcline; Bi, biotite; S, slab.

D-1042, and M-112) are nearly colorless to light brown and complexly zoned and fractured (Fig. 6). The crystals are prismatic and terminate in simple pyramids or, less commonly, in both pyramids and basal pinacoids. Both clear and opaque inclusions are common, and some grains have internal iron staining along fractures. Fission-track mapping of various splits of M-83 revealed variations in uranium within single grains and among grains by a factor of two or more, but commonly uranium is more or less uniformly distributed within grains (Fig. 6). Localized concentrations of uranium (hot spots) were noted in only a few grains.

Initially, three size fractions of clear, well-formed crystals were hand picked from D-1042 for isotopic analyses. Although the results left little doubt of the antiquity of this rock, the relatively small despersion of points on the concordia diagram (Fig. 5) prevented any precise interpretation of the lead-loss mechanism responsible for their moderate discordance. A best-fit line to the three points gave an upper intercept age of 3,600 m.y., but the large uncertainty attached to it prompted us to analyze more samples.

Separates were obtained from two additional samples—M-83, an augen gneiss identical to sample D-1042; and M-112, the banded tonalitic gneiss described earlier. Sample M-112 yielded only a small amount of zircon, and a hand-picked separate from the 100-to-150 mesh fraction was analyzed. In contrast, the yield from M-83 was much more abundant, and both size and magnetic splits were made. The clearest and most inclusion-free zircons were picked from a nonmagnetic portion of the 100-to-150 mesh size fraction. Samples of the least magnetic and most magnetic portions of the 200-to-270 mesh fraction of M-83 were also obtained. The least magnetic portion was a pure separate, and no hand picking was done. The most magnetic portion contained abundant sphene, which was removed by a Clerici-solution separation.

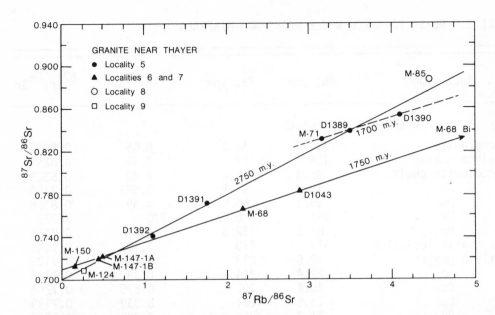

Figure 4. Rb-Sr isochron plot for mineral and whole-rock samples of the granite near Thayer. Data for D1389, D1390, D1391, and D1392 are from Sims and Peterman (1976).

Zircon from a highly cataclasized phase of the gneiss at Watersmeet, sample M-48, was also studied for evidence of multistage development. This zircon is spongy in appearance, is nearly opaque, and forms stubby crystals as long as 1 mm (Fig. 6). In spite of its apparent degraded optical properties, an X-ray diffraction pattern revealed sharp peaks suggesting minimal metamictization (Glenn Izett, 1976, personal commun.). Surprisingly, the U-Pb ages are virtually concordant at 1,755 m.y. (Table 3), in agreement with the Rb-Sr whole-rock and mineral isochron at this locality (Fig. 3). The younger Th-Pb age of 1,550 m.y. is the only indication that the zircon has not been quantitatively retentive of all daughter lead isotopes. As mentioned earlier, this

locality represents the most modified facies of the gneiss that we have seen. The major minerals are extremely fresh—all having formed by recrystallization during the shearing event. The presence of trace amounts of carbonate, fluorite, and local thin galena veinlets suggests that fluids were present when the rock recrystallized. We provisionally interpret the zircon from M-48 as a metamorphic mineral that formed 1,755 m.y. ago.

The slightly discordant leucogranite dikes that intrude the tonalitic phase of the gneiss at Watersmeet also were deformed cataclastically. A sample of leucogranite (M-45H) was collected near the center of a 2-m-wide dike for zircon analysis. Two morphologic varieties of zircon were identified—a rare coarse-

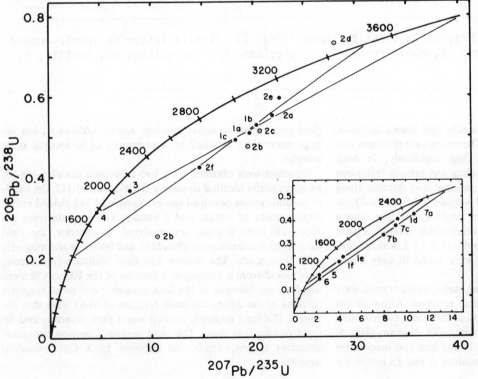

Figure 5. U-Pb concordia plot for zircons. The main diagram is for data on zircons from the tonalitic phase of the gneiss at Watersmeet. The inset shows data for the leucogranite and for the granite near Thayer. The numbers are related to samples in Table 3, and the chords are discussed in the text.

TABLE 3. URANIUM-THORIUM-LEAD ISOTOPIC AGES OF ZIRCON FROM THE MARENISCO-WATERSMEET AREA, NORTHERN MICHIGAN

| Sample[*] | Concentration (ppm) | | | Isotopic composition of lead (atom percent) | | | | Age (millions of years)[†] | | | |
|---|---|---|---|---|---|---|---|---|---|---|---|
| | U | Th | Pb | $^{204}Pb$ | $^{206}Pb$ | $^{207}Pb$ | $^{208}Pb$ | $\dfrac{^{206}Pb}{^{238}U}$ | $\dfrac{^{207}Pb}{^{235}U}$ | $\dfrac{^{207}Pb}{^{206}Pb}$ | $\dfrac{^{208}Pb}{^{232}Th}$ |
| Tonalitic augen gneiss at Watersmeet | | | | | | | | | | | |
| [1a] D1042(60-80) | 668.1 | 236.0 | 422.5 | 0.052 | 69.74 | 20.11 | 10.10 | 2647 | 3073 | 3366 | 3177 |
| [1b] D1042(100-150) | 737.2 | 276.1 | 487.0 | 0.054 | 69.42 | 19.98 | 10.55 | 2732 | 3107 | 3361 | 3252 |
| [1c] D1042(200-270) | 781.1 | 326.9 | 478.2 | 0.054 | 69.73 | 19.43 | 10.78 | 2576 | 3006 | 3309 | 2808 |
| [2a] M-83(100-150)NM | 520.2 | 170.2 | 347.8 | 0.010 | 71.06 | 20.63 | 8.301 | 2830 | 3180 | 3411 | 3402 |
| [2b] Leachable in HCl and HNO$_3$[#] | 167 | 412 | 548 | 0.880 | 23.21 | 15.85 | 60.07 | 2507 1451[§] | 3064 2487[§] | 3455 3492[§] | 7836 6549[§]' |
| [2c] Leachable in HF[#] | 669.0 | 199.9 | 416.5 | 0.013 | 71.10 | 20.90 | 7.990 | 2670 | 3122 | 3429 | 3300 |
| [2d] Residue[#] | 251.7 | 116.6 | 226.8 | 0.024 | 70.23 | 19.78 | 9.966 | 3544 | 3421 | 3351 | 3713 |
| [2e] M-83(200-270)NM | 556.6 | 188.3 | 396.9 | 0.013 | 71.95 | 20.00 | 8.038 | 3009 | 3210 | 3340 | 3359 |
| [2f] M-83(200-270)M | 815.7 | 277.8 | 419.9 | 0.039 | 71.25 | 18.39 | 10.32 | 2276 | 2797 | 3199 | 2899 |
| [3] M-112(100-150) | 820.3 | 349.4 | 394.0 | 0.209 | 68.57 | 13.44 | 17.78 | 2021 | 2243 | 2454 | 2651 |
| Highly sheared granodioritic gneiss | | | | | | | | | | | |
| [4] M-48(50-150) | 724.8 | 2033.7 | 371.5 | 0.043 | 53.24 | 6.320 | 40.40 | 1752 | 1753 | 1757 | 1550 |
| Leucogranite dike | | | | | | | | | | | |
| [1d] M-45H(50-100) | 834.4 | 356.3 | 381.5 | 0.023 | 76.62 | 12.95 | 10.41 | 2194 | 2360 | 2510 | 2218 |
| [1e] M-45H(100-150) | 2107.1 | 1080.1 | 572.1 | 0.040 | 78.13 | 10.84 | 10.99 | 1411 | 1723 | 2128 | 1130 |
| [1f] M-45H(200-270) | 2181.2 | 1839.1 | 541.8 | 0.043 | 77.61 | 10.73 | 11.61 | 1294 | 1637 | 2113 | 669 |
| Granite near Thayer | | | | | | | | | | | |
| [5] M-68(50-100) | 1754.9 | 1023.1 | 365.4 | 0.071 | 73.42 | 8.328 | 18.18 | 1041 | 1251 | 1635 | 1244 |
| [6] M-71(50-100) | 5522.5 | 3049.5 | 999.1 | 0.080 | 73.69 | 9.347 | 16.88 | 916 | 1233 | 1844 | 1033 |
| [7a] M-147(50-150) | 483.2 | 190.4 | 239.3 | 0.063 | 74.60 | 14.45 | 10.89 | 2280 | 2499 | 2685 | 2368 |
| [7b] M-147(200-270) | 792.8 | 277.0 | 317.2 | 0.107 | 73.40 | 14.03 | 12.46 | 1860 | 2229 | 2590 | 2194 |
| [7c] M-147(-270) | 905.5 | 234.2 | 394.2 | 0.085 | 75.75 | 14.32 | 9.845 | 2064 | 2349 | 2609 | 2517 |

[*]Number and letter in brackets are locality number (Fig. 2) and letters to identify samples on concordia plot (Fig. 5). Mesh size is indicated in parentheses and magnetic and nonmagnetic fractions are indicated by N and NM respectively.
[†]Isotopic composition of initial lead assumed to be $^{204}Pb:^{206}Pb:^{207}Pb:^{208}Pb$ = 1:12.1:13.8:32.1 for D1042, M-112, and M-83 (except for those values marked by [§] which are calculated for modern lead, i.e. 1:18.8:15.7:39.0), and = 1:13.5:14.6:33.3 for M-68, M-71, and M-147; = 1:13.9:14.7:33.7 for M-45; = 1:16.02:15.76:35.99 for M-48 (these values were obtained on thin galena veinlets in sheared gneiss at the M-48 locality--M. Delevaux, analyst).
[#]In the leaching experiment on M-83(100-150)NM, the fraction of total sample represented by the three analyses are: [2b] 0.6 percent, [2c] 64.0 percent, and [2d] 35.4 percent.

grained, subhedral, light-brown variety similar to those in the tonalitic gneiss; and more abundant, generally smaller, elongate, colorless to white crystals. The distinction between the populations is further enhanced by their different uranium contents and apparent ages (Table 3). The 100-to-150 and 200-to-270 mesh fractions of M-45H have high uranium contents, are quite discordant, and considered by themselves only have $^{207}Pb$-$^{206}Pb$ ages suggesting that initial crystallization took place at least 2,100 m.y. ago. An interesting possibility is that, despite their morphological differences, all three fractions are part of the same chronometric system and, therefore, can be used to construct a line intersecting concordia at about 2,600 m.y. and 800 m.y. The 2,600-m.y. age could well suggest a relationship to the major igneous and metamorphic event related to the greenstone and granite terrane to the north.

Zircons were obtained from three samples of the granite near Thayer (M-68, M-71, and M-147). The initial analyses were made of the two granitic samples (M-68 and M-71), which yielded zircons that are highly radioactive and discordant (Table 3, Fig. 5).

These results were equivocal and provided only minimum ages for the unit. Subsequently, a tonalitic phase was separated from sample M-147 which produced zircon that is much lower in uranium and much less discordant. Three size fractions (50-to-150, 200-to-270, and <270 mesh) were prepared from an abundant yield. The zircons from the 50-to-150 mesh fraction were hand picked, and the finer fractions were purified by Clerici-solution and magnetic separations. Data for the three zircon splits define a chord intersecting concordia at 2,750 m.y. Obviously, both M-71 and M-68 will not fit on an extension of this chord, but M-71 falls closer to the line (Fig. 5). When M-71 is included in the regression with points for M-147, the upper intercept with concordia is 2,745 ± 65 m.y. (95% confidence limits) and the lower intercept is 640 ± 140 m.y. Exclusion of M-71 does not affect the upper intercept appreciably. Sample M-68 is a strongly foliated flaser gneiss and is much more highly sheared than M-71. The deviation of M-68 from the chord may reflect a lead-loss pattern that is strongly influenced by the 1,750-m.y.-old cataclastic event. In other words, the zircon may have developed a discordance in response to shear-

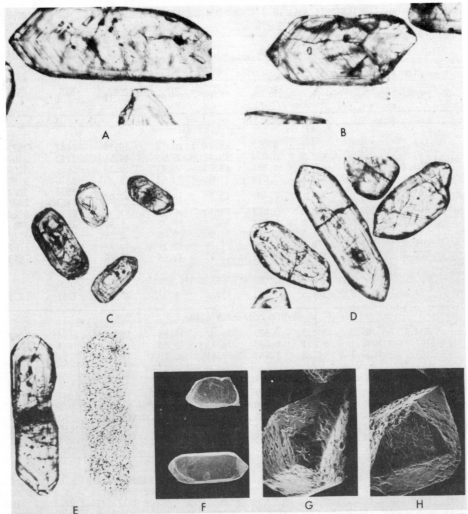

**Figure 6. Photographs of zircons. A and B are M-83(100-150)NM. C and D are M-83(200-270)NM. E is photograph and fission-track map (muscovite detector) of a grain from M-83(100-150)NM. F, G, and H are scanning electron microscope photographs. F is D1042 (60-80), and G and H are M-48 (50-150)—note spongy appearance.**

ing and a subsequent lead loss in the late Proterozoic. The age of 2,745 ± 65 m.y. for the granite near Thayer is not resolvable from the Rb-Sr isochron age of 2,650 ± 140 m.y. for the Puritan Quartz Monzonite.

We can now return to data for zircon from the tonalitic phase of the gneiss at Watersmeet to examine the discordancy patterns. Collectively, the data do not define a linear array on the concordia diagram (Fig. 5), and this fact alone implies a complex history involving two or more geologic events subsequent to the formation of the rock. Regression of all of the data points defines concordia intercepts at 3,780 m.y. and 1,740 m.y., but two points (2e and 2f in Fig. 5) deviate from the line by more than experimental error. Various other lines can be defined by selecting data points. Elimination of 2e, 3 and 4 results in a much better statistical fit, with intercepts at 3,560 m.y. and 1,200 m.y. We must note that the position of 3 (M-112) plots near the intersection of chords for tonalitic gneiss and for the leucogranite and, therefore, is equivocal. Selection of data to obtain chords are useful exercises,

but a meaningful interpretation must explain the total array of data, and certain limiting conditions are evident. The gneiss has a firm minimum age of 3,410 m.y. provided by the $^{207}Pb$-$^{206}Pb$ age of sample M-83(100–150)NM (point 2a in Fig. 5). An additional constraint is the cataclasis defined by the concordant 1,755-m.y. age for sample M-48. Rb-Sr data for whole rocks and minerals (Fig. 3) show that the effects of this event were widespread, although new zircon formed in only the most highly sheared phases of the gneiss. The positions of points for the magnetic and nonmagnetic portions of M-83(200–270) on the concordia diagram (2e and 2f in Fig. 5) indicate additional events both earlier and later than the dynamic metamorphism. The earlier event may be represented by late Archean igneous activity—the leucogranite dikes and granite near Thayer. Discordance patterns of zircon from these units (Fig. 5) suggest an episode of lead loss in the late Proterozoic.

As an additional test for a component of radiogenic lead supportive of an older age than that given by the $^{207}Pb/^{206}Pb$ ratio for

M-83(100–150)NM, we conducted a simple leaching experiment on a split of this sample. The zircon was first rinsed in hot 6 $N$ HCl followed by 6 $N$ HNO$_3$, and then partially dissolved in hot HF at atmospheric pressure to yield a soluble and residue fraction. It was anticipated that if different sites of isotopically distinct lead existed in the grains, this experiment would reveal such heterogeneity. Particular significance would be attached to a variability in the radiogenic $^{207}$Pb/$^{206}$Pb ratio; if variability was found, it would give evidence of events both older and younger than the composite $^{207}$Pb-$^{206}$Pb age. In contrast to expectations, the radiogenic $^{207}$Pb/$^{206}$Pb ratio changed very little (Table 3, Fig. 5). The experiment served only to identify a surficial "common lead" component during the rinsing step and a high uranium and low uranium—with corresponding greater discordance and less discordance—dichotomy between the HF-soluble and residue fractions. We doubt if any significance should be attached to the slight reverse discordance, and the 3,700-m.y. $^{208}$Pb-$^{232}$Th age of the residue fraction in that elemental fractionation could well have been induced during the leaching. The large mass loss in zircon effected by the HF leach seems to have occurred mainly along internal fractures withing the zircon grains. These fractures were probably caused by radiation-damage stresses and were subsequently healed by secondary silica(?) cement. The cementing material survived the HCl-HNO$_3$ rinses but rapidly dissolved in the HF, whereupon the crystals disintegrated into a myriad of fragments.

The analytical data from the leaching experiment on M-83(100–150)NM provide no evidence of a composite zircon population comprising subsystems with appreciably different ages. Rather, the linear array of points projects toward the origin of the concordia diagram; such a distribution indicates isotopic homogeneity of the radiogenic $^{207}$Pb/$^{206}$Pb ratio. At the same time, the nonlinearity of the three fractions of M-83 indicates some isotopic heterogeneity as a function of grain size and magnetic properties of the zircon; hence, the results of the leaching experiment do not rule out a multistage development of the zircon.

Even though we have generated a rather large amount of data in this study, we must conclude that a unique primary age of the tonalite gneiss at Watersmeet is not firmly established by the zircon dating. The maximum $^{207}$Pb-$^{206}$Pb age of 3,410 m.y. and the $^{208}$Pb-$^{232}$Th age of 3,400 m.y. on the same sample provide a firm estimate of the minimum age of the unit. The question that remains is how much older than 3,410 m.y. is the gneiss. The Sm-Nd determination on a sample from locality 1 (Fig. 2) by McCulloch and Wasserburg (this volume) gives a model age of 3,600 m.y., and these authors conclude that the tonalite gneiss was added to the crust at this time. The result is particularly encouraging in that the Sm-Nd system does not appear to have been disturbed by the subsequent events that affected the gneiss, and the Sm-Nd age is within the intercept limits (3,500 and 3,800 m.y.) given by the two regressions shown on Figure 5. The U-Pb systems in the zircons clearly responded to the major early Proterozoic dynamic metamorphism, and this event may have obscured the effects of an earlier discordance imposed during the late Archean. Much younger lead loss is indicated by zircon from the leucogranite and from the granite near Thayer. Possibly, the lower-uranium zircons from the tonalite gneiss also lost lead at this time, further complicating the discordance patterns.

## CONCLUSIONS

1. The precursor of the tonalitic gneiss in the Watersmeet area formed in the early Archean at least 3,400 m.y. ago, as indicated by the U-Th-Pb data on zircons. Sm-Nd dating of the gneiss gives an age of 3,600 m.y. (McCulloch and Wasserburg, this volume).

2. Leucogranite dikes (2,600 m.y. old) intruded the tonalite gneiss and the granite near Thayer (2,750 m.y. old) and represent igneous activity concurrent with the development of the major greenstone-granite terrane immediately to the north of the gneiss terrane. The leucogranite and the granite near Thayer may be highly deformed equivalents of the Puritan Quartz Monzonite dated at 2,650 ± 140 m.y. by Rb-Sr.

3. Reactivation of the gneisses of early Archean age and the granitic rocks of late Archean age during the early Proterozoic Penokean orogeny resulted in cataclasis and recrystallization, the effects of which are recorded regionally in the Rb-Sr systems and, locally, in metamorphic zircon with a set of ages concordant at 1,755 m.y. The terrane stabilized rapidly at this time, as indicated by agreement of the radiometric systems: Rb-Sr whole-rock and mineral ages, including biotite ages, and the concordant zircon age. No severe episodes of metamorphism have affected the terrane since late in the early Proterozoic. Discordance patterns in zircon ages from the leucogranite and from the granite near Thayer suggest lead loss in the late Proterozoic. This event was probably regional uplift and faulting.

4. Demonstration of an early Archean age for the tonalitic gneiss at Watersmeet provides additional evidence for an ancient sialic terrane in this region. We anticipate that more occurrences of rocks of early Archean age will be documented as further studies using methods capable of penetrating the effects of the Algoman and Penokean orogenies are completed.

## ACKNOWLEDGMENTS

We gratefully acknowledge the technical assistance of K. Futa and R. A. Hildreth in the Rb-Sr work, Margarita Gallego and Loretta Kwak in the U-Th-Pb work, and G. T. Cebula and J. W. Groen for sample preparations. C. W. Naeser prepared the zircon grain mounts and fission-track maps. We are also indebted to W. R. Van Schmus and Paul Nunes for their thoughtful comments on the manuscript, and to S. S. Goldich for many helpful discussions of the dating problem and of the geology and geochronology of the Lake Superior region in general.

## REFERENCES CITED

Baadsgaard, H., 1973, U-Th-Pb dates on zircons from the early Precambrian Amitsoq Gneisses, Godthaab district, west Greenland: Earth and Planetary Science Letters, v. 19, p. 22–28.

——1976, Further U-Pb dates on zircons from the early Precambrian rocks of the Godthaabsfjord area, west Greenland: Earth and Planetary Science Letters, v. 33, p. 261–267.

Baragar, W.R.A., and McGlynn, J. C., 1976, Early Archean basement in the Canadian Shield: A review of evidence: Canada Geological Survey Paper 76-14, 20 p.

Barton, J.M., Jr., 1975, Rb-Sr isotopic characteristics and chemistry of the 3.6 b.y. Hebron Gneiss, Labrador: Earth and Planetary Science Letters, v. 27, p. 427-435.

Catanzaro, E. J., and Kulp, J. L., 1964, Discordant zircons from the Little Belt (Montana), Beartooth (Montana), and Santa Catalina (Arizona) Mountains: Geochimica et Cosmochimica Acta, v. 28, p. 87-124.

Frith, R. A., and Doig, R., 1975, Pre-Kenoran tonalitic gneiss in the Grenville province: Canadian Journal of Earth Sciences, v. 12, p. 844-849.

Frith, Rosaline, Frith, R. A., and Doig, R., 1977, The geochronology of the granitic rocks along the Bear-Slave structural province boundary, northwest Canadian Shield: Canadian Journal of Earth Sciences, v. 14, p. 1356-1373.

Fritts, C. E., 1969, Bedrock geologic map of the Marenisco-Watersmeet area, Gogebic and Ontonagon Counties, Michigan: U.S. Geological Survey Miscellaneous Investigations Series Map I-576, scale 1:48,000.

Goldich, S. S., and Hedge, C. E., 1974, 3,800 Myr granitic gneiss in southwestern Minnesota: Nature, v. 252, p. 467-468.

Goldich, S. S., Hedge, C. E., and Stern, T. W., 1970, Age of the Morton and Montevideo Gneisses and related rocks, southwestern Minnesota: Geological Society of America Bulletin, v. 81, p. 3671-3696.

Goodwin, A. M., and West, G. F., 1974, The Superior Geotraverse Project: Geoscience Canada, v. 1, p. 21-29.

Green, D. C., and Baadsgaard, H., 1971, Temporal evolution and petrogenesis of an Archean crustal segment at Yellowknife, N.W.T., Canada: Journal of Petrology, v. 12, p. 177-217.

Hanson, G. N., and others, 1971, Age of the early Precambrian rocks of the Saganaga Lake—Northern Light Lake area, Ontario-Minnesota: Canadian Journal of Earth Sciences, v. 8, p. 1110-1124.

Hart, S. R., and Davis, G. L., 1969, Zircon U-Pb and whole-rock Rb-Sr ages and early crustal development near Rainy Lake, Ontario: Geological Society of America Bulletin, v. 80, p. 595-616.

Heimlich, R. A., and Banks, P. O., 1968, Radiometric age determinations, Bighorn Mountains, Wyoming: American Journal of Science, v. 266, p. 180-192.

Hurst, R. W., and others, 1975, 3600-m.y. Rb-Sr ages from very early Archean gneisses from Saglek Bay, Labrador: Earth and Planetary Science Letters, v. 27, p. 393-403.

Jahn, Bor-ming, and Murthy, V. Rama, 1975, Rb-Sr ages of the Archean rocks from the Vermilion district, northeastern Minnesota: Geochimica et Cosmochimica Acta, v. 39, p. 1679-1689.

Johnson, R. C., and Hills, F. A., 1976, Precambrian geochronology and geology of the Boxelder Canyon area, northern Laramie Range, Wyoming: Geological Society of America Bulletin, v. 87, p. 809-817.

Krogh, T. E., 1973, A low-contamination method for hydrothermal decomposition of zircon and extraction of U and Pb for isotopic age determinations: Geochimica et Cosmochimica Acta, v. 37, p. 485-494.

Krogh, T. E., and Davis, G. L., 1971, Zircon U-Pb ages of Archean metavolcanic rocks in the Canadian Shield: Carnegie Institution of Washington Year Book 70, p. 241-242.

Krogh, T. E., Ermanovics, I. F., and Davis, G. L., 1974, Two episodes of metamorphism and deformation in the Archean rocks of the Canadian Shield: Carnegie Institution of Washington Year Book 73, p. 573-575.

Krogh, T. E., Harris, N.B.W., and Davis, G. L., 1976, Archean rocks from the eastern Lac Seul region of the English River gneiss belt, northwestern Ontario, Part 2, Geochronology: Canadian Journal of Earth Sciences, v. 13, p. 1212-1215.

McCulloch, M. T., and Wasserburg, G. J., 1980, Sm-Nd model ages from an early Archean tonalitic gneiss, northern Michigan: Geological Society of America Special Paper 182 (this volume).

Moorbath, S., and others, 1972, Further rubidium-strontium age determinations on the very early Precambrian rocks of the Godthaab district, west Greenland: Nature, Physical Science, v. 240, p. 78-82.

Morey, G. B., and Sims, P. K., 1976, Boundary between two lower Precambrian terranes in Minnesota and its geological significance: Geological Society of America Bulletin, v. 87, p. 141-152.

Naylor, R. S., Steiger, R. H., and Wasserburg, G. J., 1970, U-Th-Pb and Rb-Sr systematics in 2700 × 10⁶ year old plutons from the southern Wind River Range, Wyoming: Geochimica et Cosmochimica Acta, v. 34, p. 1133-1150.

Nikic, Z., and others, 1975, Diatreme containing boulders of 3030 m.y. old tonalite gneiss, Con Mine, Yellowknife, Slave craton: Geological Society of America Abstracts with Programs, v. 7, p. 1213-1214.

——1980, Boulder from the basement, the trace of ancient crust?: Geological Society of America Special Paper 182 (this volume).

Nunes, P. D., and Tilton, G. R., 1971, Uranium-lead ages of minerals from the Stillwater igneous complex and associated rocks, Montana: Geological Society of America Bulletin, v. 82, p. 2231-2250.

Peterman, Z. E., and Hildreth, R. A., 1978, Reconnaissance geology and geochronology of the Precambrian of the Granite Mountains, Wyoming: U.S. Geological Survey Professional Paper 1055, 22 p.

Peterman, Z. E., Doe, B. R., and Bartel, Ardith, 1967, Data on the rock GSP-1 (granodiorite) and the isotope-dilution method of analysis for Rb and Sr, in Geological Survey Research 1967: U.S. Geological Survey Professional Paper 575-B, p. B181-B186.

Peterman, Z. E., and others, 1972, Geochronology of the Rainy Lake region, Minnesota-Ontario: Geological Society of America Memoir 135, p. 193-215.

Reed, J. C., Jr., and Zartman, R. E., 1973, Geochronology of Precambrian rocks of the Teton Range, Wyoming: Geological Society of America Bulletin, v. 84, p. 561-582.

Sims, P. K., 1976, Precambrian tectonics and mineral deposits, Lake Superior region: Economic Geology, v. 71, p. 1092-1127.

Sims, P. K., and Peterman, Z. E., 1976, Geology and Rb-Sr ages of reactivated gneisses and granite in the Marenisco-Watersmeet area, northern Michigan: U.S. Geological Survey Journal of Research, v. 4, p. 405-414.

Sims, P. K., Peterman, Z. E., and Prinz, W. C., 1977, Geology and Rb-Sr age of Precambrian W Puritan Quartz Monzonite, northern Michigan: U.S. Geological Survey Journal of Research, v. 5, p. 185-192.

Steiger, R.H., and Jager, E., 1977, Subcommission on geochronology: Convention on the use of decay constants in geo- and cosmochronology: Earth and Planetary Science Letters, v. 36, p. 359-362.

Steiger, R. H., and Wasserburg, G. J., 1969, Comparative U-Th-Pb systematics in 2.7 × 10⁹ yr plutons of different geologic histories: Geochimica et Cosmochimica Acta, v. 33, p. 1213-1232.

Stueber, A. M., and Heimlich, R. A., 1977 Rb-Sr isochron age of the Precambrian basement complex, Bighorn Mountains, Wyoming: Geological Society of America Bulletin, v. 88, p. 441-444.

Van Schmus, W. R., and Anderson, J. L., 1977, Gneisses and migmatite of Archean age in the Precambrian basement of central Wisconsin: Geology, v. 5, p. 45-48.

Wanless, R. K., and others, 1974, Age determinations and geological studies, K-Ar ages, Report 12: Canada Geological Survey Paper 74-2, 72 p.

MANUSCRIPT RECEIVED BY THE SOCIETY APRIL 25, 1979
REVISED MANUSCRIPT RECEIVED NOVEMBER 26, 1979
MANUSCRIPT ACCEPTED JANUARY 11, 1980

Geological Society of America
Special Paper 182
1980

# Early Archean Sm-Nd model ages from a tonalitic gneiss, northern Michigan

M. T. McCULLOCH
G. J. WASSERBURG
*Lunatic Asylum, Division of Geological and Planetary Sciences, California Institute of Technology, Pasadena, California 91125*

## ABSTRACT

Sm-Nd model ages on a whole-rock sample of the gneiss at Watersmeet, Michigan, show that these rocks were added to the continental crust about 3,600 m.y. ago. These model ages are calculated by assuming that the rocks were derived from a mantle reservoir with the Sm/Nd ratio of CHUR, which corresponds approximately to the chondritic ratio. These data confirm the U-Th-Pb zircon results that indicated a minimum age of 3,410 m.y. and possibly an age as old as 3,800 m.y. This study further illustrates the utility of Sm-Nd model ages in deciphering the history of complex Precambrian terranes.

## INTRODUCTION

Peterman and others (this volume) have reported an age, based on U-Th-Pb zircon systematics, of greater than 3,400 m.y. for a tonalitic gneiss in the Marenisco-Watersmeet area of northern Michigan. In an attempt to confirm this important observation, we have determined Sm-Nd model ages on a whole-rock sample of the gneiss at Watersmeet. Although dependent on the validity of the model assumptions, the Sm-Nd model ages are consistent with the early Archean ages of Peterman and others. This study further illustrates the ability of the Sm-Nd system to obtain primary crustal formation ages, despite these rocks' complex

history. The utility of this approach should enable formation ages to be determined from other rocks produced during early crustal formation events.

## ANALYTICAL PROCEDURE AND RESULTS

A 4-kg sample of the tonalitic augen gneiss from Watersmeet was provided to us by Z. Peterman, R. Zartman, and P. Sims. The plagioclase-quartz-biotite gneiss has a groundmass grain size of ∼2 mm with larger 1- to 2-cm plagioclase porphyroblasts. The sample was pulverized and ground to pass 100 mesh, and splits were taken. Splits of 0.5 and 2.5 g were dissolved and totally spiked with $^{147}$Sm and $^{150}$Nd tracers. Sm and Nd concentrations and $^{143}$Nd/$^{144}$Nd isotopic ratios were determined from these solutions through the use of procedures described by Papanastassiou and others (1977). A small difference in the Sm and Nd concentrations is present between the two splits. Although this small difference does not seriously affect the Sm-Nd model age, it does show the difficulty in obtaining representative whole-rock samples from polymetamorphic rocks where isotopic redistribution has probably occurred between mineral phases.

Sm-Nd data for a sample from locality D1042 (see Peterman and others, this volume) are given in Table 1. The model ages from two splits of the same sample are 3,620 ± 30 m.y. and 3,570 ± 50 m.y., which are the same within analytical error.

TABLE 1.   Sm-Nd DATA FOR THE GNEISS AT WATERSMEET

| Sample | Mass (g) | Sm (ppm) | Nd (ppm) | $\frac{^{147}Sm}{^{144}Nd}$ | $\frac{^{143}Nd}{^{144}Nd}$ | $T^{Nd}_{CHUR}$ (m.y.) |
|---|---|---|---|---|---|---|
| D1042 Split a | 2.56 | 5.05 | 41.38 | 0.07383 | 0.509011±37 | 3,570±50 |
| D1042 Split b | 0.54 | 5.11 | 40.50 | 0.07632 | 0.509027±22 | 3,620±30 |

## DISCUSSION

The Sm-Nd model age ($T^{Nd}_{CHUR}$) is the time in the past when the major REE fractionation in a rock occurs during its derivation from a mantle reservoir. As schematically shown in Figure 1, the $^{143}Nd/^{144}Nd$ evolves as a function of time in the mantle reservoir [$I_{CHUR}(T)$] from 4,560 m.y. ago until time $T^{Nd}_{CHUR}$, when (in this case, 3,600 m.y. ago) the magma enriched in LREE (light REE) is derived from the mantle reservoir. This LREE enrichment produces a lower Sm/Nd ratio, and the subsequent evolution of $^{143}Nd/^{144}Nd$ in the rock is retarded relative to its evolution in the mantle reservoir. The $T^{Nd}_{CHUR}$ age is then the intersection of the rock-evolution curve and mantle-evolution curve and is given by

$$T^{Nd}_{CHUR} = \frac{1}{\lambda}\ell n\left[1 + \frac{\epsilon_{Nd}(0) \times I_{CHUR}(0) \times 10^{-4}}{f_{Sm/Nd} \times (^{147}Sm/^{144}Nd)_{CHUR}}\right],$$

where

$$f_{Sm/Nd} = \left[\frac{(^{147}Sm/^{144}Nd)_{meas}}{(^{147}Sm/^{144}Nd)_{CHUR}} - 1\right]$$

and

$$\epsilon_{Nd}(0) = \left[\frac{(^{143}Nd/^{144}Nd)_{meas}}{I_{CHUR}(0)} - 1\right] \times 10^4.$$

The ratios ($^{143}Nd/^{144}Nd_{meas}$ and ($^{147}Sm/^{144}Nd)_{meas}$ are measured for the rock. The mantle reservoir (CHUR) has a Sm/Nd ratio of ($^{147}Sm/^{144}Nd)_{CHUR}$ = 0.1936 and $^{143}Nd/^{144}Nd$ ratio today (at $T$ =0) of $I^{Nd}_{CHUR}(0)$ = 0.511836.
The evolution of $^{143}Nd/^{144}Nd$ and of Sm/Nd in the mantle reservoir was first described by DePaolo and Wasserburg (1976). The model-age arguments presented here are directly dependent on the validity of the mantle-reservoir characteristics that they outlined and that have been further substantiated in Archean rocks by Hamilton and others (1977, 1978). A detailed discussion of Sm-

Nd model ages from a wide variety of crustal rock types is given by McCulloch and Wasserburg (1978) and Jacobsen and Wasserburg (1978).

In Table 2, the Sm-Nd parameters from the gneiss at Watersmeet are compared with the Amitsoq Gneiss (DePaolo and Wasserburg, 1976) and a composite sample (New Quebec) from the Superior province of the Canadian Shield (McCulloch and Wasserburg, 1978). The Canadian Shield composite, which represents a large segment of the Superior province, has a LREE enrichment ($f_{Sm/Nd} < 0$) and a negative $\epsilon_{Nd}(0)$ value. These characteristics are even more evident in the older (3,600-m.y.-old) Amitsoq Gneiss and the gneiss at Watersmeet. The $\epsilon_{Nd}(0) = -55$ from the gneiss at Watersmeet is the most primitive $^{143}Nd/^{144}Nd$ ratio so far reported. For comparison, a sample that has been Sm free during the Earth's history would today have an $\epsilon_{Nd}(0) = -114$. Thus, even without correction for $^{147}Sm$ decay, the primitive $^{143}Nd/^{144}Nd$ ratio in the gneiss shows unequivocally that this is a very ancient rock.

In a manner analogous to that discussed for Sm and Nd, Rb-Sr model ages ($T^{Sr}_{UR}$) can also be calculated (see McCulloch and Wasserburg, 1978). From the data from Peterman and others (this volume), the $T^{Sr}_{UR}$ model ages from the gneiss at Watersmeet range from 2,090 m.y. to an impossible 6,130 m.y. For the sample

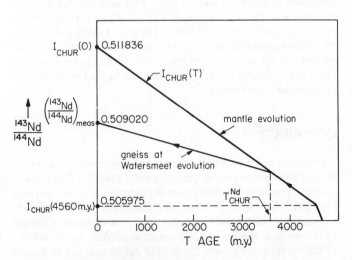

Figure 1. Evolution of $^{143}Nd/^{144}Nd$ in the gneiss at Watersmeet (to scale). The $T^{Nd}_{CHUR}$ age is given by the intersection of the sample evolution line with the mantle evolution line [$I_{CHUR}(T)$].

TABLE 2. Sm-Nd EVOLUTIONARY PARAMETERS

| Sample | $\varepsilon_{Nd}(0)$ | $f_{Sm/Nd}$ | $T^{Nd}_{CHUR}$ (m.y.) |
|---|---|---|---|
| Watersmeet Gneiss[*] | -55.0±0.5 | -0.612 | 3,600±40 |
| Amitsoq Gneiss[†] | -35.4±0.6 | -0.390 | 3,620±90 |
| New Quebec Comp.[§] | -29.4±0.3 | -0.444 | 2,660±60 |

[*]  *Average from Table 1.*
[†]  *DePaolo and Wasserburg (1976).*
[§]  *McCulloch and Wasserburg (1978).*

studied, $T^{Sr}_{UR} =\ 3,160$ m.y., which is distinctly younger than the $T^{Nd}_{CHUR}$ age. This difference is probably due to redistribution of Rb and Sr at younger times, as indicated by the wide range in $T^{Sr}_{UR}$ ages.

The gneiss at Watersmeet has had a complex history and shows evidence of cataclasis and recrystallization (Sims and Peterman, 1976). As shown by Peterman and others (this volume), this later metamorphism has resulted in highly disturbed Rb-Sr and complexly discordant U-Th-Pb zircon systems. Despite the intensive study by these workers, a definitive primary age of the gneiss at Watersmeet could not be established, although a minimum age of 3,410 m.y. was clearly indicated from the zircon study. In contrast, the Sm-Nd model age appears to indicate that the whole-rock Sm-Nd system has remained closed since 3,600 m.y. B.P. This is probably due to the geochemical coherence and relative immobility of the REE. As already pointed out, this age is dependent on model assumptions. However, a relatively large deviation in the initial $^{143}Nd/^{144}Nd$ of 1 part in $10^4$ would produce an error of only 70 m.y.

The 3,600-m.y.-old crust identified in western Greenland, Labrador, Minnesota (Moorbath and others, 1972; Baadsgaard, 1973; Barton, 1975; Hurst and others, 1975; Catanzaro, 1963; Goldich and others, 1970; Goldich and Hedge, 1974), and now in northern Michigan may represent a period of major and widespread crustal formation. Although this event is relatively poorly documented compared to the younger event at 2,600 to 2,800 m.y. B.P., the application of Sm-Nd model ages may enable identification of other rocks of this age, even in complex polymetamorphic terranes.

## ACKNOWLEDGMENTS

We thank R. Zartman, Z. Peterman, and P. Sims for the sample of the gneiss at Watersmeet from northern Michigan. Their endeavors provided the impetus for this study. This study was supported by National Science Foundation Grant PHY 76-02724 and NGL 05-002-188. Division Contribution No. 3175 (283).

## REFERENCES CITED

Baadsgaard, H., 1973, U-Th-Pb dates on zircons from the early Precambrian Amitsoq Gneisses, Godthaab district, West Greenland: Earth and Planetary Science Letters, v. 19, p. 22–28.

Barton, J. M., Jr., 1975, Rb-Sr isotopic characteristics and chemistry of the 3.6 b.y. Hebron Gneiss, Labrador: Earth and Planetary Science Letters, v. 27, p. 427–435.

Catanzaro, E. J., 1963, Zircon ages in southwestern Minnesota: Journal of Geophysical Research, v. 68, p. 2045–2049.

DePaolo, D. J., and Wasserburg, G. J., 1976, Nd isotopic variations and petrogenetic models: Geophysical Research Letters, v. 3, p. 249–254.

Goldich, S. S., and Hedge, C. E., 1974, 3,800 Myr granitic gneiss in southwestern Minnesota: Nature, v. 252, p. 467–468.

Goldich, S. S., Hedge, C. E., and Stern, T. W., 1970, Age of Morton and Montevideo gneisses and related rocks, southwestern Minnesota: Geological Society of America Bulletin, v. 81, p. 3671–3696.

Hamilton, P. J., O'Nions, R. K., and Evensen, N. M., 1977, Sm-Nd dating of Archean basic and ultrabasic volcanics: Earth and Planetary Science Letters, v. 36, p. 263–268.

Hamilton, P. J., O'Nions, R. K., and Evensen, N. M., 1978, Sm-Nd isotopic investigations of Isua supracrustals and implications for mantle evolution: Nature, v. 272, p. 41–43.

Hurst, R. W., Bridgwater, D., Collerson, K. D., and Wetherill, G. W., 1975, 3,600 m.y. Rb-Sr ages from very early Archean gneisses from Saglek Bay, Labrador: Earth and Planetary Science Letters, v. 27, p. 393–403.

Jacobsen, S. B., and Wasserburg, G. J., 1978, Interpretation of Nd, Sr and Pb isotope data from Archean migmatites in Lofoten-Vesterålen, Norway: Earth and Planetary Science Letters, v. 41, p. 245–253.

McCulloch, M. T., and Wasserburg, G. J., 1978, Sm-Nd and Rb-Sr chronology of continental crust formation: Science, v. 200, p. 1003–1011.

Moorbath, S., O'Nions, R. K., Pankhurst, R. J., and Gale, N. H., 1972, Further rubidium-strontium age determinations on the very early Precambrian rocks of the Godthaab district, west Greenland: Nature, Physical Science, v. 240, p. 78–82.

Papanastassiou, D. A., DePaolo, D. J., and Wasserburg, G. J., 1977, Rb-Sr and Sm-Nd chronology and genealogy of mare basalts from the sea of tranquility: Proceedings, Eighth Lunar and Planetary Science Conference, v. 1, p. 1639–1672.

Peterman, Z. E., Zartman, R. E., and Sims, P. K., 1978, Tonalitic gneiss of early Archean age from northern Michigan: Geological Society of America Special Paper 182 (this volume).

Sims, P. K., and Peterman, Z. E., 1976, Geology and Rb-Sr ages of reactivated gneisses and granite in the Marenisco-Watersmeet area, northern Michigan: U.S. Geological Survey Journal of Research, v. 4, p. 405–414.

MANUSCRIPT RECEIVED BY THE SOCIETY APRIL 25, 1979
REVISED MANUSCRIPT RECEIVED NOVEMBER 26, 1979
MANUSCRIPT ACCEPTED JANUARY 11, 1980

Geological Society of America
Special Paper 182
1980

# Geology and Rb-Sr Age of Lower Proterozoic granitic rocks, northern Wisconsin

P. K. SIMS
Z. E. PETERMAN
*U.S. Geological Survey, Federal Center, Denver, Colorado 80225*

## ABSTRACT

Granitic rocks ranging in composition from granite to tonalite and associated metavolcanic-metasedimentary rocks compose an east-trending belt as much as 180 km wide and 300 km long in northern Wisconsin. The granitic rocks have an initial $^{87}Sr/^{86}Sr$ of $0.7025 \pm 0.0005$ and a Rb-Sr whole-rock isochron age of $1,885 \pm 65$ m.y., which is interpreted as the time of crystallization of the granitic rocks. Rb-Sr whole-rock mineral secondary isochrons for two samples give ages of $1,655 \pm 55$ m.y. and $1,545 \pm 55$ m.y.; K-Ar ages of biotite from these samples are $1,615 \pm 55$ m.y. and $1,598 \pm 54$ m.y., respectively. These mineral ages are interpreted as resulting from isotopic resetting caused by a thermal event about 1,600 m.y. ago. The granitic rocks and associated metavolcanic-metasedimentary rocks constitute lower Proterozoic greenstone-granite complexes that are remarkably similar in pattern to the Archean greenstone-granite complexes in the Superior province of the Canadian Shield.

## INTRODUCTION

Reconnaissance geologic mapping in northern Wisconsin and concurrent geochronologic studies have delineated an extensive area underlain by granitic rocks of early Proterozoic (or Precambrian X) age. Exposed granitic rocks extend intermittently from the vicinity of Radisson and Ladysmith on the west eastward beyond Phillips (Fig. 1). The granitic rocks occur at the approximate transition from interbedded metasedimentary and metavolcanic rocks on the north to dominantly metavolcanic rocks on the south. The metasedimentary and metavolcanic rocks in this area have not been dated directly, but from regional relationships are interpreted as early Proterozoic (Sims, 1976a; Sims and others, 1978) and roughly equivalent in age to the Marquette Range Supergroup (Cannon and Gair, 1970) of the Upper Peninsula of Michigan and the adjacent Gogebic Range in Wisconsin. Because of the sparse outcrops in northern Wisconsin and the difficulty of correlating this sequence in detail with the Marquette

Range Supergroup, the metasedimentary-metavolcanic sequence in northern Wisconsin has not been formally named. Likewise, the granitic rocks have not been assigned formal names.

The main purposes of this report are to describe the distribution, general field relationships, and petrography and chemical composition of the granitic rocks and to establish the time of their emplacement. We also describe the country rocks briefly, because existing published data concerning them are sparse (Dutton and Bradley, 1970).

The samples for this study were collected by Sims and K. J. Schulz in 1976 and by Sims and M. G. Mudrey, Jr, in 1977. Descriptions of the samples are given in Table 1.

## REGIONAL GEOLOGY

The lower Proterozoic rocks in northern Wisconsin lie within the southern part of the large east-trending basin that contains the valuable sedimentary iron ores of the Lake Superior region (Bayley and James, 1973). As in northern Michigan, the sequence is strongly asymmetrical. A thin sedimentary succession on the north—as on the Gogebic Range (Fig. 1)—gives way southward to a vastly thicker succession of interbedded sedimentary and volcanic rocks, which in turn passes into a thick succession of dominantly volcanic rocks (Sims and others, 1978). The sedimentary succession mainly formed in a stable shelf environment, whereas the mixed sedimentary and volcanic rocks were deposited in a deep-water environment. The facies arrangement and contrasting thicknesses of material within the basin have been attributed to differences in stability of the Archean basement rocks during early Proterozoic time (Sims, 1976a). The shallow-water supracrustal succession was deposited on tectonically stable greenstone-granite complexes, which Morey and Sims (1976) have termed the greenstone terrane, whereas the deep-water succession was deposited on a basement of mobile gneisses designated as the gneiss terrane. The granitic rocks and associated volcanic-sedimentary rocks of concern to this report are wholly within the gneiss terrane.

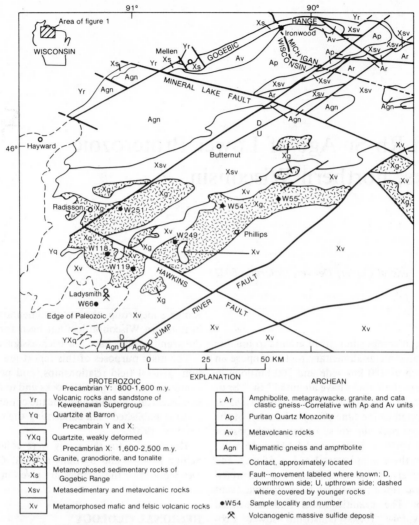

**Figure 1. Generalized geologic map of Precambrian rocks of part of northern Wisconsin (compiled from Sims and others, 1978, and W. C. Prinz, 1978, written commun.).**

The granitic rocks described herein compose the western part of an east-trending belt of granite and associated metamorphic rocks as much as 180 km wide and 300 km long that traverses northern Wisconsin. In this area (Fig. 1), the belt is bounded on the north by late Archean granitic gneiss and amphibolite. The gneisses on the south side of the belt, south of Ladysmith, are presumed to be Archean also because of their similarity in metamorphic grade and composition to gneiss and amphibolite in central Wisconsin that have been determined at two localities as being 2,800 to 3,000 m.y. old (Van Schmus and Anderson, 1977). Possibly these are metamorphic ages.

The metasedimentary and metavolcanic rocks (Fig. 1) in the northern part of the belt consist of two assemblages: (1) a medium-grade assemblage of pillowed porphyritic basalt and lesser iron formation, quartzite, slate, and mica schist, known in part from drilling (Allen and Barrett, 1915, p. 90–99; C. E. Dutton, 1975, oral commun.), which is exposed northeast of Butternut; and (2) a high-grade assemblage containing kyanite-staurolite-garnet schist, garnetiferous biotite schist, amphibolite, and local small bodies of granite and pegmatite, which is exposed

south and east of Butternut. Apparently, the two contrasting assemblages are separated by a northeast-trending, high-angle, dip-slip(?) fault that extends from east of Butternut at least 30 km to the northeast, to the vicinity of Mercer.

The succession of metamorphosed mafic and felsic volcanic rocks (Fig. 1) is more poorly exposed than the mixed succession of metasedimentary and metavolcanic rocks to the north, and its distribution has been determined (Sims and others, 1978) mainly by its aeromagnetic (Zietz and others, 1977) and gravity (Ervin and Hammer, 1974) expressions. Although metabasalt and associated subvolcanic rocks are the major rock types, felsic volcanic and subordinate volcanogenic sedimentary rocks compose substantial parts of the southern part of the volcanic succession. Where observed in outcrops, the volcanic rocks are of amphibolite grade. The Flambeau volcanogenic massive sulfide deposit south of Ladysmith (May, 1977) and a deposit along the Thornapple River, about 7 km north of Ladysmith, occur in relatively felsic parts of the volcanic pile. Galena from the Flambeau deposit has a model lead age of 1,820 ± 150 m.y. (B. R. Doe, 1976, written commun.).

TABLE 1. APPROXIMATE MODES OF GRANITIC ROCKS, NORTHERN WISCONSIN

| | W54 | W55 | W66 | W249B | W249A | W251 | W118 | W119 |
|---|---|---|---|---|---|---|---|---|
| Plagioclase | 36 | 41 | 46 | 42 | 54 | 61 | 61 | 48 |
| Quartz | 25 | 26 | 28 | 27 | 28 | 28 | 23 | 24 |
| K-feldspar | 27 | 28 | 14 | 22 | 4 | 0.5 | 4.5 | .. |
| Biotite | 9 | 5 | 10 | 6.5 | 12 | 5.5 | 10 | 6 |
| Hornblende | .. | .. | 2 | tr | .. | 5 | 1 | 22 |
| Epidote and/or clinozoisite | 1 | tr | tr | 2 | 2 | .. | tr | tr |
| Muscovite | 1.5 | tr | .. | tr | tr | .. | tr | .. |
| Accessory minerals | 0.5 | tr | tr | tr | tr | tr | 0.5 | tr |
| An content, plagioclase | 24-30 | 20-25 | 22-25 | 22-25 | 25-30 | 28-32 | 22-30 | 25-35 |
| Average grain diameter (mm) | 2 | 2.5 | 3 | 2.5 | 1 | 2 | 2 | 3 |

*Note:* Modal values in volume percent; leaders (..) indicate not present; tr, trace.

The lower Proterozoic supracrustal rocks are steeply dipping and are folded on tight, generally northeast-trending axes. The folding of the supracrustal rocks (Penokean orogeny) was accompanied by deformation of the basement gneisses, mainly by shearing and cataclasis. The shearing and cataclasis were superposed on steeply dipping folded gneisses that were deformed originally during the Archean.

## GRANITIC ROCKS

The granitic rocks in northern Wisconsin occupy an area of at least 3,000 km² (Fig. 1). The outlines of individual plutons were delineated from exposures along major rivers and interpretation of the aeromagnetic and gravity data (Sims and others, 1978); because of the sparse outcrops and uncertainties in interpreting geophysical data, they are necessarily generalized. In the following paragraphs, the granitic rocks are discussed primarily with respect to specific sample localities.

The granitic rocks range in composition from granite to tonalite (Fig. 2), as defined by Streckeisen (1976). Biotite, hornblende, or both are conspicuous, particularly in more mafic facies (Table 1), and impart a weak to moderately strong foliation to the rocks. In general, the tonalitic facies have a more prominent foliation. All the rocks contain concentrically zoned plagioclase and have other characteristics indicative of crystallization at moderate depths. Tonalite bodies in the area between Ladysmith and the Hawkins fault tend to be elliptical in outline; whereas both tonalite and granite bodies north of the fault have linear trends subparalled to the structural fabric in the country rocks. The granodiorite from a drill core south of Ladysmith, represented by sample W66 (Table 1), is reported (E. R. May, 1975, oral commun.) to intrude and truncate the Flambeau massive sulfide deposit. Inclusions of schist comparable in lithology to observed country rocks and the gross map patterns (Sims and others, 1978) also indicate that the granitic rocks intrude the adjacent metasedimentary and metavolcanic rocks. We interpret the granitic rocks as being late syntectonic relative to the regional deformation (Penokean orogeny). Cataclasis is local and sparse in the rocks examined, which indicates that tectonism during this thermotectonic event mainly ceased prior to crystallization of the rocks.

The granite, as represented by samples W54 and W55 from two localities in the body north of Phillips, is light gray or pinkish gray, medium grained, equigranular, and weakly foliated; it has a hypidiomorphic-granular texture, locally modified by weak cataclasis. Plagioclase has moderately conspicuous normal concentric zoning; adjacent to K-feldspar, it is myrmekitic. K-feldspar tends to be poikilitic, containing inclusions of plagioclase and quartz, and appears to have crystallized late paragenetically; it is locally microperthitic as a result of exsolution at subsolidus temperatures. Biotite is olive gray, commonly frayed, and slightly altered, either to muscovite and epidote or to chlorite plus sphene or iron oxide. Accessory minerals are allanite, zircon, sphene, apatite, and, rarely, opaque oxides. Typically, allanite has complete rims of clinozoisite-epidote adjacent to biotite. Near sample locality W54, at Lugerville, the granite is associated with a pinkish-gray, medium- to coarse-grained, inequigranular, migmatitic, foliated granite containing discontinuous streaks of quartz and K-feldspar and, locally, megacrysts of K-feldspar.

The tonalite, as represented by samples W118, W119, W251, and W249A, is more diverse texturally and compositionally than the granite but has many similar petrographic characteristics. Plagioclase has a conspicuous normal concentric zoning and a similar spread in anorthite content to that in granite. The sparse K-feldspar is interstitial to plagioclase and quartz. Where present, hornblende is green or bluish green. Opaque oxides are present in nearly all samples studied; other accessory minerals are allanite with clinozoisite-epidote rims, sphene, zircon, and apatite. Near locality W251 east of Radisson (Fig. 1) occur outcrops of an inequigranular, migmatitic foliated granite that resembles the migmatitic granitic rocks near Lugerville previously noted.

The granodiorite from a drill core near Ladysmith (W66 in Fig. 1) differs from other samples studied in having lath-shaped, euhedral, zoned plagioclase that is irregularly embayed and replaced by K-feldspar to yield patch microantiperthite. The difference between the mode (Table 1) and chemical composition (Table 2) of sample W66 is attributed to an irregular distribution of K-feldspar in the rock.

Chemical analyses (Table 2) reflect the modal variation from tonalite to granite (Fig. 2). $SiO_2$ ranges from about 60% (W119) to 71% (W249B); as it increases, total Fe, MgO, and CaO decrease. Total alkalies range from 4.3% to 8.0%, the major variation being in $K_2O$. The samples represent a calc-alkalic trend, but cogenetic relationships have not been demonstrated.

Rb ranges from 19 to 131 ppm and increases generally with increasing $SiO_2$ and $K_2O$. Except for the two low and high extremes

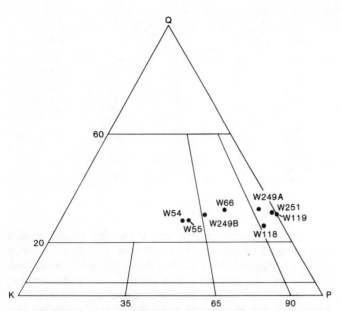

Figure 2. Triangular diagram showing modal composition of analyzed granitic rocks, northern Wisconsin (modified from Streckeisen, 1976).

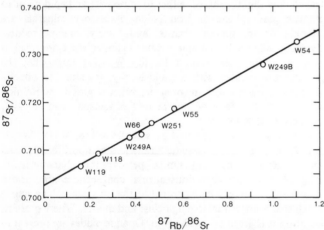

Figure 3. Whole-rock Rb-Sr isochron for tonalites, granodiorites, and granites from northern Wisconsin. The isochron corresponds to an age of 1,885 ± 65 m.y. with an initial $^{87}Sr/^{86}Sr$ ratio of 0.7025 ± 0.0005.

## LOCALITY DESCRIPTIONS FOR FIGURES 2 AND 3

**W54** Collected from large outcrop on south bank of South Fork of Flambeau River, near Lugerville, Price County, Wisconsin; SW¼, NW¼, NW¼ sec. 20, T. 38 N., R. 1 W., lat 45°46′N, long 90°32′W. Pinkish-gray, medium-grained, weakly foliated, equigranular, biotite granite.

**W55** Collected from blasted roadcut on U.S. Forest Service road 136; Price County, Wisconsin; NW¼ sec. 6, T. 38 N., R. 3 E., lat 45°49′N, long 90°10′W. Light-gray, medium-grained, weakly foliated, equigranular, biotite granite.

**W66** From drill core HT-4, depth 15 m, of Bear Creek Mining Co., south of Ladysmith, Rusk County, Wisconsin; SE¼ sec. 19, T. 34 N., R. 6 W., lat 45°24′N, long 91°09′W. Reddish-brown, medium-grained, equigranular, hornblende-biotite tonalite. Has strong foliation and lineation.

**W249B** Collected from large outcrop, east bank at Little Falls on Flambeau River, Sawyer County, Wisconsin; sec. 27, T. 37 N., R. 3 W.; lat 45°40′N, long 90°44′W. Pinkish-gray, medium- to coarse-grained, foliated biotite granodiorite.

**W249A** Collected from same locality as W249B. Light olive-green, medium-grained tonalite. This facies crosscuts and is subordinate to that represented by sample W249B.

**W251** Collected from outcrop east of bridge over Brunet River, Sawyer County, Wisconsin; NW¼ sec. 20, T. 38 N., R. 5 W., lat 45°46′N, long 91°02′W. Medium-gray, medium-grained, foliated tonalite.

**W118** Collected from blasted roadcut, east side Wisconsin highway 27, 13 km north of Ladysmith, Rusk County, Wisconsin; SW¼ NW¼ sec. 22, T. 36 N., R. 6 W.; lat 45°36′N, long 91°07′W. Medium light-gray, medium-grained, weakly foliated biotite tonalite.

**W119** Collected from large outcrops south of Big Falls Dam, west side Flambeau River, Rusk County, Wisconsin; SW¼ SW¼ sec. 35, T. 36 N. R. 5 W.; lat 45°33′N, long 90°57′30″W. Medium-gray, medium-to coarse-grained, foliated, biotite-hornblende tonalite. Tonalite is cut by thin dikes of leucogranite.

## ANALYTICAL DATA AND INTERPRETATION OF AGES

Whole-rock and mineral samples were analyzed for Rb and Sr (Table 2), using conventional isotope-dilution methods similar to those described by Peterman and others (1967), except that a high-purity $^{84}Sr$ spike was used. The program of McIntyre and others (1966) was used to regress the data and to define the statistics of the isochrons (Figs. 3, 4); uncertainties on the ages and intercepts are at the 95% confidence level. The ages are calculated using the isotope ratios and decay constants recommended by the IUGS Subcommission on Geochronology (Steiger and Jäger, 1977). Estimated analytical errors for $^{87}Rb/^{86}Sr$ and $^{87}Sr/^{86}Sr$, both at 1$\delta$ are given in Table 3.

Apatite, microcline, and biotite were separated from two samples, W55 and W249B (Table 2), to evaluate postcrystallization events that have affected the rocks. The two samples were selected because they contain fresh biotite and microcline and lack significant cataclasis. Both suites identify later events that resulted in internal redistribution of Sr 200 to 300 m.y. after the rocks crystallized.

The results of the regressions of the whole-rock and mineral ages (and initial $^{87}Sr/^{86}Sr$ ratios) are as follows: whole-rock samples, 1,885 ± 65 m.y (0.7025 ± 0.0005); W55 (whole rock and minerals), 1,545 ± 55 m.y. (0.7063 ± 0.0008); and W249B (whole rock and minerals), 1,655 ± 55 m.y. (0.7046 ± 0.0009).

indicated by W251 and W119, K/Rb ratios are in the normal range for granitic rocks of this composition. Rb/Sr ratios are characteristically low and contrast with those of granitic rocks of similar age in northeastern Wisconsin. The Athelstane Quartz Monzonite and Hoskin Lake Granite from northeastern Wisconsin have Rb/Sr in the range 0.37 to 1.43 (Van Schmus and others, 1975), whereas samples from northern Wisconsin, with significantly higher Sr and lower Rb, have ratios between 0.055 and 0.38. These are pronounced differences and must reflect differences in source materials, degrees of partial melting, or differentiation patterns.

U and Th, determined by delayed-neutron activation analyses (Table 2), are generally in the normal range of granitic rocks although W66 has high U, and W54 and W55 have higher than average Th contents. Th/U ratios are variable: the most extreme value, 17.7, is shown by W249A.

TABLE 2. CHEMICAL ANALYSES OF GRANITIC ROCKS

| | W66 | W118 | W119 | W251 | W249A | W249B | W54 | W55 |
|---|---|---|---|---|---|---|---|---|
| $SiO_2$ | 68.5 | 69.7 | 60.2 | 65.7 | 69.3 | 71.2 | 70.0 | 71.0 |
| $Al_2O_3$ | 14.6 | 15.3 | 16.7 | 16.4 | 15.2 | 14.6 | 14.5 | 15.1 |
| $Fe_2O_3$ | 1.0 | 1.2 | 1.8 | 0.53 | 0.85 | 0.62 | 1.3 | 0.62 |
| FeO | 2.6 | 1.6 | 4.7 | 3.2 | 2.6 | 2.3 | 1.7 | 1.5 |
| MgO | 1.3 | 0.89 | 2.7 | 1.6 | 0.87 | 0.92 | 0.68 | 0.44 |
| CaO | 2.2 | 3.4 | 6.4 | 4.3 | 3.3 | 2.5 | 1.9 | 2.0 |
| $Na_2O$ | 3.1 | 3.8 | 2.7 | 4.0 | 3.4 | 3.1 | 3.0 | 4.1 |
| $K_2O$ | 4.2 | 2.0 | 1.6 | 1.7 | 2.2 | 3.8 | 4.4 | 3.9 |
| $H_2O^+$ | 0.29 | 0.48 | 1.2 | 0.76 | 0.61 | 0.64 | 0.64 | 0.47 |
| $H_2O^-$ | 0.18 | 0.08 | 0.08 | 0.13 | 0.15 | 0.10 | 0.07 | 0.15 |
| $TiO_2$ | 0.27 | 0.22 | 0.32 | 0.35 | 0.31 | 0.30 | 0.27 | 0.25 |
| $P_2O_5$ | 0.12 | 0.10 | 0.10 | 0.09 | 0.15 | 0.10 | 0.09 | 0.09 |
| MnO | 0.07 | 0.03 | 0.10 | 0.07 | 0.04 | 0.06 | 0.04 | 0.02 |
| $CO_2$ | 0.16 | 0.02 | 0.08 | 0.08 | 0.03 | 0.06 | 0.00 | 0.08 |
| Total | 99 | 99 | 99 | 99 | 99 | 100 | 99 | 100 |
| K/Rb | 472 | 399 | 699 | 190 | 347 | 300 | 279 | 274 |
| Rb/Sr | 0.146 | 0.081 | 0.055 | 0.161 | 0.128 | 0.329 | 0.379 | 0.196 |
| U | 8.0 | 1.2 | 0.3 | 2.6 | 0.6 | 2.2 | 2.7 | 3.8 |
| Th | 11.6 | 5.9 | 1.4* | 12.0 | 11.0 | 19.6 | 31.8 | 28.4 |
| Th/U | 1.4 | 4.9 | 4.3 | 4.6 | 17.7 | 8.8 | 11.7 | 7.4 |

*Note:* Oxide values in percent. Major elements analyzed by Z. A. Hamlin and H. Smith, U.S. Geological Survey; U and Th (in parts per million) analyzed by H. T. Millard, Jr., C. McFee, C. Bliss, and C. M. Ellis, U.S. Geological Survey. Sample locality descriptions given in Table 1; sample localities shown in Figure 1.
*The coefficient of variation exceeds 30% for this analysis.

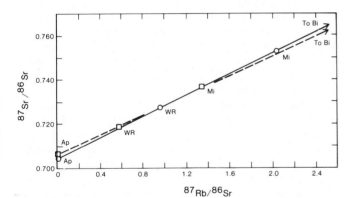

**Figure 4. Whole-rock and mineral isochrons for samples W55 (squares) and W249B (circles). Letter designations are: Ap = apatite; WR = whole rock; Mi = microcline, Bi = biotite. The isochrons represent ages and initial ratios of 1,545 ± 55 m.y. and 0.7063 ± 0.0008 for W55, and 1,655 ± 55 m.y. and 0.7046 ± 0.0009 for W249B. The biotite points are included in the regressions but not plotted because of their extremely high $^{87}Rb/^{86}Sr$ ratios (Table 2).**

TABLE 3. RB-SR ANALYTICAL DATA

| Sample no. | Material | Rb (ppm) | Sr (ppm) | $^{87}Rb/^{86}Sr$ | $^{87}Sr/^{86}Sr$ |
|---|---|---|---|---|---|
| W119 | Whole rock | 19 | 344 | 0.160 | 0.7066 |
| W118 | ---do---- | 41.6 | 514 | 0.234 | 0.7092 |
| W249A | ---do---- | 52.6 | 411 | 0.371 | 0.7125 |
| W66 | ---do---- | 73.8 | 505 | 0.423 | 0.7132 |
| W251 | ---do---- | 74.1 | 459 | 0.467 | 0.7156 |
| W55 | ---do---- | 118 | 603 | 0.565 | 0.7186 |
| W55 | Apatite | 0.86 | 458 | 0.0054 | 0.7064 |
| W55 | Microcline | 325 | 704 | 1.341 | 0.7366 |
| W55 | Biotite | 828 | 19 | 174 | 4.5384 |
| W249B | Whole rock | 105 | 319 | 0.951 | 0.7277 |
| W249B | Apatite | 0.50 | 210 | 0.0070 | 0.7044 |
| W249B | Microcline | 261 | 374 | 2.028 | 0.7529 |
| W249B | Biotite | 677 | 12.9 | 234 | 6.1894 |
| W54 | Whole rock | 131 | 346 | 1.102 | 0.7327 |

*Note:* Uncertainties (1 σ) are ±1.0% of the $^{87}Rb/^{86}Sr$ ratio and ±0.04% of the $^{87}Sr/^{86}Sr$ ratio. Sr-isotope ratios are relative to a value of 0.7080 for the E and A standard.

TABLE 4. K-Ar AGES OF BIOTITE

| Sample no. | $K_2O$ (%) | $^{40}Ar_{rad}$ ($\times 10^{-10}$ mol/g) | $^{40}Ar_{rad}$ (%) | $^{40}Ar_{rad}/^{40}K$ | Age ± 2σ (m.y.) |
|---|---|---|---|---|---|
| W55 | 9.04 | 334.3 | 99 | 0.149 | 1,598 ± 54 |
| W294B | 8.61 | 323.7 | 99 | 0.152 | 1,615 ± 55 |

*Note:* Analyzed by R. F. Marvin, H. H. Mehnert, and V. M. Merritt, U.S. Geological Survey.

The data for all three suites are nearly within estimated analytical error in defining the isochrons. The residual variances (MSWD as defined by McIntyre and others, 1966) are 3.83 for the whole-rock samples, 1.61 for W55, and 3.09 for W249B. Because of the small number of samples and the latitude available in choosing analytical errors used in the regression, these values are not interpreted as indicating excessive scatter beyond experimental error.

K-Ar ages of 1,598 ± 54 m.y. and 1,615 ± 55 m.y. were obtained on biotite from samples W55 and W249B, respectively (Table 4). Within analytical error, these ages are in agreement with the Rb-Sr mineral isochron ages that include data for biotite from these samples.

Geologic considerations are more critical in interpreting the whole-rock isochron than the small scatter in the data points. Because of the sparse, scattered outcrops of granitic rocks in northern Wisconsin, it was necessary to use widely spaced samples from separate bodies to develop an isochron. Implicit in the isochron technique is the assumption that all the samples are of the same age and crystallized with a common initial $^{87}Sr/^{86}Sr$. Although we cannot prove conclusively that this assumption is correct, the data support it. For example, data from three sample localities in the body near Phillips (Fig. 1) define an isochron indistinguishable from that obtained using all samples. The same argument can be made for samples from the granitic body near Radisson, although they are less radiogenic than the others. The sample from near Ladysmith (W66) is separated from the main bodies, but deleting this point from the isochron would have negligible effect on the slope and intercept. We conclude that the samples conform to the assumptions in the Rb/Sr isochron method and that these bodies are 1,885 ± 65 m.y. old. The age is interpreted as the time of crystallization of the granitic rocks. This age is indistinguishable from several zircon ages of other igneous rocks in Wisconsin, as discussed in the following section.

The initial $^{87}Sr/^{86}Sr$ ratio of 0.7025 ± 0.0005 defined by the isochron is relatively precise because of the colinearity of the data and the characteristically low Rb/Sr ratios of the samples. This low ratio is within the field of mantle evolution defined by tonalites and trondhjemites of various ages (Peterman, 1979). Thus, the northern Wisconsin granitic rocks were derived from an isotopically primitive source with a low Rb/Sr ratio. Probably the granitic rocks represent juvenile additions to the crust, although the possibility that they were derived from older tonalites or other rock types with low Rb/Sr ratios cannot be precluded. Clearly, however, these rocks cannot have been derived from 3,500-m.y.-old tonalite or 2,700-m.y.-old tonalite and granite similar to those exposed in the Marenisco-Watersmeet area (Peterman and others, this volume) immediately to the north in Michigan. The $^{87}Sr/^{86}Sr$ ratios in these rocks at 1,885 m.y. ago would have been substantially higher on the average than 0.7025 (Sims and Peterman, 1976; Sims and others, 1977). Limited data for the Archean tonalites south of Mellen in northern Wisconsin (see Fig. 1) suggest that these have lower Rb/Sr ratios than those in northern Michigan, but even these would have generated $^{87}Sr/^{86}Sr$ ratios of 0.705 to 0.710 by 1,885 m.y. ago.

The Rb-Sr data for whole-rock, apatite, microcline, and biotite samples for W55 and W249B define precise isochron ages that are significantly younger than the whole-rock age (Figs. 3, 4) but in agreement with the K-Ar biotite ages. The regressions are strongly influenced by data for biotite. However, data for the whole-rock, apatite, and microcline samples are colinear and define similar ages independently of the biotite points. Thus these samples, and presumably the region as a whole, experienced a moderately severe thermal event of sufficient magnitude to cause isotopic redistribution of $^{87}Sr$ and loss of $^{40}Ar$ from biotite about 1,600 m.y. ago. The precise ages defined by the Rb-Sr mineral isochrons and the agreement of these with the K-Ar biotite ages suggest that the region stabilized (cooled) rapidly at 1,600 m.y. ago in order to maintain concordancy in these different radiometric and mineral systems. This could have been accomplished by relatively rapid uplift from a depth where the temperature exceeded approximately 300 °C at 1,600 m.y. ago or by a sharp thermal event at this time. Similar metamorphic Rb-Sr ages are common in a variety of rock types in central and eastern Wisconsin (Van Schmus, 1976).

## DISCUSSION

The radiometric dating of lower Proterozoic granitic rocks in northern Wisconsin extends the known occurrence of these rocks in Wisconsin. Previously, Banks and Cain (1969) established U-Pb concordia-intercept ages between 1,860 and 1,830 m.y. for four granitic bodies in northeastern Wisconsin, and Van Schmus (this volume) determined by the same method ages of 1,840 to 1,820 m.y. for scattered bodies of granitic and tonalitic rocks in central and north-central Wisconsin. Among the granitic rocks dated by Van Schmus is a mesozonal granite that intrudes metavolcanic rocks in the Wausau area, central Wisconsin (Sims and others, 1978). These metavolcanic rocks as well as those in northeastern Wisconsin (Banks and Rebello, 1969) have been dated in the range 1,850 to 1,875 m.y. These data indicate that approximately 1,850-m.y.-old granitic plutons are abundant and widespread in Wisconsin. They compose a wide belt across northern Wisconsin between lat 45° and 46° N, a distance of 300 km, and underlie a substantial area in central Wisconsin. The volcanic-plutonic activity preceded and in part was contemporaneous with the main peak of lower Proterozoic deformation and metamorphism (Penokean orogeny).

The association of abundant lower Proterozoic granitic plutons with metavolcanic and metasedimentary rocks of approximately the same age in northern Wisconsin and the pattern of occurrence are remarkably similar to those of the greenstone-granite complexes of Archean age in northern Minnesota (Sims, 1976b; Goldich, 1972). The Archean and apparently the lower Proterozoic volcanic-plutonic complexes were formed during relatively short time spans, perhaps 50 to 100 m.y.

A lower Proterozoic volcanic-plutonic episode that is younger than both the Penokean orogeny and the intrusion of the about 1,850-m.y.-old granitic rocks has been recognized in south-central Wisconsin. This episode is represented mainly by supracrustal rhyolite and cogenetic granophyric granite that occur in small windows through Paleozoic cover rocks (Smith, 1978), over an area of about 2,000 km² in southern Wisconsin. Both the rhyolite and the granite have U-Pb zircon ages of 1,765 ± 10 m.y. (Van Schmus, this volume). These rocks underlie quartzite in the Baraboo area in the same region; both the rhyolite and the quartzite are mildly folded and faulted. In northern Wisconsin, a coarse leucogranite near Monico, which is 95 km east of Phillips, and a coarse porphyritic granite from near Radisson (Fig. 1) also have U-Pb concordia-intercept ages of 1,765 m.y. (Van Schmus, this volume). The porphyritic granite at Radisson is included in the area of granitic rocks in Figure 1; its areal extent is not known.

## REGIONAL GEOLOGIC RELATIONSHIPS

Data from the Lake Superior region indicate that the syntectonic, mesozonal granitic rocks ($\sim$1,850 m.y. old) of the type described herein probably are mainly confined to the lower Proterozoic metavolcanic belt in the northern part of Wisconsin, whereas the younger (1,765 m.y. old) posttectonic, epizonal suite probably is more widespread. With respect to the mesozonal suite, tonalitic rocks of this age intrude Archean gneisses in central Wisconsin (Van Schmus, this volume) on the south margin of the lower Proterozoic structural basin. Outside of Wisconsin, a small pluton (granite of section 28) in the southern Minnesota has a comparable age ($\sim$1,830 m.y.; Goldich and others, 1970). It intrudes Archean gneisses that have ages in the range 3,500 to 3,800 m.y. (Goldich and Hedge, 1974). Except for a few small bodies of granitic rocks (James and others, 1961), as yet undated, northern Michigan lacks rocks of either granitic suite. However, a small composite mafic intrusive body (Peavy Pond Complex; Bayley, 1959) in the southeastern part of the Upper Peninsula has an age of about 1,875 m.y. (Banks and Van Schmus, 1972).

The approximately 1,600-m.y.-old event recorded by Rb-Sr whole-rock mineral ages in northern Wisconsin apparently reflects a widespread tectonothermal event that affected lower Proterozoic and older rocks in much of Wisconsin and the eastern part of the Upper Peninsula of Michigan. With the exception of granitic rocks in northern Wisconsin discussed in this report, the Rb-Sr whole-rock systems in all granitic rocks from Wisconsin that have been analyzed are severely disturbed and give secondary, rotated isochrons (Van Schmus and others, 1975). Van Schmus (this volume) has suggested that this event possibly represented epeirogenic uplift of this part of the crust about 1,600 m.y. ago; alternatively, it might represent the episode of folding and faulting that is recorded in the 1,765-m.y.-old rhyolite and granite in southern Wisconsin and in scattered platform quartzite bodies between 1,500 and 1,765 m.y. old (Van Schmus, 1976) in central and northern Wisconsin.

So far as known, the multiple events recorded by the lower Proterozoic rocks in Wisconsin—two distinct episodes of plutonic activity ($\sim$1,850 and 1,765 m.y. ago) and apparently several pulses of tectonic activity—are confined to that part of the Lake Superior region that is underlain by Archean gneiss basement rocks (gneiss terrane of Morey and Sims, 1976). The more stable greenstone terrane in the northern parts of this and adjacent areas in Michigan and Minnesota (Sims, 1976a) was not affected by these disturbances in early Proterozoic time.

## ACKNOWLEDGMENTS

Whole-rock samples and mineral separates were prepared by G. T. Cebula and J. W. Groen. The Rb-Sr chemistry and some of the mass spectrometry was done by K. Futa. K-Ar biotite ages were determined by R. F. Marvin, H. H. Mehnert, and V. M. Merritt. All analysts are at the U.S. Geological Survey. The geology of the Mercer, Wisconsin, area was discussed with Richard Black and Charles Guidotti of the University of Wisconsin, Madison.

## REFERENCES CITED

Allen, R. C., and Barrett, L. P., 1915, Contributions to the pre-Cambrian geology of northern Michigan and Wisconsin: Michigan Geological Survey Publication 18, Geologic Series 15, p. 13–164.

Banks, P. O., and Cain, J. A., 1969, Zircon ages of Precambrian granitic rocks, northeastern Wisconsin: Journal of Geology, v. 77, p. 208–220.

Banks, P. O., and Rebello, D. P., 1969, Zircon age of a Precambrian rhyolite, northeastern Wisconsin: Geological Society of America Bulletin, v. 80, p. 907–910.

Banks, P. O., and Van Schmus, W. R., 1972, Chronology of Precambrian rocks of Iron and Dickinson Counties, Michigan [abs.]: Institute on Lake Superior Geology, 18th, Houghton, Michigan, Proceedings, pt. 2, Paper 23.

Bayley, R. W., 1959, Geology of the Lake Mary quadrangle, Iron County, Michigan: U.S. Geological Survey Bulletin 1077, 112 p.

Bayley, R. W., and James, H. L., 1973, Precambrian iron-formations of the United States: Economic Geology, v. 68, p. 934–959.

Cannon, W. F., and Gair, J. E., 1970, A revision of stratigraphic nomenclature for middle Precambrian rocks in northern Michigan: Geological Society of America Bulletin, v. 81, p. 2843–2846.

Dutton, C. E., and Bradley, R. E., 1970, Lithologic, geophysical., and mineral commodity maps of Precambrian rocks in Wisconsin: U.S. Geological Survey Miscellaneous Geologic Investigations Map I-631, Scale 1:1,000,000.

Ervin, C. P., and Hammer, S., 1974, Bouguer anomaly gravity map of Wisconsin: Wisconsin Geological and Natural History Survey, scale 1:500,000.

Goldich, S. S., 1972, Geochronology in Minnesota, in Sims, P. K., and Morey, G. B., eds., Geology of Minnesota—A centennial volume: Minnesota Geological Survey, p. 27–37.

Goldich, S. S., and Hedge, C. E., 1974, 3,800-Myr granitic gneiss in south-western Minnesota: Nature, v. 252, no. 5483, p. 467–468.

Goldich, S. S., Hedge, C. E., and Stern, T. W., 1970, Age of the Morton and Montevideo gneisses and related rocks, southwestern Minnesota: Geological Society of America Bulletin, v. 81, p. 3671–3695.

Goldich, S. S., and others, 1961, The Precambrian geology and geochronology of Minnesota: Minnesota Geological Survey Bulletin 41, 193 p.

James, H. L., and others, 1961, Geology of central Dickinson County, Michigan: U.S. Geological Survey Professional Paper 310, 176 p.

May, E. R., 1977, Flambeau—A Precambrian supergene enriched massive sulfide deposit: Wisconsin Geological and Natural History Survey, Geoscience Wisconsin, v. 1, p. 1–26.

McIntyre, G. A., and others, 1966, The statistical assessment of Rb-Sr isochrons: Journal of Geophysical Research, v. 71, no. 22, p. 5459–5468.

Morey, G. B., and Sims, P. K., 1976, Boundary between two Precambrian W terranes in Minnesota and its geologic significance: Geological Society of America Bulletin, v. 87, p. 141–152.

Peterman, Z. E. 1979, Sr-Isotope geochemistry of tonalites and trondhjemites, late Archean to late Cretaceous in Barker, Fred, ed., Trondhjemites, dacites, and related rocks: Amsterdam Elsevier Scientific Publishing Company, p. 133–147.

Peterman, Z. E., Doe, B. R., and Bartel, Ardith, 1967, Data on rock GSP-1 (granodiorite) and the isotope-dilution method of analysis for

Rb and Sr, *in* Geological Survey research 1967: U.S. Geological Survey Professional Paper 575-B, p. B181-B186.

Peterman, Z. E., Zartman, R. E., and Sims, P. K., 1980, Tonalitic gneiss of early Archean age from northern Michigan: Geological Society of America Special Paper 182 (this volume).

Sims, P. K., 1976a, Precambrian tectonics and mineral deposits, Lake Superior region: Economic Geology, v. 71, p. 1092–1118.

——1976b, Early Precambrian tectonic-igneous evolution in the Vermilion district, northeastern Minnesota: Geological Society of America Bulletin, v. 87, p. 379–389.

Sims, P. K., and Peterman, Z. E., 1976, Geology and Rb-Sr ages of reactivated Precambrian gneisses and granite in the Marenisco-Watersmeet area, northern Michigan: U.S. Geological Survey Journal of Research, v. 4, p. 405–414.

Sims, P. K., Cannon, W. F., and Mudrey, M. G., Jr., 1978, Preliminary geologic map of Precambrian rocks in part of northern Wisconsin: U.S. Geological Survey Open-File Report 78-318.

Sims, P. K., Peterman, Z. E., and Prinz, W. C., 1977, Geology and Rb-Sr age of Precambrian W Puritan Quartz Monzonite, northern Michigan: U.S. Geological Survey Journal of Research, v. 5, p. 185–192.

Smith, E. I., 1978, Precambrian rhyolites and granites in south-central Wisconsin—Field relations and geochemistry: Geological Society of America Bulletin, v. 89, p. 875–890.

Steiger, R. H., and Jäger, E., 1977, Subcommission on Geochronology—Convention on the use of decay constants in geo- and cosmochronology: Earth and Planetary Science Letters, v. 36, p. 359–362.

Streckeisen, A. L., 1976, To each plutonic rock its proper name: Earth-Science Reviews, v. 12, p. 1–33.

Van Schmus, W. R., 1976, Early and middle Proterozoic history of the Great Lakes area, North America: Royal Society of London Philosophical Transactions, ser. A280, no. 1298, p. 605–628.

——1980, Chronology of igneous rocks associated with the Penokean orogeny in Wisconsin: Geological Society of America Special Paper 182 (this volume).

Van Schmus, W. R., and Anderson, J. L., 1977, Gneiss and migmatite of Archean age in the Precambrian basement of central Wisconsin: Geology, v. 5, p. 45–48.

Van Schmus, W. R., Thurman, E. M., and Peterman, Z. E., 1975, Geology and Rb-Sr chronology of middle Precambrian rocks in eastern and central Wisconsin: Geological Society of America Bulletin, v. 86, p. 1255–1265.

Zietz, Isidore, Karl, J. H., and Ostrom, M. E., 1977, Preliminary aeromagnetic map covering most of the exposed Precambrian terrane in Wisconsin: U.S. Geological Survey Miscellaneous Field Studies Map MF-888, scale 1:250,000.

Manuscript Received by the Society April 25, 1979
Revised Manuscript Received November 26, 1979
Manuscript Accepted January 11, 1980

Geological Society of America
Special Paper 182
1980

# Penokean deformation in central Wisconsin

**R. S. MAASS**
**L. G. MEDARIS, JR.**
*Department of Geology and Geophysics, University of Wisconsin, Madison, Wisconsin 53706*

**W. R. VAN SCHMUS**
*Department of Geology, University of Kansas, Lawrence, Kansas 66045*

## ABSTRACT

An Archean, early Proterozoic, and middle Proterozoic sequence of metamorphic and plutonic rocks is exposed along and near the Wisconsin River in central Wisconsin, between Stevens Point and Wisconsin Rapids. Major lithologic units, from oldest to youngest, consist of (1) banded tonalitic gneiss with interlayered amphibolite, (2) 2,800-m.y.-old migmatite, (3) diorite gneiss, (4) 1,840-m.y.-old foliated tonalite, (5) a series of 1,820-m.y.-old lineated tonalites, (6) granitic aplites and pegmatites, (7) amphibolite dikes, and (8) diabase dikes. All units except the diabase dikes have been metamorphosed under amphibolite-facies conditions and possess a steeply plunging, penetrative mineral lineation.

Deformation during the Penokean orogeny produced three steeply plunging sets of folds ranging from isoclinal to open in style. $F_1$ folding is interpreted as culminating prior to 1,840 m.y. ago, but also extending beyond this time. Foliated tonalite appears to have been emplaced during the waning stages of $F_1$ deformation. $F_2$ and $F_3$ folding occurred after the intrusion of 1,820-m.y.-old lineated tonalite, but younger limits on the age of folding are not yet available. All three fold sets are coaxial and parallel to the penetrative mineral lineation. Although not conclusive, there is evidence that suggests that an older deformation, perhaps Archean in age, affected the terrane prior to the Penokean orogeny.

## INTRODUCTION

Study of the Precambrian terrane in Wisconsin is hampered by lack of exposure due to an extensive cover of Pleistocene glacial deposits and Paleozoic sedimentary rocks. Despite this, recent field investigations, combined with isotopic age determinations, have provided comprehensive regional syntheses of the Precambrian geology of the Lake Superior region (Van Schmus, 1976; Sims, 1976) and a relatively complete picture of the later Proterozoic history of Wisconsin, involving such rocks as the Ke-

weenawan lavas (Craddock, 1972), Wolf River batholith (Van Schmus and others, 1975a; Anderson and Cullers, 1978), rhyolites and granites of south-central Wisconsin (Smith, 1978), and Baraboo quartzite (Dott and Dalziel, 1972). Although a general framework for the Archean and early Proterozoic history of Wisconsin has been established (Van Schmus, 1976, and this volume; Van Schmus and Anderson, 1977; Sims, 1976), the details for this portion of the geologic record are still poorly understood in Wisconsin.

The present investigation is part of a long-term project designed to determine, by means of coordinated field and isotopic studies, the regional extent and petrologic and structural characteristics of the early Proterozoic Penokean orogeny and a probable Archean orogeny in Wisconsin.

## LOCALITY AND METHODS

Although outcrops of Precambrian rocks are sparsely distributed in Wisconsin, good exposures of limited extent occur locally along streams and in quarries. Our approach has been to examine such exposures in detail in order to establish a relative sequence of events on the basis of geologic evidence and then to correlate this relative sequence with isotopic ages obtained from U-Pb measurements on separated zircons. This approach should eventually provide a fairly complete chronicle of Archean and Proterozoic events for Wisconsin, even though traditional geologic mapping on a regional scale cannot be carried out.

This investigation deals with Precambrian rocks exposed along and near the Wisconsin River between Stevens Point and Wisconsin Rapids (Fig. 1). These rocks were first described by Weidman (1907) and are now known to include one of the southernmost occurrences of confirmed Archean basement in the Southern province (Linwood Township quarry, described by Anderson, 1972, and Van Schmus and Anderson, 1977). The extensive exposures at Conants Rapids and Biron Dam have been mapped by plane table at scales of 30 and 50 ft to the inch, respectively (Maass, 1977), and the results are illustrated in Figures 2 and 3.

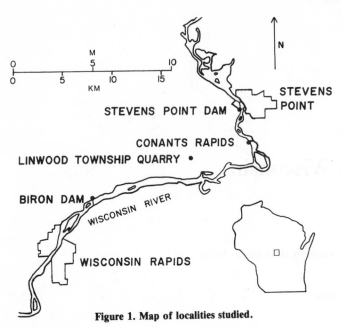

Figure 1. Map of localities studied.

## LITHOLOGIC UNITS

### Banded gneiss

The oldest lithologic unit in the area is a fine-grained tonalitic banded gneiss composed of oligoclase, quartz, biotite, microcline, and hornblende that crops out at Conants Rapids (Fig. 2) and at Stevens Point Dam. The banding is due to alternating mafic-rich and mafic-poor layers 5 cm or less (typically 1 to 2 cm) in thickness. Fine-grained amphibolite composed of hornblende and andesine occurs as layers as much as 30 cm thick, concordant with layering in the gneiss. An early generation of pegmatite dikes as much as 30 cm thick and exhibiting pinch-and-swell structure is also concordant with layering in the gneiss. Both amphibolite and pegmatite have been included with banded gneiss in the map shown in Figure 2.

Regionally, the banded gneiss occurs as small, irregular masses surrounded and invaded by younger intrusive bodies having steep contacts. The nature of the protolith from which the gneiss was derived is unknown, because any primary structures or textures that might have been present have been completely obliterated by metamorphism and deformation. From compositional considerations, the protolith could have been a graywacke, if sedimentary, or andesite or tonalite, if igneous.

### Migmatite

Exposure of "veined" migmatite (Van Schmus and Anderson, 1977) occurs in a quarry in Linwood Township (NW¼SE¼ sec. 15, T 23 N., R. 7 E.). The migmatite consists of fine-grained, gray, banded melanosomes interlayered with medium-grained, white or pink leucosomes. The melanosomes and leucosomes range as much as 10 cm in width and are in sharp contact with each other. Concordantly interlayered amphibolite is similar in appearance and composition to amphibolite in the banded gneiss. The tonalitic melanosomes contain oligoclase, quartz, biotite, and microcline, and the granitic leucosomes contain oligoclase,

quartz, and microcline. Some leucosomes are homogeneous, whereas others are segregated into microcline-rich cores and oligoclase-rich walls. The bulk composition of the leucosomes remains approximately the same, however, whether or not segregation has occurred (Anderson, 1972).

On the basis of field relationships, petrology, and chemistry, Anderson (1972) suggested that the migmatite is an injection gneiss in which the melanosomes represent unaltered banded gneiss. The leucosomes are anatectic in composition, corresponding to experimental melts produced at 5- to 6-kb partial pressure of water.

### Diorite Gneiss

Black, fine-grained, massive diorite gneiss, composed of oligoclase-andesine, hornblende, biotite, and quartz, crops out as irregularly shaped patches less than 5 m in width at Stevens Point Dam. The unit is not in exposed contact with banded gneiss or migmatite, thus its age relative to these units is based on structural evidence. Foliated tonalite has intruded and contains inclusions of diorite gneiss.

### Foliated Tonalite

Gray, foliated tonalite is exposed at Stevens Point Dam, Conants Rapids, and Biron Dam. Porphyroclastic grains of oligoclase-andesine, from 1 to 6 mm in length, are surrounded by a fine-grained matrix of oligoclase-andesine, quartz, biotite, hornblende, and microcline (Fig. 4). Small lenticular dioritic inclusions, elongated parallel or subparallel to the foliation, are common in foliated tonalite and probably represent diorite gneiss brought up from depth. Their consistent and abundant presence suggests that distribution of diorite gneiss may be extensive at depth.

At Stevens Point Dam foliated tonalite is intrusive into banded gneiss and diorite gneiss and contains inclusions of both. Anderson (1972) reported that foliated tonalite has intruded migmatite along one of the small streams in the area and that the contact is discordant to layering in the migmatite.

### Lineated Tonalites

A series of fine-grained lineated tonalites, composed of oligoclase, quartz, biotite, microcline, and hornblende (Fig. 5), occurs as dikes and irregular bodies throughout the terrane. The series varies systematically in color and mafic mineral content with age: the older members are dark gray and mafic-rich, the younger members are light gray, white, or even pink and contain fewer mafic minerals. A single exception is the youngest tonalite of the series, which is very fine grained and dark gray. Two members of the series are actually granodioritic, containing slightly more than 10% microcline, but have been included under the designation "tonalite" as a matter of convenience. For mapping purposes the series has been separated into five divisions on the basis of color (Fig. 2), each division containing more than one member of the series (except the "fine-grained dark gray"). The relationships between individual members of the series are best displayed at Conants Rapids, where crosscutting veinlets and small inclusions of the various tonalite members may be observed.

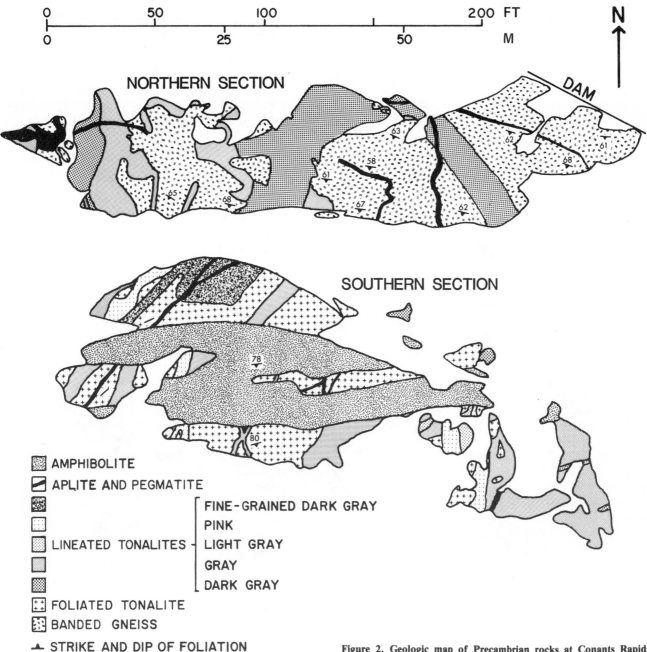

Figure 2. Geologic map of Precambrian rocks at Conants Rapids, Wisconsin (NW¼SW¼ sec. 8, T. 23 N., R. 8 E.); selected segments of larger outcrop.

Lineated tonalites have intruded banded gneiss and foliated tonalite at Conants Rapids, migmatite at the Linwood Township quarry, and foliated tonalite at Biron Dam.

### Aplites and Pegmatites

Granitic aplite and pegmatite, containing oligoclase, quartz, microcline, and biotite, are present in every outcrop as discordant dikes and veins. Maximum width is 4 m; average width is 0.3 m. The majority were emplaced between the time of formation of lineated tonalite and the amphibolite dikes, although minor amounts of aplite and pegmatite were emplaced between the time of intrusion of foliated tonalite and lineated tonalite, and a few, after the amphibolite dikes.

### Amphibolite Dikes

Amphibolite dikes, ranging in width from less than 1 up to 25 m, occur at Conants Rapids and Biron Dam. They are black and fine grained, composed of a schistose assemblage of andesine, hornblende, and biotite, and probably represent metamorphosed diabase. Two generations closely spaced in time have been

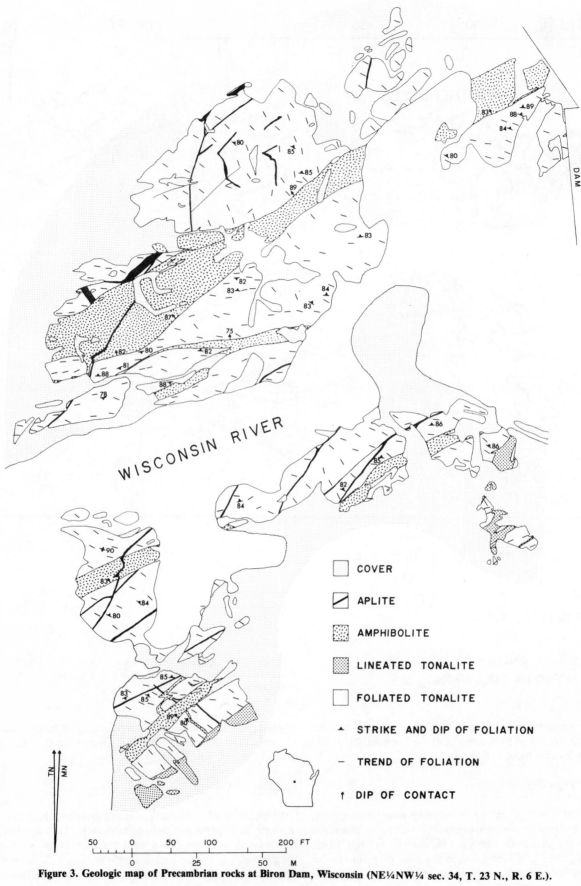

**COVER**

**APLITE**

**AMPHIBOLITE**

**LINEATED TONALITE**

**FOLIATED TONALITE**

▲ STRIKE AND DIP OF FOLIATION

— TREND OF FOLIATION

↑ DIP OF CONTACT

WISCONSIN RIVER

DAM

TN  MN

50   0   50   100        200 FT

0        25        50 M

**Figure 3. Geologic map of Precambrian rocks at Biron Dam, Wisconsin (NE¼NW¼ sec. 34, T. 23 N., R. 6 E.).**

**Figure 4.** Photomicrograph of foliated tonalite, *bc* plane. Porphyroclastic grains of oligoclase-andesine at right center and left center. Other major constituents are quartz, biotite, and hornblende. Field of view is 6 mm in width; crossed polarizers.

**Figure 5.** Photomicrograph of lineated tonalite, lineation in plane of thin section. Major constituents are oligoclase, quartz, biotite, and hornblende. All minerals define the lineation, but to varying degrees. Field of view is 6 mm in width; crossed polarizers.

recognized. At Conants Rapids they are distinguished by a slight difference in grain size, and at Biron Dam, by the presence or absence of inclusions of foliated tonalite.

The dikes trend east and northeast at Conants Rapids (Fig. 2), and northeast at Biron Dam (Fig. 3). In both localities their contacts are nearly vertical, and the dikes have transected foliated and lineated tonalites.

### Diabase Dike

The youngest lithologic unit in the area is a diabase dike 25 m wide that occurs at the southern end of the Biron Dam outcrop (beyond the map area in Fig. 3), where it has intruded foliated tonalite and lineated tonalite. The black, fine-grained diabase is composed of labradorite, augite, iron oxides, and minor olivine and displays well-developed subophitic texture. The fresh igneous texture of the diabase stands in contrast to the recrystallized textures of all the older igneous rocks in the area.

### ISOTOPIC AGES

In the course of a major regional geochronologic study of Penokean rocks (Van Schmus, this volume), radiometric ages have been obtained on a number of tonalitic units. These units are summarized in Table 1 and are important because they represent rocks intruded at various stages in the deformational history of the region. Two samples (11 and 15) are not from the Wisconsin River region, but they appear to be related to the intrusives of this region.

The ages have been determined on separated zircon fractions from the rocks concerned. In all but one case (sample 11) the zircons have euhedral morphologies, are relatively clear, and show little to no evidence of older cores. For sample 11 the zircons are rounded and subhedral but otherwise show little evidence of older history. All indications are that the zircons analyzed represent igneous crystallization times of the units concerned. There is no reason to believe they are metamorphic products. Hence, the ages

obtained represent times of intrusion of these units during the Penokean orogeny.

The analytical results are shown in Figure 6. One major advantage of the U-Pb method for zircons is that it can yield results precise to $\pm 0.5\%$ or better and thus can be used to resolve events occurring over relatively short intervals, even though absolute ages may exceed 2,000 m.y. This is very well demonstrated by the rocks represented in Figure 6. Those that are more strongly deformed (for example, foliated tonalites) appear to be distinctly older than those that are less deformed (for example, lineated tonalite and an undeformed dacite dike), although the results are indistinguishable at the $2\sigma$ uncertainty level. Thus, we conclude that the tonalite igneous activity began about 1,840 m.y. ago and lasted for about 20 m.y., until about 1,820 m.y. ago. Furthermore, these ages clearly limit a portion of the deformation to the interval 1,820 to 1,840 m.y. ago, which we regard as representing

TABLE 1.   SPECIMENS RADIOMETRICALLY DATED BY THE U-Pb METHOD ON ZIRCON

| Sample no.* | Rock type and location | Age[†] (m.y.) |
|---|---|---|
| 9 | Foliated tonalite, Biron Dam | 1,842 ± 10 |
| 11 | Foliated tonalite, Eau Claire | |
| 15 | Dacite dike, Pittsville Quarry | |
| 20 | Lineated tonalite, Conants Rapids | 1,824 ± 25 |
| 22 | Lineated tonalite, Conants Rapids | |

*From Van Schmus (this volume). Refer to that source for detailed locations and analytical data.

[†]From Figure 6. Ages based on least-squares fit for data from foliated suite and from lineated suite and represent concordia intercepts at the 95% confidence level. Ages based on decay constants of Jaffey and others (1971).

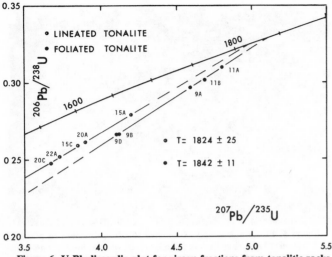

Figure 6. U-Pb discordia plot for zircon fractions from tonalitic rocks having varying degrees of Penokean deformation. The foliated suite defines an age of 1,842 ± 10 m.y., whereas the younger, crosscutting rocks define an age of 1,824 ± 25 m.y. Ages are defined by least-squares fits to the data and are quoted at the 95% confidence level. Although the ages overlap at the quoted uncertainties, the systematic separation of the data in this plot strongly suggests a real age difference of about 20 m.y. See text for further discussion.

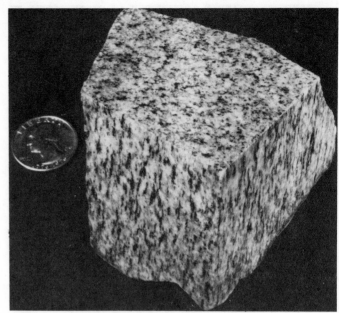

Figure 7. Hand specimen of lineated tonalite. In hand specimen, the lineation is best defined by the orientation of biotite and hornblende. A very weak foliation is visible on the upper surface of the specimen, trending from left to right.

the younger part of the Penokean orogeny (deformation prior to 1,840 m.y. may have occurred over an indefinite time interval).

## STRUCTURAL GEOLOGY

### Mesoscopic Elements

All lithologic units in the terrane except the diabase have been metamorphosed to amphibolite facies and exhibit a penetrative mineral lineation defined by the orientation of elongate grains of hornblende and biotite. Individual quartz and feldspar grains are only slightly elongated, although aggregates of these minerals are elongated parallel to the mineral lineation and enhance it.

In contrast to the pervasive development of a mineral lineation, not all lithologic units are foliated. A penetrative foliation is well developed in banded gneiss and migmatite, less so in diorite gneiss and foliated tonalite, and generally absent from lineated tonalite, a sample of which is illustrated in Figure 7.

Amphibolite dikes are foliated, but foliation is parallel to walls of the dikes and appears to have developed during and shortly after intrusion. The relationships are best shown at Biron Dam, where amphibolite dikes locally transect foliation in foliated tonalite at a high angle, but within the dikes, foliation and schlieren of foliated tonalite are aligned parallel to walls of the amphibolite dikes. Furthermore, an aplite dike that cuts across one of the amphibolite dikes (SW¼ of Fig. 3) has been tightly folded, with an axial plane parallel to the walls of the amphibolite dike, attesting to late-stage differential movements within the amphibolite dikes parallel to the walls.

Some of the aplite and pegmatite dikes in the terrane contain a foliation that is parallel to the margins of the dikes rather than to the regional foliation. In certain cases the foliation is a relict

primary structure; in others it is a later cataclastic feature. Late stresses in the region appear to have been preferentially relieved by movement within some of the aplite and pegmatite dikes parallel to their walls. This has resulted in cataclasis, which is most intense in the center of the dikes and grades out toward the margins.

Three sets of folds have been recognized in the terrane. The earliest set, $F_1$, is characterized in banded gneiss and migmatite by tight to isoclinal folds of similar style (Fig. 8) and in diorite gneiss and foliated tonalite by open to tight folds of similar or concentric style. Fold geometry and field relationships indicate that the diorite gneiss and foliated tonalite were emplaced during the waning stages of $F_1$ folding. Contacts of foliated tonalite transect isoclinal folds in banded gneiss, but at Stevens Point Dam, veins

Figure 8. Isoclinal $F_1$ fold in banded gneiss, Conants Rapids. The core of the fold consists of amphibolite.

Figure 9. This specimen depicts the relationship between units that were subjected to the entire period of F₁ deformation and those which were intruded during the late stages of F₁ deformation. In this example, a felsic vein injected into banded gneiss during the waning stages of F₁ folding crosscuts isoclinal folds and has been tightly folded along the same axial planes. Veins of foliated tonalite that crosscut isoclinally folded banded gneiss at Stevens Point Dam have been folded in a similar manner.

Figure 10. Isoclinal F₁ fold deformed by F₂, Z-style fold, in banded gneiss, Conants Rapids. Penetrative F₁ axial plane (S₁) and nonpenetrative F₂ axial plane indicated by dashed lines.

of foliated tonalite, which have intruded isoclinally folded banded gneiss at moderate to high angles to F₁ axial planes, have been tightly folded along the same axial planes. An analogous situation is illustrated in Figure 9, in which an aplite vein that intruded isoclinally folded banded gneiss has been tightly folded.

Two inclusions of banded gneiss in foliated tonalite provide further evidence that the most intense phase of F₁ folding occurred prior to the emplacement of foliated tonalite, but that F₁ deformation continued subsequent to intrusion. Rotation of the inclusions during intrusion positioned the axial planes of isoclinal folds within the inclusions approximately normal to foliation in foliated tonalite. The axial planes of the isoclinal folds were then tightly folded. The axial planes of the tight folds are parallel to foliation in the surrounding foliated tonalite, demonstrating that the tight folds are associated with deformation of foliated tonalite. Open folding of the diorite gneiss-foliated tonalite contact also attests to F₁ deformation of foliated tonalite.

The axial planes of F₁ folds define a penetrative foliation, S₁, which is parallel to compositional layering, S₀, in banded gneiss and migmatite, except in the hinges of folds, where S₀ is transected by S₁. Transposition of compositional layering during F₁ folding, as evidenced by the occurrence of intrafolial folds and rootless fold hinges in banded gneiss and migmatite, has resulted in the development of S₁. Where banded gneiss or migmatite are in contact with foliated tonalite, S₁ passes uninterrupted from one unit across the contact into the other.

The second set of folds, F₂, are disharmonic, intrafolial, S- and Z-style folds (Fig. 10), which occur in all lithologic units older than the amphibolite dikes. The axial planes, S₂, of F₂ folds are nonpenetrative features generally oblique to S₁, which wraps around F₂ fold hinges. F₂ folds are scarce, lack a consistent sense of rotation, and are similar or concentric in style. The folds rarely exceed 20 cm in wavelength or amplitude.

The youngest fold set, F₃, has affected all lithologic units in the area except for the diabase. F₃ deformation is characterized by broad, open, disharmonic folds of concentric style, with amplitudes ranging from centimetres to tens of metres, and wavelengths generally ten or more times the fold amplitude. F₃ axial planes are nonpenetrative features oriented approximately perpendicular to S₁, which wraps around the hinges of F₃ folds.

Geometric relationships among the various mesoscopic structural elements are best displayed at Conants Rapids, where exposure is extensive enough for a significant number of structural measurements to be taken. In banded gneiss, poles to S₁ are distributed in a girdle that defines a fold axis, β, bearing S22°E and plunging 61° (Fig. 11A). All three sets of folds (F₁ through F₃) are coaxial, and fold axes are parallel to the statistically defined fold axis, β (Fig. 11B). In addition, mineral lineations and crenulations in banded gneiss are parallel to fold axes and β (Fig. 11C).

Mineral lineations in lineated tonalites at Conants Rapids are parallel to the linear elements in banded gneiss, although greater scatter is exhibited by the tonalites (Fig. 11D). Part of this scatter is due to inhomogeneity on the scale of the map area at Conants Rapids; measurements of lineations in lineated tonalites were collected over the entire map area, whereas measurements in banded gneiss were taken only from the northernmost outcrops. However, part of the scatter may reflect the fact that lineations could not be measured directly in the tonalites but had to be obtained by plotting the trace of lineation on three nonparallel surfaces on a stereonet. In some cases the surfaces involved were very irregular, and the resulting triangles of error in locations of the lineation were rather large.

Mineral lineations in amphibolite dikes at Conants Rapids are close to vertical, falling within the field for mineral lineations in lineated tonalite. Measurements of lineation in amphibolite dikes were collected only at the southern end of Conants Rapids, where the lineation in lineated tonalite is also nearly vertical.

At Biron Dam, poles to foliation (S₁) in foliated tonalite define a fold axis, β, bearing N70°E and plunging 80° (Fig. 12A). Mineral lineations in lineated tonalite, amphibolite, and aplite are parallel to β (Fig. 12B). It was difficult to determine the orientation of mineral lineations in foliated tonalite because of the steep

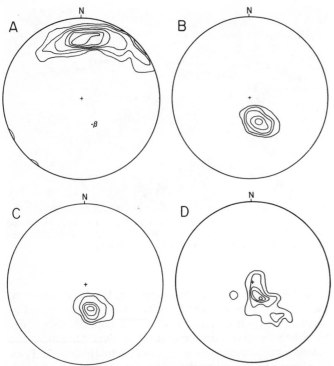

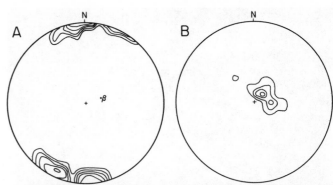

Figure 12. (A) Poles to foliation (S₁) in foliated tonalite, Biron Dam; 31 measurements; contours in percent per 1% area: 3, 6, 10, 15, 20. (B) Mineral lineation in lineated tonalite, amphibolite dikes, and aplite, Biron Dam; 11 measurements; contours in percent per 1% area: 9, 18, 27, 36.

Figure 11. (A) Poles to S₁ in banded gneiss, Conants Rapids; 125 measurements; contours in percent per 1% area: 1, 2.5, 5, 10, 15, 20. (B) F₁, F₂, and F₃ fold axes in banded gneiss, Conants Rapids; 44 measurements; contours in percent per 1% area: 2, 10, 20, 40, 60. (C) Mineral lineation and crenulations in banded gneiss, Conants Rapids; 51 measurements; contours in percent per 1% area: 2, 10, 30, 50, 70. (D) Mineral lineation in lineated tonalite, Conants Rapids; 18 measurements; contours in percent per 1% area: 5.6, 11.1, 16.6, 22.2, 27.7. Mesoscopic and microscopic structrual diagrams; all are lower-hemisphere projections.

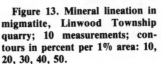

Figure 13. Mineral lineation in migmatite, Linwood Township quarry; 10 measurements; contours in percent per 1% area: 10, 20, 30, 40, 50.

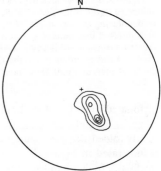

dip of foliation in the tonalite and the low topographic relief of tonalite outcrops, but in the few places where mineral lineations could be measured, they were found to be parallel to mineral lineations in the other lithologic units.

Although detailed analysis of structures at the Linwood Township quarry has not yet been completed, it is clear that geometric relations there are more complex than at Conants Rapids or Biron Dam. Isoclinal fold axes in migmatite plunge at divergent angles, ranging from shallow to the northwest, through vertical, to shallow to the southeast. In contrast, mineral lineations in migmatite (Fig. 13) plunge consistently to the southeast (S30°E, 65°), parallel to mineral lineations in lineated tonalite at the quarry, and to mineral lineations in banded gneiss and lineated tonalite at Conants Rapids. Where isoclinal fold axes plunge at shallow angles, steeply plunging mineral lineations pass through the fold hinges; therefore, the mineral lineations are evidently younger than some isoclinal folds in migmatite.

## MICROSCOPIC ANALYSIS

Similar orientations of penetrative mesoscopic structures exist in the different lithologic units investigated here, and petrofabric analyses, utilizing a Zeiss four-axis universal stage, have been per-

formed on selected samples to test whether similarities exist on the microscopic scale as well. In subsequent figures, *a, b,* and *c* refer to mesoscopic fabric axes, *b* being parallel to lineation, *c,* normal to foliation, and *a,* normal to *b* in the plane of foliation. In each of the samples depicted, the fabric axes are clearly defined; however, in lineated tonalite the *a* and *c* fabric axes may be located only approximately if foliation is weakly developed, or in the absence of foliation the *a* and *c* fabric axes lose any physical significance.

Foliation in aplite, as described previously, is parallel to walls of the aplite dike and represents a late, cataclastic foliation, rather than regional foliation.

### Quartz

The orientation of [0001] in quartz has been measured in samples of banded gneiss, foliated tonalite, and lineated tonalite from Conants Rapids (Fig. 14A, 14B, 14C), and in a sample of aplite from Biron Dam (Fig. 14D), which has intruded an amphibolite dike and is one of the youngest deformed units in the terrane. There were 600 grains measured in the sample of banded gneiss, and 300 grains in the samples of foliated tonalite, lineated tonalite, and aplite. All of the petrofabric diagrams are similar and feature a girdle normal to the *b* fabric axis and exhibit monoclinic symmetry. Petrofabric analyses of four additional samples of banded gneiss and lineated tonalite (not illustrated here) have yielded similar patterns.

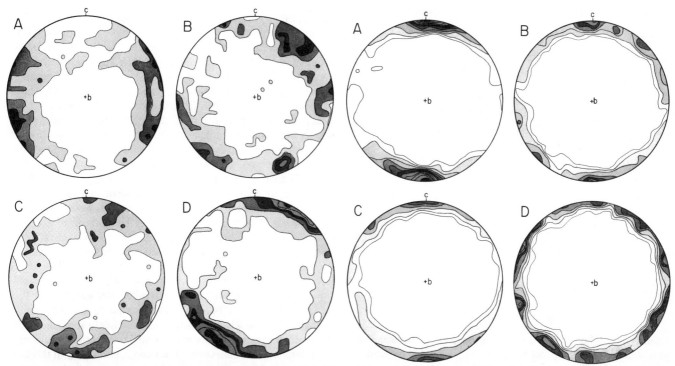

Figure 14. (A) [0001] of quartz in banded gneiss, Conants Rapids; 600 measurements; contours in percent per 1% area: 1, 2, 3. (B) [0001] of quartz in foliated tonalite, Conants Rapids; 300 measurements; contours in percent per 1% area: 1, 2, 3, 4. (C) [0001] of quartz in lineated tonalite, Conants Rapids; 300 measurements; contours in percent per 1% area: 1, 2, 3. (D) [0001] of quartz in aplite, Biron Dam; 300 measurements; contours in percent per 1% area: 1, 2, 3, 4, 5, 6, 7.

Figure 15. (A) Poles to (001) of biotite in banded gneiss, Conants Rapids; 200 measurements; contours in percent per 1% area: 1, 2, 4, 6, 8, 10, 12, 14, 16, 18, 20. (B) Poles to (001) of biotite in foliated tonalite, Conants Rapids; 200 measurements; contours in percent per 1% area: 1, 2, 4, 6, 8, 10, 12. (C) Poles to (001) of biotite in lineated tonalite, Conants Rapids; 200 measurements; contours in percent per 1% area: 1, 2, 4, 6, 8, 10; early mafic-rich member of lineated tonalite series. (D) Poles to (001) of biotite in lineated tonalite, Conants Rapids; 200 measurements; contours in percent per 1% area: 1, 2, 3, 4, 5, 6, 7; intermediate-aged, less mafic-rich member of lineated tonalite series.

The preferred orientation of quartz in aplite (Fig. 14D) is much stronger than that in the other samples, possibly due to the greater amount of quartz in aplite (35%) compared to banded gneiss, foliated tonalite, and lineated tonalite (<25% in each case), but more likely related to the preferential relief of late stresses in aplite and pegmatite dikes, with resulting cataclastic deformation of the dikes.

## Biotite

Poles to (001) of biotite have been measured in four samples in order (1) to document the decrease in degree of foliation in passing from banded gneiss to foliated tonalite to lineated tonalite, (2) to test whether foliation is truly absent in some members of the lineated tonalite series, and (3) to ascertain whether the monoclinic symmetry exhibited by quartz is shared by biotite.

The degree of foliation in banded gneiss (Fig. 15A) is greater than that in foliated tonalite (Fig. 15B) and that in a mafic-rich, early member of the lineated tonalite series (Fig. 15C). Foliation is absent from a less mafic-rich lineated tonalite (Fig. 15D) of intermediate age in the tonalite series. Late, mafic-poor members of the lineated tonalite series do not contain sufficient biotite for analysis but also appear to lack foliation, both in hand specimen and thin section. The mesoscopic fabric of most lineated tonalites resembles the mesoscopic fabric of the specimen used to generate Figure 15D. Like the petrofabric diagrams for [0001] in quartz, all four diagrams for (001) in biotite display a girdle with monoclinic symmetry oriented normal to the b fabric axis.

## DISCUSSION

Our interpretation of the sequence and timing of intrusive and deformational events that have affected the Precambrian terrane between Stevens Point and Wisconsin Rapids, based on a correlation of isotopic ages and outcrop relations, is summarized in Table 2.

Two possibilities exist for the timing of $F_1$ folding that produced steeply plunging, tight to isoclinal folds in banded gneiss and migmatite. $F_1$ may have occurred during Archean time, shortly after injection of leucosomal melt into migmatite, or alternatively, during the Penokean orogeny, just prior to intrusion of foliated tonalite. We prefer to assign $F_1$ folding to the Penokean orogeny owing to the parallelism of linear and planar $F_1$ structures in banded gneiss and migmatite with linear and planar structures contained in Penokean intrusives. $F_1$ fold axes are parallel to $F_2$ and $F_3$ fold axes and to the pervasive mineral lineation found in Penokean intrusives. $F_1$ axial-plane foliation in banded gneiss and migmatite is parallel to foliation in 1,840-m.y.-old foliated

TABLE 2.  GEOLOGIC HISTORY OF THE PRECAMBRIAN
TERRANE IN THE WISCONSIN RIVER REGION

| Rock units | Folding events |
|---|---|
| Late Proterozoic(?) | |
| Diabase | |
| Early Proterozoic | |
| | $F_3$, Broad open folds |
| Aplites | |
| Amphibolite dikes | |
| Pegmatites and aplites | |
| | $F_2$, S- and Z-style folds |
| Lineated tonalites (1,820 m.y.) | |
| Pegmatites and aplites | Tight to open folds |
| Foliated tonalite (1,840 m.y.) | $F_1$ |
| Diorite gneiss | Isoclinal to tight folds |
| Archean | |
| Migmatite (>2,800 m.y.) | Isoclinal folds(?) |
| Pegmatites | |
| Banded gneiss and associated amphibolite | |

tonalite, and where banded gneiss or migmatite are in contact with foliated tonalite, foliation passes through the contact uninterrupted.

If $F_1$ folding is interpreted as an Archean event, it becomes necessary to explain the parallelism of structures that formed during orogenies separated in time by 1 b.y. or more. The possibility that Penokean structures were controlled by preexisting Archean structures has been considered; however, the terrane is composed almost entirely of Penokean intrusive rocks, and therefore, there is very little Archean material to control later structures.

We have assigned $F_2$ and $F_3$ folding to the Penokean orogeny because of the parallelism of $F_1$, $F_2$, and $F_3$ fold axes and the outcrop relations among $F_2$ and $F_3$ folds and the various intrusive rocks. However, at present there is no way to set a younger limit on the age of $F_2$ and $F_3$ folds, other than to say that $F_2$ is older than the amphibolite dikes and $F_3$ is older than the diabase dike at Biron Dam (for which an age has not yet been determined). The fresh mineralogy and igneous texture of the diabase suggest that it is probably Keweenawan in age (about 1,100 m.y. old).

It is possible, however, that $F_2$ and $F_3$ folds developed after the Penokean orogeny. Throughout the Great Lakes region Rb-Sr whole-rock and mineral ages have been reset at about 1,650 to 1,700 m.y. ago, including Rb-Sr whole-rock ages for tonalites of the Wisconsin River area (Van Schmus and others, 1975b). Possible structural manifestations of this middle Proterozoic event are recorded in folded 1,765-m.y.-old volcanic rocks in south-central Wisconsin (Smith, 1978). Perhaps $F_2$ and $F_3$ folds are physical expressions of this event, resulting from differential vertical (?) block movements of crustal segments (P. K. Sims, 1978, personal commun.). However, we feel that $F_2$ and $F_3$ folds probably are not correlative with folding in the 1,765-m.y.-old volcanic rocks because of significant differences in style and orientation of folding. Additional isotopic work is needed to bracket more precisely the age of $F_2$ and $F_3$ folds, and paleostrain analyses should be undertaken to establish the strain history for this terrane.

Archean banded gneiss and migmatite and lower Proterozoic

intrusive rocks in this terrane have been pervasively recrystallized under amphibolite-facies metamorphic conditions. This metamorphism is thought to have occurred during the Penokean orogeny because mineral lineations produced during this event are parallel to the fold axes ($F_1$, $F_2$, and $F_3$), and the crystallographic orientations of quartz and biotite grains are systematically related to these mineral lineations (b fabric axes) and fold axes.

In the Wisconsin River region, deformation of Archean age has not yet been established unequivocally. However, at the Linwood Township quarry, steeply plunging Penokean mineral lineations transect the axes of gently plunging isoclinal folds, signifying that the folds are older. Furthermore, the gentle plunge of these folds is at variance with the consistent Penokean pattern of steeply plunging structures, suggesting that these folds were not formed during early Proterozoic time and may be Archean in age. The possibility that an Archean deformational episode has affected this terrane is supported by reconnaissance studies 30 to 80 km to the west along the Yellow and Black Rivers. Along these rivers isoclinally folded Archean gneisses, migmatites, and iron formation are intruded by undeformed to mildly deformed lower Proterozoic (1,820 to 1,842-m.y.-old) granitic to tonalitic rocks. Along the Yellow River the intrusives are undeformed; along the Black River they vary from undeformed to slightly granulated.

In summary, we believe that the Penokean orogeny in this part of Wisconsin began with isoclinal folding of an Archean gneiss terrane, continued with less intense deformation that accompanied the early Proterozoic intrusion of diorite, tonalite, aplite, pegmatite, and diabase (now amphibolite), and concluded with broad, open folding. All fold sets are characterized by steeply plunging axes. Amphibolite-facies metamorphism accompanied deformation and resulted in recrystallization of Penokean igneous rocks as well as Archean gneiss. In this region both Archean and lower Proterozoic rocks were folded during the Penokean orogeny, in contrast with the situation in the Upper Peninsula of Michigan, where the Archean basement behaved as rigid blocks during Penokean folding of superincumbent lower Proterozoic volcanic and sedimentary rocks (Cannon, 1973; Klasner, 1978).

Further studies of this type should provide a basis for establishing the extent and character of the Penokean orogeny in Wisconsin and provide a better understanding of the tectonic evolution of the Lake Superior region during early and middle Proterozoic times.

## ACKNOWLEDGMENTS

We thank Paul Sims and Cam Craddock for their perceptive and constructive reviews of this paper. Financial support was provided by the Wisconsin Alumni Research Foundation of the University of Wisconsin (to Maass and Medaris) and National Science Foundation Grants GA-15951, GA-43426, and EAR-15007 (to Van Schmus).

## REFERENCES CITED

Anderson, J. L., 1972, Petrologic study of a migmatite-gneiss terrain in central Wisconsin and the effect of biotite-magnetite equilibria on partial melts in the granite system [M.S. thesis]: Madison, University of Wisconsin, 103 p.

Anderson, J. L., and Cullers, R. L., 1978, Geochemistry and evolution of the Wolf River batholith, a late Precambrian rapakivi massif in north Wisconsin, U.S.A.: Precambrian Research, v. 7, p. 287–324.

Cannon, W. F., 1973, The Penokean orogeny in northern Michigan, *in* Young, G. M., ed., Huronian stratigraphy and sedimentation: Geological Association of Canada Special Paper 12, p. 251–271.

Craddock, C., 1972, Keweenawan geology of east-central and southeastern Minnesota, *in* Sims, P. K., and Morey, G. B., eds., Geology of Minnesota: A centennial volume: Minnesota Geological Survey, p. 416–424.

Dott, R. H., and Dalziel, I.W.D., 1972, Age and correlation of the Precambrian Baraboo Quartzite of Wisconsin: Journal of Geology, v. 80, p. 552–568.

Klasner, J. S., 1978, Penokean deformation and associated metamorphism in the western Marquette Range, northern Michigan: Geological Society of America Bulletin, v. 89, p. 711–722.

Maass, R. S., 1977, Structure and petrology of an early and middle Precambrian gneiss terrane between Stevens Point and Wisconsin Rapids, Wisconsin [M.S. thesis]: Madison, University of Wisconsin, 128 p.

Sims, P. K., 1976, Precambrian tectonics and mineral deposits, Lake Superior region: Economic Geology, v. 71, p. 1092–1118.

Smith, E. I., 1978, Precambrian rhyolites and granites in south-central Wisconsin: Field relations and geochemistry: Geological Society of America Bulletin, v. 89, p. 875–890.

Van Schmus, W. R., 1976, Early and middle Proterozoic history of the Great Lakes area, North America: Royal Society of London Philosophical Transactions, sec. A, v. 280, p. 605–628.

——1980, Chronology of igneous rocks associated with the Penokean orogeny in Wisconsin, *in* Morey, G. B., and Hanson, G. N., eds., Selected studies of Archean gneisses and lower Proterozoic rocks, southern Canadian Shield: Geological Society of America Special Paper 182 (this volume).

Van Schmus, W. R., and Anderson, J. L., 1977, Gneiss and migmatite of Archean age in the Precambrian basement of central Wisconsin: Geology, v. 5, p. 45–48.

Van Schmus, W. R., Medaris, L. G., and Banks, P. O., 1975a, Geology and age of the Wolf River batholith, Wisconsin: Geological Society of America Bulletin, v. 86, p. 907–914.

Van Schmus, W. R., Thurman, M. E., and Peterman, Z. E., 1975b, Geology and Rb-Sr chronology of middle Precambrian rocks in eastern and central Wisconsin: Geological Society of America Bulletin, v. 86, p. 1255–1265.

Weidman, S., 1907, The geology of north-central Wisconsin: Wisconsin Geological and Natural History Survey Bulletin, v. 16, 697 p.

MANUSCRIPT RECEIVED BY THE SOCIETY APRIL 25, 1979
REVISED MANUSCRIPT RECEIVED NOVEMBER 26, 1979
MANUSCRIPT ACCEPTED JANUARY 11, 1980

Geological Society of America
Special Paper 182
1980

# Chronology of igneous rocks associated with the Penokean orogeny in Wisconsin

W. R. VAN SCHMUS

*Department of Geology, University of Kansas, Lawrence, Kansas 66045*

## ABSTRACT

U-Pb analyses have been carried out on 36 zircon fractions representing 23 lower Proterozoic plutonic and volcanic units from throughout Wisconsin. These data, combined with published results from northeastern Wisconsin, allow detailed definition of the chronology of igneous events in Wisconsin associated with the Penokean orogeny. Two distinct pulses can be recognized. The older one began with mafic to felsic volcanism 1,859 ± 20 m.y. ago and was followed immediately by tonalitic to granitic plutonism 1,840 to 1,820 m.y. ago. Structural studies indicate that these rocks were emplaced during the main phase of the orogeny. The second pulse consisted predominantly of rhyolite and granophyric granite and occurred about 1,760 ± 10 m.y. ago. No Proterozoic igneous units with zircon ages in excess of 1,860 m.y. have been found to date in Wisconsin, nor have any been found with ages on the order of 1,615 to 1,630 m.y., the time of widespread alteration of Rb-Sr geochronologic systems in the region.

## INTRODUCTION

The general Precambrian geology of the western Great Lakes region has been reviewed recently by Card and others (1972), Van Schmus (1976), and Sims (1976a). The area of interest in this paper comprises the exposed or near-surface Precambrian rocks of Wisconsin (Fig. 1), a region that represents the south-central part of the Penokean fold belt of the Lake Superior region. The Penokean orogeny, as redefined by Goldich and others (1961), is a major tectonic and petrogenetic event in this region and was originally placed by these authors in the time interval 1,600 to 1,800 m.y. ago. Subsequent studies (Goldich, 1972a, 1972b; Van Schmus, 1972) placed the age closer to 1,800 to 1,900 m.y. ago, and recent work (Van Schmus, 1974, 1976; Van Schmus and others, 1975b) has shown that much of the igneous terrane of Wisconsin belongs to the Penokean orogeny.

Because of the limited distribution of outcrops in this region, correlation of geologic units based on field relationships is difficult. Consequently it is necessary to use radiometric

geochronology extensively in order to establish the basic time framework for the geologic history of the region and to correlate precisely specific events, such as the igneous activity associated with the Penokean orogeny. However, previous studies, based largely on Rb-Sr methods, have shown that many rock units in Wisconsin cannot be accurately dated by that method (Van Schmus and others, 1975b).

For the past several years I have been doing U-Pb analyses of zircons separated from many of the igneous rocks associated with the Penokean orogeny. At present, sufficient data have been obtained to show that the igneous events associated with the main pulse of the Penokean orogeny occurred about 1,860 to 1,820 m.y. ago, and that this was followed by a later pulse of more felsic igneous activity about 1,760 m.y. ago. This represents a significant refinement in our knowledge of the time framework for the Penokean orogeny. The basis for these conclusions is presented in this paper, along with additional details on the correlation of rock types with ages. In addition to providing a sound framework for the chronology of the Penokean orogeny, these results will also be invaluable in deciphering the complex history of Archean gneiss and migmatite that occur throughout central Wisconsin (Van Schmus and Anderson, 1977).

## GEOLOGIC SETTING

The generalized Precambrian geology of Wisconsin (Fig. 1) is modified from Dutton and Bradley (1970), Van Schmus and others (1975a, 1975b), Van Schmus (1976), Sims (1976a), and Van Schmus and Anderson (1977). The oldest rocks are Archean (> 2,500 m.y.) and can be roughly divided into two regions. One is the northwestern region where granitic, metavolcanic, and gneissic rocks (unit 7, Fig. 1) underlie the Gogebic Range and extend southeastward for at least 40 km. These rocks are for the most part probably about 2,600 to 2,800 m.y. old. This estimate is based on unpublished data by S. Chaudhuri of Kansas State University and me and correlation of these rocks with similar rocks in Upper Michigan (Sims and others, 1977). Apparently these rocks belong to the greenstone-granite terrane of Morey and Sims (1976), although the boundary between this terrane and their

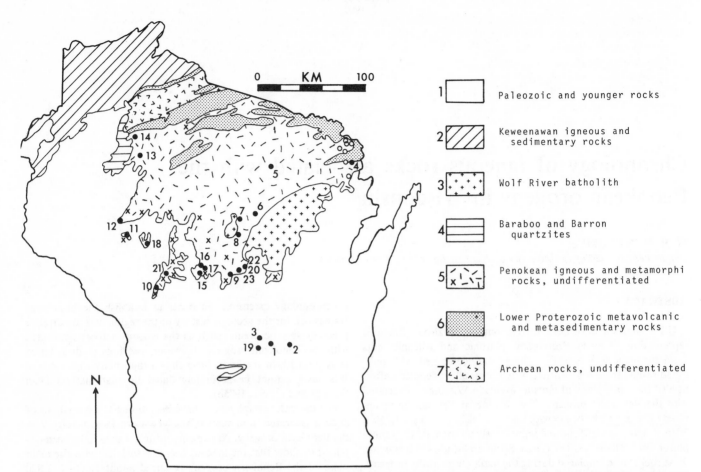

**Figure 1. Generalized geologic map of Wisconsin showing major Precambrian terranes and distribution of units for which U-Pb analyses on separated zircons are available. Sample localities summarized in Table 1 and Appendix 1; open circles from Banks and Cain (1969). Unit 1, undifferentiated Paleozoic and younger rocks; unit 2, Keweenawan igneous and sedimentary rocks, undifferentiated; unit 3, Wolf River batholith, 1,500 m.y. old; unit 4, Baraboo and Barron quartzites; unit 5, undifferentiated igneous and metamorphic rocks, contains Penokean plutonic complexes, remnants of middle Precambrian metavolcanic and metasedimentary rocks, and remnants or windows of Archean gneiss and migmatite (denoted by "x" where known); unit 6, middle Precambrian metavolcanic and metasedimentary rocks 1,850 to 1,950 m.y. old; unit 7, Archean basement rocks.**

gneiss-migmatite terrane to the south may lie within this area (Sims, 1976a).

The other region containing Archean rocks is central and west-central Wisconsin. In this region the Archean rocks consist predominantly of felsic gneiss, migmatite, and amphibolitic gneiss (Weidman, 1907; Myers, 1974; Dutton and Bradley, 1970). Field and geochronologic studies (Anderson, 1972; Sherwood, 1976; Van Schmus and Anderson, 1977; Maass and others, this volume) indicate that the Archean rocks have been extensively intruded by Penokean plutonic rocks and remetamorphosed and deformed during the Penokean orogeny. In fact, it is not clear whether this terrane can be best described as one consisting of an Archean basement with many Penokean plutons or as a Penokean plutonic complex with many remnants of Archean crust. Thus, although data available (Van Schmus and Anderson, 1977) indicate that these rocks are possibly an eastward extension of the ancient gneiss-migmatite terrane in Minnesota (Morey and Sims, 1976; Sims, 1976a), their extensive reworking during the Penokean orogeny has made deciphering of their original character very difficult. Therefore, a complete understanding of the Penokean history of these rocks is necessary in order to isolate pre-Penokean characteristics.

Most of the exposed Precambrian rock in Wisconsin consists of metavolcanic, metasedimentary, and plutonic rocks of early Proterozoic age and can be separated into three general suites. The oldest of these consists of metavolcanic and lesser metasedimentary rocks (unit 6, Fig. 1). These occur extensively in northern Wisconsin, where the metavolcanic rocks range from basalt through rhyolite in composition and contain the massive sulfide deposits recently discovered in Wisconsin (Sims, 1976a). Scattered occurrences of metavolcanic rock also exist throughout the state, but are too small to indicate on Figure 1.

The second suite consists of plutonic rocks that range in composition from tonalite through quartz monzonite. These rocks intrude the lower Proterozoic metavolcanic and metasedimentary rocks and the Archean gneissic terrane and occur throughout the entire Precambrian basement of Wisconsin. The first and second suites comprise the rocks most definitely associated with the main phase of the Penokean orogeny in Wisconsin (Van Schmus and others, 1975b; Van Schmus, 1976).

The third suite of rocks consists of rhyolitic volcanic rocks and associated granites that are exposed principally as inliers in southern Wisconsin (Asquith, 1964; Van Schmus and others, 1975b; Smith, 1978) and in the Baraboo area (Dalziel and Dott,

1970). Ages recently obtained on these rocks (Van Schmus, 1978, and this report) indicate that they are about 80 m.y. younger than the main orogenic suites mentioned above.

Subsequent to the rhyolitic volcanism 1,760 m.y. ago, extensive deposits of sandstone (now quartzite) were formed (Dott and Dalziel, 1972). These were subsequently deformed, along with the underlying rhyolite, some time prior to the emplacement of the Wolf River batholith, 1,500 m.y. ago (Van Schmus and others, 1975a). However, there was also a significant event 1,615 to 1,630 m.y. ago that reset Rb-Sr systems in many of the older units (Van Schmus and others, 1975b; Van Schmus, 1978), and if this resetting occurred during the deformation of the sandstones, then these rocks are older than 1,630 m.y. The youngest igneous activity was Keweenawan and occurred about 1,100 m.y. ago (Silver and Green, 1972). For the most part these post-Penokean events have had little to no effect on zircon U-Pb age interpretations for the Penokean units.

## ANALYZED SAMPLES

The rocks reported on here represent a wide range of lithology and tectonic setting and are from widely distributed localities throughout the state (Fig. 1; App. 1). They were collected over several field seasons and represent both reconnaissance sampling of isolated occurrences of igneous rocks and of specific units selected on the basis of regional geology. This suite of rocks can be considered generally representative of early Proterozoic igneous rocks in Wisconsin with reference to ages, compositions, and tectonic settings. As such, it provides an excellent basis for outlining the history of igneous activity in Wisconsin associated with the Penokean orogeny.

## ANALYTICAL PROCEDURES

The analyses reported in this paper were obtained in the geochronology laboratories of the University of Kansas. A Wilfley table was used to separate heavy mineral concentrates from 50- to 100-kg samples of crushed and ground rock. Zircon was separated from the concentrates and purified with the use of heavy liquids, a magnetic separator, and acid washing. A Franz isodynamic separator was used to split the purified zircon separates into various magnetic fractions, and the four least magnetic (hence, generally most concordant) fractions were reserved for analysis. Some of the magnetic fractions were also sieved into +100, 100 to 200, and −200 mesh-size fractions. In a few instances, insufficient zircon was obtained for splitting into fractions, and the total population was retained as a single sample.

After final purification by hand picking, 5 to 15 mg of a zircon fraction were dissolved and U and Pb were separated using the procedures of Krogh (1973). Isotopic analyses were carried out on a 6-in. Nier-type mass spectrometer equipped with programmable magnetic field scanning and digital output. Analytical blanks were typically low ($2 \times 10^{-9}$ g total Pb and $1 \times 10^{-9}$ g total U or less) and represented less than a few percent of the total nonradiogenic Pb in most samples. All Pb data were corrected using (204):(206):(207):(208) ratios of (1.0):(18.6):(15.8):(39.0) for

blank Pb and an appropriate original common Pb composition based on the two-stage model of Stacey and Kramers (1975) and the age of the sample concerned. In all cases the radiogenic $^{206}$Pb and $^{207}$Pb concentrations and, particularly, the radiogenic $^{207}$Pb/$^{206}$Pb ratio (and hence the age) were not significantly dependent (± 0.1% or less) on reasonable variations of the nonradiogenic Pb composition used.

Total analytical precision for the data presented is estimated at ± 2% to 3% for U/Pb ratios and ± 0.5% to 1% for radiogenic $^{207}$Pb/$^{206}$Pb ratios, and results in a corresponding uncertainty of 0.5% to 1% for individual Pb-Pb ages obtained. Where appropriate, concordia intercept ages were determined by least-squares fitting of a line to the data using the method of York (1966). Uncertainties for these cases are given at the 95% confidence level. Ages were calculated using the decay constants of Jaffey and others (1971): $\lambda$ ($^{238}$U) = $1.5513 \times 10^{-10}$ yr$^{-1}$ and $\lambda$ ($^{235}$U) = $9.8485 \times 10^{-10}$ yr$^{-1}$. For internal consistency any Rb-Sr ages quoted in discussion have been recalculated using $\lambda$ ($^{87}$Rb) = $1.42 \times 10^{-11}$ yr$^{-1}$ (Steiger and Jäger, 1977). Consequently, ages discussed in this paper will be 1.5% to 2% younger than those given in previous reports on these and other rocks in the Great Lakes area (Goldich, 1968; Van Schmus, 1976; Sims, 1976a).

## RESULTS

As mentioned above, several suites of rock can be recognized throughout Wisconsin. In the presentation below, the results are broadly grouped according to similarities in age and petrology, but this should not be taken to mean that all of the individual units within each group are closely related petrogenetically.

### Tonalite-Trondhjemite Suite

Throughout the state the older Penokean plutonic complexes range in composition from quartz-monzonite through tonalite. Because of the close geographic association of the more sodic rocks with the gneiss-migmatite terrane in central Wisconsin (Anderson, 1972; Maass and others, this volume), these rocks have been extensively analyzed and are represented by samples 8, 9, 11, 12, 13, 15, 20, 22, and 23 (Figs. 1, 2; Tables 1, 2). Also included is the Marinette Quartz Diorite in northeastern Wisconsin (Aldrich and others, 1965; "MQD" in Fig. 2). Within this grouping, three subgroupings can be recognized: medium-grained foliated tonalite; fine-grained "lineated tonalite"; and leucotonalite or trondhjemite.

From Figure 2 it can be seen that ages defined by zircons from the tonalitic rocks appear to range from about 1,820 to 1,840 m.y. This spread in ages is considered to be real, first because it appears to be distinctly outside analytical uncertainty, and second because the relative ages so obtained agree with field relationships and structural data presently available. In particular, Anderson (1972) and Maass and others (this volume) have concluded that there are two main tonalite suites in central Wisconsin, which are best exposed along and near the Wisconsin River from Stevens Point to Wisconsin Rapids. The older of these is medium grained and well foliated (almost gneissic) and has been referred to as "foliated tonalite" or "blastic tonalite." This unit generally occurs as large bodies and is represented by sample 9, collected at

TABLE 1. SUMMARY OF SAMPLE DESCRIPTIONS AND LOCATIONS

| Sample no. | Original field no. | Rock type | Approximate location |
|---|---|---|---|
| 1 | VS70-7 | Quartz porphyry rhyolite | Noble Quarry, Green Lake County |
| 2 | VS73-1 | Quartz porphyry rhyolite | Utley Quarry, Green Lake County |
| 3 | VS73-2 | Granophyric granite | Montello Quarry, Marquette County |
| 4 | VS73-8 | Amberg Quartz Monzonite | Quarry near Amberg, Marinette County |
| 5 | VS73-11 | Granite near Monico | South of Monico, Oneida County |
| 6 | VS73-16 | Kalinke Quartz Monzonite | Northeast of Wausau, Marathon County |
| 7 | VS73-17 | Rhyolite porphyry | Wausau, Marathon County |
| 8 | VS73-18 | Tonalite | Near Mosinee, Marathon County |
| 9 | VS73-21 | Foliated tonalite | Biron Dam, Wood County |
| 10 | VS73-22 | Granite | Black River Falls, Jackson County |
| 11 | VS73-25 | Tonalite | Little Falls, Eau Claire County |
| 12 | VS73-26 | Leucotonalite | Chippewa Falls, Chippewa County |
| 13 | VS73-34 | Leucotonalite | North of Ladysmith, Rusk County |
| 14 | VS73-37 | Porphyritic granite | Radisson, Sawyer County |
| 15 | VS74-2 | Dacite dike | Quarry near Pittsville, Wood County |
| 16 | VS74-6 | Granite | North of Pittsville, Wood County |
| 17 | VS74-9 | Quartz Monzonite | North of Pittsville, Wood County |
| 18 | VS74-10 | Quartz porphyry felsite | Eau Claire River, Eau Claire County |
| 19 | VS74-13 | Quartz porphyry rhyolite | Observatory Hill, Marquette County |
| 20 | VS75-8 | Tonalite dike | Conant's Rapids, Portage County |
| 21 | VS76-26 | Granite | Arbutus Lake, Clark County |
| 22 | VS75-9E | Tonalite | Near Conant's Rapids, Portage County |
| 23 | VS75-7 | Tonalite dike | Linwood Quarry, Portage County |

Figure 2. U-Pb discordia plot for zircons from tonalitic-trondhjemitic rocks in Wisconsin. Sample numbers as in Figure 1 and Table 1. Cross is for zircon from the Marinette Quartz Diorite in northeastern Wisconsin (Aldrich and others, 1965). Chords shown are least-squares fit to data from relatively older, foliated tonalite (1,842 ± 10 m.y.) and younger, cross-cutting dikes (1,824 ± 25 m.y.). Ages are quoted at the 95% uncertainty level. See text for further discussion.

Biron Dam along the Wisconsin River. Sample 11, from Eau Claire County, is similar in general lithologic character. Zircons from these two samples yield a concordia intercept age of 1,842 ± 10 m.y. (Fig. 2).

The younger suite is generally fine grained, less deformed, and commonly occurs as dikes cutting older rocks, including the "foliated tonalite." The younger suite is generally referred to as "lineated tonalite" in the Wisconsin River area and is represented by samples 20 and 22; sample 15 is a dacite dike from farther west, near Pittsville. Zircons from these samples yield a concordia intercept age of 1,824 ± 25 m.y. (Fig. 2), consistent with the fact that these rocks postdate the "foliated tonalite." Because only 1 to 3 zircon fractions have been analyzed from each rock, it is not possible to define precisely the true difference in the ages; they are indistinguishable at the 2σ level of uncertainty.

However, zircons from the younger, cross-cutting rocks consistently plot to the left (or to the young side) of those from the older, more deformed rocks. Thus, in spite of overlapping uncertainty, the apparent 20- to 30-m.y. range in age is considered real. Final resolution of this question will have to await additional and more precise analyses.

The more leucocratic sodic rocks range from leucotonalite to trondhjemite and are represented by samples 8, 12, 13, and 23. These units occur throughout the region and often appear to be discrete plutonic masses. Their precise ages are currently not well defined, but the results presented in Figure 2 suggest ages intermediate to the main tonalite suites for several of these units.

## Older Granitic Suite

Typical mesozonal plutonic rocks ranging from granite to granodiorite, including quartz monzonite, occur extensively throughout central and northern Wisconsin. These plutons commonly occur associated with metavolcanic rocks that are predominantly mafic in character but range to rhyolite in com-

position, particularly in northeastern Wisconsin (Banks and Cain, 1969; Van Schmus and others, 1975b) and in central Wisconsin near Wausau (LaBerge and Myers, 1973; Van Schmus and others, 1975b). In the southern portions of the exposed Precambrian, the plutons are primarily associated with Archean gneiss and migmatite, which they intrude (Sherwood, 1976; Van Schmus and Anderson, 1977). Samples 10, 16, 17, and 21 intrude Archean rocks; sample 6 is from the Wausau area and intrudes Penokean rocks. No analyses are reported here for zircons from rocks of this suite from northeastern Wisconsin, since they have been analyzed by Banks and Cain (1969).

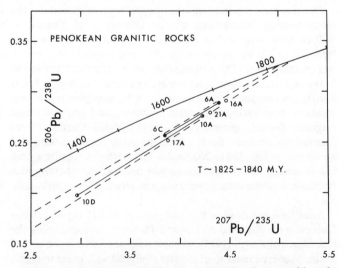

Figure 3. U-Pb discordia plot for zircons from Penokean granitic rocks in Wisconsin. Only samples analyzed by the author are shown; solid lines connect cogenetic zircon pairs. Dashed lines represent age limits from tonalitic rocks (Fig. 2). All chords for zircons from granitic rocks in northeastern Wisconsin (Banks and Cain, 1969) would also plot within the dashed lines. See text for further discussion.

TABLE 2. ANALYTICAL DATA ON SEPARATED ZIRCON FRACTIONS

| No.[*] | Frac.[†] | U (ppm) | Pb (ppm) | 204/206 | 207/206 | 208/206 | 206/238 | 207/235 | 207/206 |
|---|---|---|---|---|---|---|---|---|---|
| 1 | total | 847 | 274 | 0.00217 | 0.1360 | 0.1834 | 0.2748 (1,565) | 4.031 (1,640) | 0.1064 (1,739) |
| 2 | A | 469 | 158 | 0.00215 | 0.1368 | 0.1824 | 0.2854 (1,618) | 4.228 (1,680) | 0.1075 (1,757) |
| 3 | A | 471 | 137 | 0.00198 | 0.1333 | 0.1991 | 0.2445 (1,410) | 3.583 (1,546) | 0.1063 (1,737) |
|  | B | 758 | 199 | 0.00330 | 0.1495 | 0.2357 | 0.2081 (1,219) | 2.996 (1,407) | 0.1044 (1,704) |
| 4 | A | 904 | 225 | 0.00079 | 0.1156 | 0.1245 | 0.2298 (1,334) | 3.322 (1,486) | 0.1048 (1,711) |
|  | D | 1,543 | 306 | 0.00185 | 0.1268 | 0.1886 | 0.1699 (1,011) | 2.376 (1,235) | 0.1014 (1,651) |
| 5 | A | 1,197 | 272 | 0.00094 | 0.1168 | 0.1305 | 0.2081 (1,219) | 2.981 (1,403) | 0.1039 (1,695) |
|  | B | 2,051 | 321 | 0.00137 | 0.1189 | 0.1528 | 0.1398 (844) | 1.926 (1,090) | 0.0999 (1,622) |
| 6 | A | 720 | 237 | 0.00157 | 0.1316 | 0.1575 | 0.2893 (1,638) | 4.400 (1,712) | 0.1103 (1,805) |
|  | C | 1,011 | 282 | 0.00086 | 0.1206 | 0.1261 | 0.2567 (1,473) | 3.853 (1,604) | 0.1089 (1,781) |
| 7 | A | 433 | 146 | 0.00036 | 0.1171 | 0.1460 | 0.3083 (1,732) | 4.772 (1,780) | 0.1123 (1,836) |
|  | D | 466 | 147 | 0.00041 | 0.1170 | 0.1580 | 0.2849 (1,616) | 4.377 (1,708) | 0.1114 (1,823) |
| 8 | A | 313 | 89 | 0.00018 | 0.1126 | 0.1011 | 0.2729 (1,555) | 4.143 (1,663) | 0.1101 (1,801) |
| 9 | A | 348 | 121 | 0.00072 | 0.1221 | 0.2275 | 0.2965 (1,674) | 4.592 (1,748) | 0.1123 (1,838) |
|  | B | 477 | 153 | 0.00083 | 0.1236 | 0.2602 | 0.2659 (1,520) | 4.120 (1,658) | 0.1124 (1,839) |
|  | D | 472 | 153 | 0.00091 | 0.1241 | 0.2689 | 0.2663 (1,522) | 4.102 (1,658) | 0.1117 (1,828) |
| 10 | A | 1,112 | 347 | 0.00086 | 0.1229 | 0.1755 | 0.2756 (1,569) | 4.226 (1,679) | 0.1112 (1,819) |
|  | D | 1,978 | 458 | 0.00141 | 0.1276 | 0.2002 | 0.1980 (1,165) | 2.962 (1,398) | 0.1085 (1,774) |
| 11 | A | 197 | 64 | 0.00035 | 0.1172 | 0.1009 | 0.3098 (1,740) | 4.801 (1,785) | 0.1124 (1,839) |
|  | B | 264 | 82 | 0.00007 | 0.1135 | 0.0877 | 0.3018 (1,700) | 4.685 (1,765) | 0.1126 (1,842) |
| 12 | total | 453 | 119 | 0.00058 | 0.1170 | 0.0885 | 0.2506 (1,442) | 3.772 (1,587) | 0.1091 (1,785) |
| 13 | total | 1,075 | 270 | 0.00044 | 0.1113 | 0.1017 | 0.2392 (1,382) | 3.472 (1,521) | 0.1053 (1,719) |
| 14 | A1 | 986 | 310 | 0.00276 | 0.1433 | 0.2337 | 0.2534 (1,456) | 3.687 (1,569) | 0.1055 (1,724) |
|  | B | 1,090 | 321 | 0.00332 | 0.1493 | 0.2631 | 0.2289 (1,329) | 3.283 (1,477) | 0.1040 (1,697) |
| 15 | A | 795 | 241 | 0.00015 | 0.1112 | 0.1493 | 0.2789 (1,586) | 4.200 (1,674) | 0.1092 (1,786) |
|  | C | 1,058 | 299 | 0.00007 | 0.1086 | 0.1588 | 0.2590 (1,485) | 3.846 (1,602) | 0.1077 (1,761) |
| 16 | A | 428 | 138 | 0.00041 | 0.1170 | 0.1596 | 0.2911 (1,647) | 4.472 (1,726) | 0.1114 (1,823) |
| 17 | A | 439 | 128 | 0.00095 | 0.1245 | 0.1965 | 0.2524 (1,451) | 3.883 (1,610) | 0.1116 (1,825) |
| 18 | A | 1,023 | 313 | 0.00008 | 0.1143 | 0.0351 | 0.3094 (1,738) | 4.829 (1,791) | 0.1132 (1,851) |
|  | B | 1,360 | 391 | 0.00023 | 0.1155 | 0.0416 | 0.2881 (1,632) | 4.469 (1,725) | 0.1125 (1,840) |
| 19 | A | 352 | 112 | 0.00063 | 0.1159 | 0.1258 | 0.2945 (1,664) | 4.357 (1,704) | 0.1073 (1,754) |
|  | B1 | 386 | 127 | 0.00149 | 0.1271 | 0.1581 | 0.2900 (1,642) | 4.271 (1,688) | 0.1068 (1,746) |
|  | C | 484 | 151 | 0.00110 | 0.1223 | 0.1531 | 0.2795 (1,589) | 4.135 (1,661) | 0.1073 (1,754) |
| 20 | A | 1,633 | 462 | 0.00031 | 0.1126 | 0.1414 | 0.2609 (1,494) | 3.896 (1,613) | 0.1083 (1,772) |
|  | C | 1,699 | 461 | 0.00035 | 0.1122 | 0.1527 | 0.2477 (1,427) | 3.669 (1,565) | 0.1074 (1,756) |
| 21 | A | 437 | 139 | 0.00054 | 0.1195 | 0.1940 | 0.2787 (1,585) | 4.310 (1,695) | 0.1121 (1,835) |
| 22 | A | 3,166 | 829 | 0.00059 | 0.1153 | 0.0824 | 0.2518 (1,448) | 3.725 (1,577) | 0.1073 (1,754) |
| 23 | A | 596 | 187 | 0.00042 | 0.1170 | 0.0947 | 0.2985 (1,684) | 4.582 (1,746) | 0.1113 (1,821) |
|  | C | 872 | 242 | 0.00062 | 0.1184 | 0.1056 | 0.2604 (1,492) | 3.946 (1,623) | 0.1099 (1,798) |

Columns grouped as: Concentrations (U ppm, Pb ppm); Measured Pb Isotope Ratios[§] (204/206, 207/206, 208/206); U/Pb, Pb/Pb Ratios[#] (206/238, 207/235, 207/206).

[*] Sample numbers as given in Table 1 and figures.

[†] A = least magnetic fraction; B, C, D refer to more magnetic fractions. A1, B1 denote 100-200 mesh seive fractions for respective magnetic splits.

[§] Corrected for analytical blank (see text).

[#] Corrected for non-radiogenic Pb (see text). Numbers in parentheses are corresponding apparent ages based on the uranium decay constants of Jaffey and others (1971).

From Figure 3 it can be seen that rocks of this suite have the same general age range as obtained for the tonalitic suite. The data from Banks and Cain (1969) have not been plotted in Figure 3, but they would also fall within the limits defined by the tonalitic suite. Furthermore, both the data shown in Figure 3 and the discordia lines obtained by Banks and Cain also show the same general tendency to diverge in the direction of increasing discordance as shown by data from the tonalitic suite (Fig. 2).

**Older Volcanic Suite**

There are two distinctly different volcanic suites in Wisconsin (Van Schmus and others, 1975b). The younger, discussed in the next section, consists almost exclusively of rhyolite. In contrast, the older consists primarily of basaltic rocks, but it includes compositions ranging to and including rhyolite. Unfortunately, only in a few instances is it possible to obtain samples of this suite that will yield zircons for analyses. Only three such localities have been studied in Wisconsin: the Quinnesec rhyolite in northeastern Wisconsin (Banks and Rebello, 1969), a rhyolite near Wausau (sample 7), and a rhyolite from Eau Claire County (sample 18). The data for these rocks are shown in Figure 4 and indicate an age of 1,859 ± 20 m.y. This is fully consistent with a Pb-Pb model age of 1,830 ± 150 m.y. obtained from sulfide deposits associated with this volcanic suite in the Rhinelander area (Sims, 1976b). These rocks are probably analytically indistinct from some of the older units in the tonalitic and granitic suites, and it is reasonable to conclude that all three are genetically related to each other and to the Penokean orogeny.

**Rhyolite-Granite Suite**

Throughout southern Wisconsin are several inliers of Precambrian rhyolite and granite (see Van Schmus and others, 1975b, and Smith, 1978, for general summaries and earlier references). The volcanic rocks consist almost exclusively of rhyolite, and the

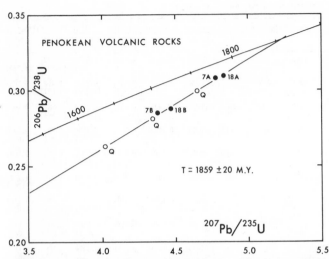

**Figure 4. U-Pb discordia plot for zircons from Penokean volcanic rocks in Wisconsin that predate intrusion of granitic to tonalitic plutonic rocks. Solid circles from this study; open circles from rhyolite of the Quinnesec Formation in northeastern Wisconsin (Banks and Rebello, 1969). A least-squares fit chord through these data intercepts concordia at 1,859 ± 20 m.y. (95% confidence limit).**

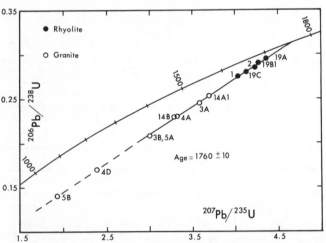

**Figure 5. U-Pb plot of zircon data from the rhyolite-granite suite of southern Wisconsin (samples 1, 2, 3, 19; Table 2) and other apparently coeval plutonic rocks from northern Wisconsin (samples 4, 5, 14). The chord defined by data from the rhyolite and related granophyric granite intersects concordia at 1,760 ± 10 m.y. (least-squares fit at 95% confidence levels).**

granite is primarily epizonal granophyre. Recent chemical data (Smith, 1978) indicate that the rhyolite and granite are genetically closely related, if not in fact comagmatic. Zircons have been separated from several of these units (samples 1, 2, 3, and 19); they define an age of 1,760 ± 10 m.y. (Fig. 5), about 100 ± 30 m.y. younger than the more mafic volcanic rocks of central and northern Wisconsin. Thus, this suite represents a distinct, separate igneous event, both in terms of its chemical character and its age.

Included in Figure 5 are zircon data from three plutonic units in northern Wisconsin (samples 4, 5, and 14). These units are mesozonal in character and intermediate in composition; the only direct link with the southern rhyolite-granophyre suite is their ap-

parent equality of age. Thus, these northern units could represent deeper-seated equivalents of the southern units.

## Comparison with Rb-Sr Data

Many of the units for which zircon U-Pb ages are reported above have also been analyzed by Rb-Sr whole-rock methods (Van Schmus and others, 1975b). The Rb-Sr ages are typically much younger (about 1,615 to 1,630 m.y. old) than the zircon ages obtained on the same units. Van Schmus and others concluded that the younger Rb-Sr ages were due to a later metamorphic event and that the zircon ages (then based on much fewer data) represented the true ages of these igneous rocks. The more extensive data reported here support the earlier interpretation. In particular, the zircons from the 1,760-m.y.-old rhyolitic rocks are clear, euhedral, show no evidence of an inherited population, and are only slightly discordant. Thus, there is little doubt that the 1,760-m.y. age represents a time of extrusion, even though Rb-Sr data yield an age of 1,630 ± 40 m.y. (Van Schmus and others, 1975b).

## Significance of Lower Intercepts

The question often arises in presentations of U-Pb data for zircons as to the significance of lower intercepts of discordia lines. At present there are two basic models applicable to most single-stage history zircons (for example, no high-grade metamorphic alteration): (1) continuous diffusion loss of radiogenic lead with either a constant diffusion parameter (Tilton, 1960) or with a continually increasing diffusion parameter (Wasserburg, 1963); or (2) the dilatancy model of Goldich and Mudrey (1972), which invokes partial episodic loss during regional uplift. The zircon discordia lines obtained for the several units discussed above have lower intercepts that range from a low of approximately 50 m.y. for the foliated tonalite to a high of approximately 400 m.y. for the lineated tonalite suite with intermediate values for the older volcanic suite and the rhyolite-granophyre suite. The apparent spread in lower intercept ages is difficult to reconcile with exclusive adherence to the dilatancy model, and it appears more reasonable at present to accept continuous loss models as dominant, with the time dependence of the diffusion parameter being different for different suites of rock. Thus, the samples with older

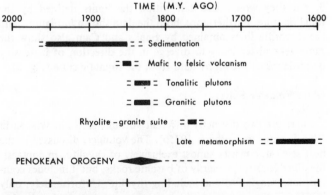

**Figure 6. Representation of the timing of petrogenetic events associated with the Penokean orogeny in Wisconsin.**

lower intercepts would more closely obey the Tilton (1960) model with nearly constant $D/a^2$, while those with younger lower intercepts would more closely follow the Wasserburg (1963) model with $D/a^2$ increasing with age.

However, the spread in lower intercept ages may be more apparent than real. First, the extrapolations are long for the suites of zircon studied. Second, part of the discordance may have been introduced by leaching during the acid washes used to remove pyrite from the zircon concentrates. Therefore, the conclusions reached above can only be considered tentative.

## DISCUSSION

The principal information gained from this study is the time framework of igneous activity during the Penokean orogeny and further insight into the details of the orogeny itself. The main chronology is summarized in Figure 6.

An essential aspect of the nature of the Penokean orogeny is that of the identity of the crust on and through which Penokean rocks were formed. Although earlier reports (Van Schmus and others, 1975b; Van Schmus, 1976) inferred the absence of Archean continental crust, recent hypotheses (Morey and Sims, 1976) and results (Van Schmus and Anderson, 1977) now clearly show that much, if not all, of the exposed Penokean fold belt is underlain by Archean continental basement rocks (see Sims, this volume, for a detailed review).

Imminent to the Penokean orogeny, the Lake Superior region was the site of deposition of the thick, lower Proterozoic sedimentary sections of Minnesota, northern Wisconsin, and Upper Michigan. In the upper parts of the section in Michigan, the sedimentation was accompanied by mafic volcanism, and in Wisconsin the exposed supracrustal rocks are predominantly mafic volcanic rocks. There is considerable evidence for mild regional deformation prior to the extrusion of the volcanic rocks (James, 1954; Cannon, 1973), but most of the intense deformation apparently postdated the formation of the basaltic volcanic assemblages. These volcanic units are $1,859 \pm 20$ m.y. old (Fig. 4), which is taken as the onset of the main pulse of the Penokean orogeny.

After extrusion of the volcanic rocks, and perhaps in part contemporaneous with it, the southern part of the Lake Superior region was invaded by numerous plutonic masses of tonalite to granite, similar in many ways to plutonic suites associated with more recent orogenic belts. These plutons were emplaced over a period of about 20 to 30 m.y ., 1,850 to 1,820 m.y. ago, and are considered to be related to the main peak of orogenic activity. Recent studies have shown that earlier formed plutons in the sequence tend to be more deformed than younger ones (Maass and others, this volume). Continued structural and geochronologic studies will allow the establishment of a detailed history of the deformational styles.

There is an absence of rocks having ages between about 1,820 m.y. and 1,760 m.y. Thus, the next major event was the formation of a granite-rhyolite suite of rocks $1,760 \pm 10$ m.y. ago. These rocks are presently best preserved and exposed in southern Wisconsin (Smith, 1978). However, the presence of contemporaneous mesozonal plutons in northern Wisconsin suggests that the granite-rhyolite suite may have been much more widespread,

but that uplift to the north and subsequent erosion have removed the supracrustal and epizonal rocks.

The rocks of this event are petrologically different from the main orogenic suites of predominantly mafic volcanic rocks and associated calcalkaline plutons, and it is not clear whether they should be considered genetically as a direct product of late-stage Penokean activity or as a distinct, later event affecting the earlier formed Penokean belt (Anderson and others, 1975). Considering that many Phanerozoic orogenies span up to 100 m.y. or more, the rhyolite-granite suite is tentatively thought to be a late-stage event related to the Penokean orogeny directly. The reason for the 60- to 80-m.y. gap between the two igneous pulses is still a matter of conjecture. It should be noted, however, that a similar bimodality of igneous activity occurs in many other middle Precambrian orogenic belts (Van Schmus, 1976) and may be a manifestation of some fundamentally common petrogenetic environment 1,600 to 2,000 m.y. ago.

In conclusion, two additional observations resulting from this study should be noted. First, sampling of plutonic rocks in Wisconsin has not been limited to only known lower Proterozoic plutons. Despite this, no well-defined plutonic masses, particularly in the gneiss-migmatite terrane, have yielded any ages older than 1,850 m.y. Thus, not only is there no evidence of Archean magmatism subsequent to formation of the Archean gneiss in most of Wisconsin, but also there is no evidence for magmatic activity in Wisconsin related to a possible 2,100-m.y.-old metamorphic event in the Lake Superior region (Van Schmus, 1976; Van Schmus and Anderson, 1977).

Second, the same sampling has yielded *no* igneous rocks with ages of about 1,615 to 1,630 m.y., corresponding to the time of major resetting of Rb-Sr systems in the southern Lake Superior region (Van Schmus and others, 1975b; Van Schmus and Woolsey, 1975; Van Schmus, 1976). Therefore, the event represented by this age may have been primarily deformational, accompanied by low-grade metamorphism, and responsible for the folding of the post-Penokean quartzites. It is possible, however, that this deformation and mild metamorphism was the northern fringe of a more major event recorded in the buried basement to the south, in the midcontinent region, but at present there is not much evidence for such an event.

## ACKNOWLEDGMENTS

In a very direct sense this study has benefited substantially by the work of S. S. Goldich and his students and colleagues over the past years.

Specific thanks are due to my many colleagues in Wisconsin who have helped immeasurably over the past several years in bringing good localities to my attention and, in some cases, helping to physically carry samples to more convenient modes of transport. Notable among these are J. L. Anderson, G. L. LaBerge, L. G. Medaris, P. E. Myers, and E. I. Smith. Laboratory assistance has been provided by J. F. DuBois, K. L. Harrower, and R. T. Laney. Financial support was provided principally through NSF grants GA-15951, GA-43426, and EAR 75-15007. The Wisconsin Geological and Natural History Survey provided valuable cooperation throughout the project.

# APPENDIX 1. DESCRIPTIONS AND LOCATIONS OF SAMPLES

1. (VS70-7)                                    lat 43°44.7'N, long 89°10.1'W
Quartz porphyry rhyolite. Collected from old Noble quarry, SE¼NE¼ sec. 34, T. 15 N., R. 11 E., Green Lake County, Wisconsin (Randolph 15-minute quadrangle).

2. (VS73-1)                                    lat 43°43.9'N, long 88°53.8'W
Quartz porphyry rhyolite. Collected from east end of old Utley quarry, NE¼SE¼NW¼ sec. 36, T. 15 N., R. 13 E, Green Lake County, Wisconsin (Fox Lake 15-minute quadrangle).

3. (VS73-2)                                    lat 43°47.6'N, long 89°19.6'W
Pink granophyric granite. Collected from waste pile at quarry in Montello, NW¼SW¼ sec. 9, T. 15 N., R. 10 E., Marquette County, Wisconsin (Montello 15-minute quadrangle).

4. (VS73-8)                                    lat 45°31.8'N, long 87°59.0'W
Medium-grained gray granite (quartz monzonite). Collected from quarry near Amberg. NW¼SE¼ sec. 3, T. 35 N., R. 20 E., Marinette County, Wisconsin (Pembine 15-minute quadrangle).

5. (VS73-11)                                    lat 45°32.5'N, long 89°9.8'W
Coarse-grained biotite-bearing granite from near Monico. Collected from outcrop east side of U.S. Highway 45, SW¼SE¼ sec. 6, T. 35 N., R. 11 E., Oneida County, Wisconsin (Monico 7.5-minute quadrangle).

6. (VS73-16)                                    lat 45°1.9'N, long 89°23.3'W
Kalinke Quartz Monzonite. Coarse-grained hornblende-bearing quartz monzonite collected from small exposure along north side of road, SE¼SE¼SE¼ sec. 34 T. 30 N., R. 9 E., Marathon County, Wisconsin (Doering 15-minute quadrangle).

7. (VS73-17)                                    lat 44°58.5'N, long 89°36.4'W
Quartz-feldspar rhyolite porphyry from Wausau. Collected from outcrop on north side of Wisconsin Highway 52 south, side SW¼ sec. 19, T. 29 N., R. 8 E., Marathon County, Wisconsin (Wausau 15-minute quadrangle).

8. (VS73-18)                                    lat 44°48.4'N, long 89°40.6'W
Medium- to coarse-grained tonalite from near Mosinee. Collected from outcrop on west side of U.S. Highway 51, NW¼SE¼ sec. 21, T. 27 N., R. 7 E., Marathon County, Wisconsin (Wausau 15-minute quadrangle).

9. (VS73-21)                                    lat 44°26.2'N, long 89°46.9'W
Medium-grained foliated tonalite from Biron Dam. Collected from rubble pile near base of power-line pylon, NW¼NW¼ sec. 34, T. 23 N., R. 6 E., Wood County, Wisconsin (Wisconsin Rapids 15-minute quadrangle).

10. (VS73-22)                                    lat 44°17.8'N, long 90°50.8'W
Medium-grained pink leucocratic granite from Black River Falls. Collected from outcrop on west side of river, NE¼SE¼ sec. 15, T. 21 N., R. 4 W., Jackson County, Wisconsin (Black River Falls 15-minute quadrangle).

11. (VS73-25)                                    lat 44°48.6'N, long 91°17.0'W
Medium-grained tonalite collected from small quarry near Eau Claire River, SW¼NW¼ sec. 19, T. 27 N., R. 7 W., Eau Claire County, Wisconsin (Chippewa Falls 15-minute quadrangle).

12. (VS73-26)                                    lat 44°55.9'N, long 91°23.4'W
Medium-grained leucotonalite or trondhjemite from Chippewa Falls. Collected from outcrop below dam, SE¼SE¼ sec. 6, T. 28 N., R. 8 W., Chippewa County, Wisconsin (Chippewa Falls 15-minute quadrangle).

13. (VS73-34)                                    lat 45°35.6'N, long 91°6.8'W
Medium-grained biotite-bearing leucotonalite or trondhjemite from north of Ladysmith. Collected from outcrop on east side of Wisconsin Highway 27, NW¼NW¼ sec. 22, T. 36 N., R. 6 W., Rusk County, Wisconsin (Exeland 15-minute quadrangle).

14. (VS73-37)                                    lat 45°45.7'N, long 91°13.3'W
Coarse-grained porphyritic biotite granite from Radisson. Collected from fresh boulders at west end of dam, NW¼NW¼SE¼ sec. 22, T. 38 N., R. 7 W., Sawyer County, Wisconsin (Radisson 15-minute quadrangle).

15. (VS74-2)                                    lat 44°25.6'N, long 90°11.9'W
Quartz porphyry dacite dike cutting Archean gneiss and migmatite near Pittsville. From small quarry to west of Turner Road, SE¼SE¼ sec. 36, T. 23 N., R. 2 E., Wood County, Wisconsin (Pittsville 15-minute quadrangle).

16. (VS74-6)                                    lat 44°31.7'N, long 90°12.2'W
Red, coarse-grained granite from north of Pittsville. Collected from small quarry pit on side of hill, north of road, SW¼SE¼ sec. 25, T. 24 N., R. 2 E., Wood County, Wisconsin (Marshfield 15-minute quadrangle).

17. (VS74-9)                                    lat 44°30.1'N, long 90°7.9'W
Coarse-grained quartz monzonite from north of Pittsville. Collected from outcrop on east bank of Yellow River, NE¼SW¼ sec. 3, T. 23 N., R. 3 E., Wood County, Wisconsin (Marshfield 15-minute quadrangle).

18. (VS74-10)                                    lat 44°43.8'N, long 90°59.5'W
Quartz porphyry rhyolite from eastern Eau Claire County. Collected from outcrop on north side of Eau Claire River, east edge SW¼ sec. 16, T. 26 N., R. 5 W., Eau Claire County, Wisconsin (Fairchild 15-minute quadrangle).

19. (VS74-13)                                    lat 43°42.2'N, long 89°20.7'W
Quartz porphyry rhyolite collected from outcrop on southwest slope of Observatory Hill, SW¼SW¼ sec. 8, T. 14 N., R. 10 E., Marquette County, Wisconsin (Portage 15-minute quadrangle).

20. (VS75-8)                                    lat 44°29.3'N, long 89°34.7'W
Fine-grained tonalite dike cutting Archean gneiss at west end of dam (Conant's Rapids), NW¼SE¼ sec. 8, T. 23 N., R. 8 E., Portage County, Wisconsin (Whiting 15-minute quadrangle).

21. (VS76-26)                                    lat 44°27.7'N, long 90°40.7'W
Coarse-grained granite from Lake Arbutus. Collected from outcrop at base of eastern, abandoned bridge abutment, NW¼NW¼ sec. 19, T. 23 N., R. 2 W., Clark County, Wisconsin (Hatfield 15-minute quadrangle).

22. (VS75-9E)                                    lat 44°29.3'N, long 89°34.9'W
Fine-grained granodioritic variety of "lineated tonalite" series (R. Maass, personal commun.). Collected from construction site of new water treatment plant near Conant's Rapids, NE¼SW¼ sec. 8, T. 23 N., R. 8 E., Portage County, Wisconsin (Whiting 15-minute quadrangle).

23. (VS75-7)                                   lat 44°28.3'N, long 89°39.3'W
Fine-grained tonalite dike cutting Archean migmatitic gneiss at east side of quarry, NW¼SE¼ sec. 15, T. 23 N., R. 7 E., Portage County, Wisconsin (Whiting 15-minute quadrangle).

## REFERENCES CITED

Aldrich, L. T., Davis, G. L., and James, H. L., 1965, Ages of minerals from metamorphic and igneous rocks near Iron Mountain, Michigan: Journal of Petrology, v. 6, p. 445–472.

Anderson, J. L., 1972, Petrologic study of a migmatite-gneiss terrain in central Wisconsin and the effect of biotite-magnetite equilibria on partial melts in the granite system [M.S. thesis]: Madison, University of Wisconsin, 103 p.

Anderson, J. L., Van Schmus, W. R., and Medaris, L. G., 1975, Proterozoic granitic plutonism in the Lake Superior region and its tectonic implications [abs.]: EOS (American Geophysical Union Transactions), v. 56, p. 603.

Asquith, G. B., 1964, Origin of the Precambrian Wisconsin rhyolites: Journal of Geology, v. 72, p. 835–847.

Banks, P. O., and Cain, J. A., 1969, Zircon ages of Precambrian granitic rocks, northeastern Wisconsin: Journal of Geology, v. 77, p. 208–220.

Banks, P. O., and Rebello, D. P., 1969, Zircon ages of a Precambrian rhyolite, northeastern Wisconsin: Geological Society of America Bulletin, v. 80, p. 907–910.

Cannon, W. R., 1973, The Penokean orogeny in northern Michigan, in Young, G. M., ed., Huronian stratigraphy and sedimentation: Geological Association of Canada Special Paper 12, p. 251–271.

Card, K. D., and others, 1972, The southern province, in Price, R. A., and Douglas, R.J.W., eds., Variations in tectonic styles in Canada: Geological Association of Canada Special Paper 11, p. 381–433.

Dalziel, I.W.D., and Dott, R. H., Jr., 1970, Geology of the Baraboo district, Wisconsin — A description and field guide incorporating structural analysis of the Precambrian rocks and sedimentologic studies of the Paleozoic strata (with summaries by R. F. Black and J. H. Zimmerman): Wisconsin Geological and Natural History Survey Information Circular 14, 164 p.

Dott, R. H., Jr., and Dalziel, I.W.D., 1972, Age and correlation of the Precambrian Baraboo quartzite of Wisconsin: Journal of Geology, v. 80, p. 552–568.

Dutton, C. E., and Bradley, R. E., 1970, Lithologic, geophysical, and mineral commodity maps of Precambrian rocks in Wisconsin: U.S. Geological Survey Miscellaneous Geological Investigations Map I-631, scale 1:500,000, with accompanying report, 15 p.

Goldich, S. S., 1968, Geochronology in the Lake Superior region: Canadian Journal of Earth Sciencs, v. 5, p. 715–724.

———1972a, Geochronology in Minnesota, in Sims, P. K., and Morey, G. B., eds., Geology of Minnesota: Minnesota Geological Survey, p. 27–37.

———1972b, The Penokean orogeny [abs.]: 18th Annual Institute on Lake Superior Geology, Houghton, Michigan (Department of Geology, Michigan Technological University), paper 25.

Goldich, S. S., and Mudrey, M. G., Jr., 1972, Dilatancy model for discordant U-Pb zircon ages, in Tugarinov, A. I., ed., Contributions to recent geochemistry and analytical chemistry: Moscow, Nauka Publication Office, p. 415–418 (original in Russian).

Goldich, S. S., and others, 1961, The Precambrian geology and geochronology of Minnesota: Minnesota Geological Survey Bulletin 41, 193 p.

Jaffey, A. H., and others, 1971, Precision measurement of half-lives and specific activities of $^{235}U$ and $^{238}U$: Physical Review C., v. 4, p. 1889–1906.

James, H. L., 1954, Sedimentary facies of iron-formation: Economic Geology, v. 49, p. 235–293.

Krogh, T. E., 1973, A low-contamination method for hydrothermal decomposition of zircon and extraction of U and Pb for isotopic age determination: Geochimica et Cosmochimica Acta, v. 37, p. 485–494.

LaBerge, G. L., and Myers, P. E., 1973, Precambrian geology of Marathon County, in Guidebook to the Precambrian geology of northeastern and northcentral Wisconsin: Wisconsin Geological and Natural History Survey, Madison, p. 31–36.

Maass, R. S., Medaris, L. G., Jr., and Van Schmus, W. R., 1980, Penokean deformation in central Wisconsin: Geological Society of America Special Paper 182 (this volume).

Morey, G. B., and Sims, P.K., 1976, Boundary between two Precambrian W terranes in Minnesota and its geologic significance: Geological Society of America Bulletin, v. 87, p. 141-152.

Myers, P. E., 1974, General geology, in 38th Annual Tri-State Geological Field Conference Guidebook: Eau Claire, University of Wisconsin (Department of Geology), p. 1-3.

Sherwood, E. S., 1976, Study of a Precambrian terrane in central Wisconsin near Pittsville [M.S. thesis]: Madison, University of Wisconsin, 64 p.

Silver, L. T., and Green, J. C., 1972, Time constants for Keweenawan igneous activity [abs.]: Geological Society of America Abstracts with Programs, v. 4, p. 665.

Sims, P. K., 1976a, Precambrian tectonics and mineral deposits, Lake Superior region: Economic Geology, v. 71, p. 1092, 1118.

———1976b, Middle Precambrian age of volcanogenic massive sulfide deposits in northern Wisconsin [abs.]: 22nd Annual Institute on Lake Superior Geology, St. Paul, Minnesota (Minnesota Geological Survey), p. 57.

———1980, Boundary between Archean greenstone and gneiss terranes in northern Wisconsin and Michigan: Geological Society of America Special Paper 182 (this volume).

Sims, P. K., Peterman, Z. E., and Prinz, W. C., 1977, Geology and Rb-Sr age of Precambrian W Puritan Quartz Monzonite, northern Michigan: Journal of Research of the U.S. Geological Survey, v. 5, P. 185-192.

Smith, E. I., 1978, Precambrian rhyolites and granites in south-central Wisconsin: Geological Society of America Bulletin, v. 89, p. 875-890.

Stacey, J. S., and Kramers, J. D., 1975, Approximation of terrestrial lead isotope evolution by a two-stage model: Earth and Planetary Science Letters, v. 26, p. 207-221.

Steiger, R. H., and Jager, E., 1977, Subcommission on geochronology: Convention on the use of decay constants in geo- and cosmochronology: Earth and Planetary Science Letters, v. 36, p. 359-362.

Tilton, G. R., 1960, Volume diffusion as a mechanism for discordant lead ages: Journal of Geophysical Research, v. 65, p. 2933-2945.

Van Schmus, W. R., 1972, Geochronology of Precambrian rocks in the Penokean fold belt subprovince of the Canadian Shield [abs.]: 18th Annual Institute on Lake Superior Geology, Houghton, Michigan (Department of Geology, Michigan Technological University), paper 32.

———1974, Chronology of Precambrian events in Wisconsin [abs.]: EOS (American Geophysical Union Transactions), v. 55, p. 465.

———1976, Early and middle Proterozoic history of the Great Lakes area, North America: Philosophical Transactions of the Royal Society of London, ser. A, v. 280, p. 605-628.

———1978, Geochronology of the southern Wisconsin rhyolites and granites: Geoscience Wisconsin, v. 2, p. 19-24. (Wisconsin Geological and Natural History Survey, Madison).

Van Schmus, W. R., and Anderson, J. L., 1977, Gneiss and migmatite of Archean age in the Precambrian basement of central Wisconsin: Geology, v. 5, p. 45-48.

Van Schmus, W. R., Medaris, L. G., and Banks, P. O., 1975a, Geology and age of the Wolf River batholith, Wisconsin: Geological Society of America Bulletin, v. 86, p. 907–914.

Van Schmus, W. R., Thurman, M. E., and Peterman, Z. E., 1975b, Geology and Rb-Sr chronology of middle Precambrian rocks in eastern and central Wisconsin: Geological Society of America Bulletin, v. 86, p. 1255–1265.

Van Schmus, W. R., and Woolsey, L. L., 1975, A Rb-Sr geochronologic study of the Republic metamorphic node, Marquette County, Michigan: Canadian Journal of Earth Science, v. 12, p. 1723–1733.

Wasserburg, G. J., 1963, Diffusion processes in lead-uranium systems: Journal of Geophysical Research, v. 68, p. 4823–4846.

Weidman, S., 1907, The geology of north-central Wisconsin: Wisconsin Geological and Natural History Survey Bulletin 16, 697 p.

York, D., 1966, Least-squares fitting of a straight line: Canadian Journal of Physics, v. 44, p. 1079–1086.

MANUSCRIPT RECEIVED BY THE SOCIETY APRIL 25, 1979
REVISED MANUSCRIPT RECEIVED NOVEMBER 26, 1979
MANUSCRIPT ACCEPTED JANUARY 11, 1980

Geological Society of America
Special Paper 182
1980

# Boulders from the basement, the trace of an ancient crust?

**Z. NIKIC***
*Cominco Ltd., Pine Point, N.W.T., Canada*

**H. BAADSGAARD**
**R.E. FOLINSBEE**
**J. KRUPICKA**
**A. PAYNE LEECH***
*Department of Geology, University of Alberta, Edmonton, Alberta T6G 2E3, Canada*

**A. SASAKI**
*Mineral Deposits Department, Geological Survey of Japan, 8 Kawada-Cho, Shinjuku-Ku, Tokyo 162, Japan*

## ABSTRACT

A tabular diatreme encountered on the 1,450-, 1,800-, and 3,500-ft (442-, 549-, and 1,067-m) levels of the Con gold mine cuts steeply dipping Yellowknife pillow basalts almost at right angles and lies parallel to the gold-bearing Campbell and Con shear zones. Rounded gray gneissic tonalite boulders 6 in. to 2 ft (15 to 61 cm) in diameter predominate and are encased in a dark, pyritized, carbonatized, highly altered hornblende-biotite schist matrix, rich in apatite, which appears to have been a fluidized phase at the time of diatreme emplacement. Sulfur-isotope measurements deviate little from the meteoritic standard; they range between $-9.5^o/_{oo}$ $^{34}S$ from pyrite in a tonalite boulder to $+7.2^o/_{oo}$ $^{34}S$ from a pyrite-free anhydrite portion of the vein material. Zircons from the tonalite are highly metamict, as they are rich in uranium and lead. Zircons from one boulder yield the oldest $^{207}Pb$-$^{206}Pb$ date of 3,210 m.y. B.P.; three fractions from the same boulder yield younger dates of 3,030 to 3,040 m.y. B.P.; a denser fourth fraction, possibly annealed, gives a date of 2,900 m.y. B.P. Zircons from a quartz monzonite boulder in the diatreme yield a typical Kenoran date of 2,570 m.y. B.P., and this rock may be derived from the hood zone of the nearby diapiric Stock Lake quartz monzonite that was intruded closely following the extrusion of silicic phases of the Yellowknife volcanics 2,600 m.y. ago. The diatreme is believed to have transported, from a deep crustal source, fragments of a previously unrecognized basement gneiss of early Archean age into the 11,000-m-thick overlying Yellowknife volcanic pile.

---

*Present address: Nickic, Cominco Ltd., 409 Granville Street, Vancouver, British Columbia V6C 1T8, Canada; Leech, Gulf Canada Resources Inc., Calgary, Alberta T2P 2H7, Canada.

## INTRODUCTION

For more than 20 years, continuing radiometric studies of the Yellowknife district on the north shore of Great Slave Lake, N.W.T., have failed to demonstrate the existence of the elusive pre-Kenoran basement, although its geologic presence was often suspected (for example, by Baragar, 1966; Robertson and Cumming, 1968; Henderson, 1941; Folinsbee and others, 1968; Henderson, 1972; McGlynn and Henderson, 1972). Elsewhere in the Canadian Shield, radiometric dates of 2,900 to 3,622 m.y. B.P. from both granitoid and metavolcanic rocks have been reported (and summarized by Baragar and McGlynn, 1976), and recent preliminary Rb-Sr isochron data show that 3,000-m.y.-old tonalitic gneisses also exist in the Slave craton (Frith and others, 1974). However, the zircons from boulders discovered in a diatreme encountered at depth in a Yellowknife gold mine (Fig. 1) yield the oldest dates so far obtained, 3,210 m.y. B.P., and confirm the existence there of a previously undated basement gneiss of early Archean age.

## REGIONAL GEOLOGY

The Slave province is the second largest area of Archean rocks in the Canadian Shield. It is bordered on the east and southeast by the Churchill structural province, on the west by the Bear province, and on the southwest by Paleozoic cover. Volcanic piles occurring as narrow belts of subaqueously extruded mafic rocks between the major plutons are overlain by metasedimentary units.

The general geology of the Yellowknife area is described in detail in a number of publications, the best known of which are by Jolliffe (1942, 1946), Boyle (1961), and Henderson and Brown (1966). In summary, 11,000 m of Archean volcanics are overlain

**Figure 1. Photo of diatreme, 3,500-ft (1,067-m) level, Con mine; scale is 6.5 in. or 16.5 cm.**

by graywackes, intruded by granites and a series of diabase dikes, and cut by a system of gold-bearing chlorite schist zones and late faults. Henderson and Brown (1966) subdivided the stratigraphy into an upper (possibly unconformable) sequence (division B) of conglomerates, arkoses, and sandstones and a lower sequence (division A) of massive and pillowed basaltic lavas with a minor sequence of dacitic rocks. Both sequences are intruded by quartz porphyry dikes, porphyritic gabbro dikes and sills, large homogeneous granodiorite masses with a number of satellite stocks, and late mafic dikes. The regional structure is interpreted as a south-plunging syncline, the axial-plane trace of which lies along Yellowknife Bay (Fig. 2, after Boyle, 1961, and Cumming and Tsong, 1975). The steeply dipping volcanics of the greenstone belt yield Rb-Sr and U-Pb dates of about 2,650 m.y. B.P.; dates from the younger granitoid rocks range from 2,640 to 2,575 m.y. B.P. (Folinsbee and others, 1968; Green and others, 1968; Green and Baadsgaard, 1971).

Gold has been mined since 1938 at the Con-Rycon mine, from the shear zones developed in the volcanics. On the 1,450-, 1,800-, and 3,500-ft (442-, 549-, and 1,067-m) levels of this mine, geologists encountered a tabular, dikelike fragmental rock unit cutting the greenstones and lying parallel to the gold-bearing Campbell and Con shear zones (Figs. 3, 4), striking N7°E and dipping steeply westward.

Angular to subrounded gneissoid boulders as much as 2 ft (61 cm) in diameter are surrounded by a dark, carbonatized, fine-grained, more-mafic matrix. Contacts with the volcanic rocks are sharp, there are few boulders of the enclosing greenstones, and boulder population varies from high to low. This intriguing rock unit generated much speculation about its origin—was it a diatreme, an explosive breccia, a tillite, or an ordinary conglomerate?—but field observations convinced Nikic, then chief geologist at the mine, that the unit was an intrusive breccia emplaced as a diatreme in a low-temperature fluidized state.

## THIN-SECTION STUDY

The matrix of the diatreme appears somewhat grayer and coarser textured than the greenish, fine-grained volcanics. In thin section, the matrix is a massive, fine-grained metatuffite or metadiorite, inequigranular and granoblastic in texture and crosscut by carbonate and quartz-carbonate veinlets. There is little or no quartz, and albite or oligoclase-albite is the main felsic mineral; it occurs in individual grains or encloses all other minerals. Hornblende is totally or partially altered and replaced by greenish biotite and actinolite, clinozoisite-epidote may be scattered throughout the rock, and unmixed sphene sometimes accompanies the biotite. There is an abundance of carbonate. Accessory minerals include apatite, zircon, magnetite, ilmenite, and pyrite.

The subangular, conspicuously banded boulders are all fine- to medium-grained, more or less altered (carbonatized and albitized) metatonalites. Quartz occurs in elongated pockets and grains, not uniformly distributed and partly recrystallized after deformation. Albite, in equant grains, is crowded with grains of fine clinozoisite plus Fe-poor epidote and some calcite, biotite, and

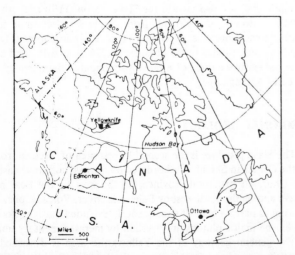

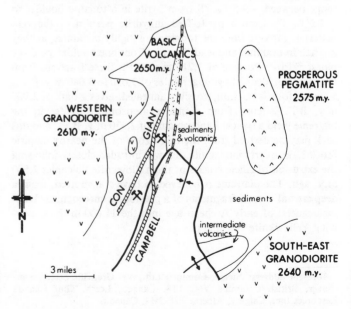

**Figure 2. Location map (above) adapted from Boyle (1961, Fig. 1). Schematic geologic map (right) of the Yellowknife area, with the displacement on late faults removed.**

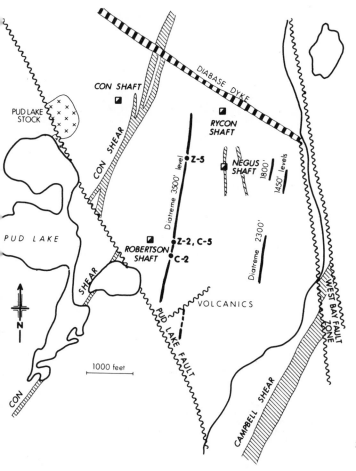

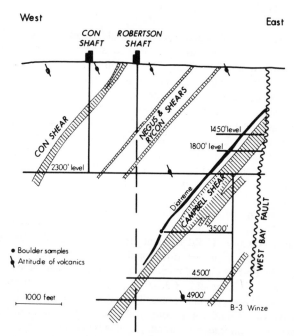

Figure 3 (left). Composite map of the Con-Rycon mine, Yellowknife, showing sample locations.

Figure 4. Cross section of the Con mine, Yellowknife, showing the gold-bearing shear zones and the diatreme cutting the volcanics.

chlorite. Former Ca-rich mafic minerals (hornblende, pyroxene?) have been altered to clusters of Fe-rich epidote and actinolite, and secondary greenish biotite with enclosed sphene and ilmenite may be associated with the Fe-rich epidote. Pyrite, zircon, and apatite are accessory, and carbonate is once more abundant and uniformly distributed.

Thin sections prepared from samples of vein material showed that it consists mostly of carbonate, but that anhydrite and brucite also exist separately and scattered throughout the carbonate. Pyrite occurs only in the carbonate.

Examination of the boulder-matrix boundary of one sample (Z-8) shows that the boundary is well defined but not sharp, that there is no shearing or development of parallel texture along the boundary, and that no thermal contact effects are evident (that is, no chilled margins and no grain-size changes). In sum, neither of the original premetamorphic parent rocks intruded the other, as a melt or highly mobile matter, and the boulders were most likely mechanically incorporated during emplacement of the mafic magma in the diatreme.

## CHEMICAL ANALYSIS

Samples of the Yellowknife boulders were analyzed chemically by Grenville Holland of Durham University using X-ray fluorescence methods; the results (Table 1), though somewhat variable, show that the boulders are indeed typical tonalites, although Krupicka (1975) warned that the oldest rocks of the sialic crust are not always tonalitic or granodioritic in composition. Previously published analyses from the Prosperous Lake granite (Green and others, 1968) and the southeast granodiorite (Folinsbee and others, 1968) are included in Table 1 for comparison. The tonalite analyses from the Yellowknife area may also be compared (Table 2) with those from the Morton Gneiss of Minnesota (Goldich and others, 1970), the Ancient Gneiss Complex in the eastern Transvaal and Swaziland (Hunter, 1974), and the tonalite diapirs in the Barberton region of South Africa (Anhaeusser, 1971; Condie and Hunter, 1976). The tonalites have in common relatively low amounts of K and $SiO_2$, but high amounts of Na, Ca, Fe, and Mg—a result predictable from the thin-section study.

In the Barberton greenstone belt, the four major tonalite diapirs form a distinct chemical group with $K_2O/Na_2O$ ratios less than 0.5, which distinguishes them from other Archean granitic plutons in the same region. The Yellowknife tonalites share this same major-element characteristic, but trace-element geochemical comparisons break down: the Yellowknife boulders contain one-half to one-third as much Rb, but three to six times more Sr and nearly ten times as much Ba on the average. The Rb/Sr, K/Rb, and Ba/Sr ratios thus reflect the effect of the carbonate-rich diatreme solution, which provided additional Sr and Ba for incorporation during alteration and removed some Rb; therefore it is difficult to do Rb-Sr dating work and impossible to speculate on crustal models based on Rb/Sr ratios (for example, see Condie and Hunter, 1976).

## U-Pb DATING ON ZIRCONS

Zircons were obtained from four boulder samples (Con 2, Con 5, Z-2, and Z-5) and further subdivided into size and gravity fractions, where possible. Except for the Con 5 zircon sample, the zircons are highly metamict and show increasing metamictization with increasing U content. The degree of discordance does not correlate with U content, however. The U-Pb analyses were carried out by the method of Krogh (1973) with the total Pb-blank = 1.2 ng and the U-blank $\leqslant$1 ng. The analytical results are given in Table 3 and plotted on a concordia diagram in Figure 5.

The zircon from the Con 5 sample falls on the discordia obtained by Green and Baadsgaard (1971) for zircons separated from the 2,600-m.y.-old southeast granodiorite and for the boulders in the conglomerate on the Sub Islands, Yellowknife Bay. The other zircon samples are from boulders definitely older than 2,600 m.y. The discordant array in Figure 5 cannot be attributed to simple or diffusional Pb-loss discordance or to episodic Pb-loss discordance. It is quite possible that these zircons are from rocks as old as 3,600 m.y. and have lost Pb both by "normal" and episodic means. The pattern of discordance is not unlike that found for zircons from the ~3,600-m.y.-old Uivak gneisses of Labrador (according to Baadsgaard).

In any event, the zircons indicate material as old as 3,200 m.y., and probably somewhat older, was (is?) present in the Yellowknife area, N.W.T.

## SULFUR–ISOTOPE ANALYSIS

Sulfur-isotope analysis of whole-rock boulder and matrix samples, pyrite separates, and the carbonate vein material produced the results shown in Table 4. In general, the per mil values deviate little from the zero meteoritic standard, ranging between $-9.5°/oo$ and $+7.2°/oo$. Comparison of our results with sulfur-isotope distribution in nature (see Thode, 1963) shows that (1) the range of values obtained from igneous rocks or reworked sediments is very broad, (2) direct comparisons would be difficult because of the mode of emplacement as a breccia in a diatreme, and (3) per mil values may reflect the complex history involving possible influence from alteration.

The whole-rock samples are consistently enriched in the heavier sulfur isotope compared to the corresponding pyrite separates, a result compatible with thin-section observation: the rocks are shot through with the anhydrite-brucite-carbonate vein material and contain some sulfate as well as pyrite.

In their study of sulfur isotopes of the Yellowknife gold deposits, Wanless and others (1960) analyzed 140 sulfide samples, only 8 of which had ratios equal to or lighter than the meteoritic standard (these were from silicic tuffs and pyrite or chalcopyrite in quartz in late faults and fractures); most of the results were between $+0.5°/oo$ and $+4.0°/oo$. Sulfide minerals from the metamorphosed greenstones ($+0.6°/oo$ to $+2.2°/oo$) contained less $S^{34}$ than those from the western granodiorite and quartz-

TABLE 1.   CHEMICAL COMPOSITIONS OF ARCHEAN GRANITOID ROCKS, YELLOWKNIFE, NWT

|  | Con 2 | Tonalite boulders* Con 5 | Tonalite boulders* Z - 2 | Tonalite boulders* Z - 5 | Southeast granodiorite[†] | Prosperous Lake granite[§] |
|---|---|---|---|---|---|---|
| SiO2 | 72.15 | 68.93 | 66.95 | 63.74 | 62.68 | 74.09 |
| Al2O3 | 13.37 | 13.20 | 12.04 | 15.69 | 17.68 | 14.30 |
| Fe2O3 | 1.47 | 3.18 | 4.02 | 4.43 | 0.61 | 0.18 |
| FeO | - | - | - | - | 4.29 | 0.99 |
| MgO | 1.26 | 1.19 | 1.66 | 2.11 | 1.60 | 0.31 |
| CaO | 5.06 | 4.16 | 7.63 | 3.60 | 4.28 | 0.82 |
| Na2O | 5.32 | 5.76 | 5.29 | 5.38 | 4.30 | 3.61 |
| K2O | 1.05 | 1.45 | 0.73 | 2.67 | 2.42 | 5.08 |
| TiO2 | 0.22 | 0.60 | 0.45 | 0.66 | 0.66 | 0.14 |
| MnO | 0.04 | 0.03 | 0.06 | 0.04 | 0.06 | 0.02 |
| P2O5 | 0.01 | 0.10 | 0.24 | 0.26 | 0.19 | 0.26 |
| S | 0.05 | 1.41 | 0.93 | 1.46 | - | - |
| TOTAL | 100.00 | 100.01 | 100.00 | 100.04 | 98.77 | 99.80 |
| Ba | 622 | 4773 | 2239 | 3750 | | |
| Nb | 2 | 6 | 1 | 6 | | |
| Zr | 131 | 223 | 254 | 275 | | |
| Y | 3 | 6 | 8 | 8 | | |
| Sr | 613 | 1404 | 2438 | 1391 | | |
| Rb | 24 | 19 | 9 | 43 | | |
| Zn | 24 | 17 | 15 | 37 | | |
| Cu | 15 | 47 | 39 | 181 | | |
| Ni | 11 | 11 | 23 | 17 | | |

Note: Major elements in wt %, trace elements in ppm.
*Analyses from this paper.
[†]Average of two, from Folinsbee and others (1968).
[§]Green and others (1968).

TABLE 2.   CHEMICAL COMPOSITIONS OF SELECTED ARCHEAN GRANITIC ROCKS, NORTH AMERICA AND SOUTHERN AFRICA

|  | Yellowknife boulders* >3,210 m.y. | Morton Gneiss[†] >3,240 m.y. | Ancient Gneiss Complex[§] 3,400 m.y. | Kaap Valley Tonalite diapir[#] 3,300 m.y. |
|---|---|---|---|---|
| SiO2 | 67.94 | 69.1 | 70.4 | 64.6 |
| Al2O3 | 13.58 | 15.6 | 14.6 | 15.4 |
| Fe2O3 | 3.28 | 0.90 | 3.92 | 4.84 |
| FeO | - | 3.42 | - | - |
| MgO | 1.56 | 0.86 | 1.44 | 2.58 |
| CaO | 5.11 | 3.14 | 3.55 | 4.33 |
| Na2O | 5.44 | 5.17 | 4.45 | 4.72 |
| K2O | 1.48 | 1.38 | 1.32 | 1.55 |
| TiO2 | 0.48 | 0.39 | 0.30 | 0.51 |
| MnO | 0.04 | 0.04 | - | - |
| P2O5 | 0.35 | 0.09 | - | - |
| S | 0.96 | - | - | - |
| Rb | 24 | - | 68 | 59 |
| Sr | 1461 | - | 233 | 552 |
| Ba | 2846 | - | 183 | 450 |
| K2O/Na2O | 0.27 | 0.27 | 0.29 | 0.33 |
| K/Rb | 547 | - | 161 | 219 |
| Rb/Sr | 0.02 | - | 0.29 | 0.11 |
| Ba/Sr | 2.01 | - | 0.79 | 0.82 |

*Average of 6 samples, Yellowknife boulders.
[†]Sample 389 D from Goldich and others (1970).
[§]Ancient Gneiss Complex, South Africa, Condie and Hunter (1976).
[#]Tonalite diapirs, South Africa, Condie and Hunter (1976).

TABLE 3.     U-Pb ANALYTICAL RESULTS AND DATES ON THE BOULDER ZIRCONS
CON MINE, YELLOWKNIFE, NWT

|  | Measured Pb ratios | | | | | Dates, m.y.* | | |
| SAMPLE | 206/204 | 207/204 | 205/204 | U, ppm | $Pb^{206}$, ppm | 206/238 | 207/235 | 207/206 |
|---|---|---|---|---|---|---|---|---|
| CON 2, >120 mesh | 2567 ± 20 | 0.2318 ± 1 | 0.0663 ± 1 | 2193 | 1041 | 2819 | 2945 | 3030 |
| CON 2, 120-230 | 3844 ± 30 | 0.2318 ± 3 | 0.06260 ± 7 | 1740 | 755 | 2620 | 2866 | 3040 |
| CON 2, >230 | 4150 ± 70 | 0.2294 ± 1 | 0.06710 ± 7 | 1306 | 585 | 2688 | 2887 | 3030 |
| CON 2, "sink" | 5300 ±100 | 0.21106 ± 7 | 0.090.3 ± 4 | 584 | 261 | 2685 | 2807 | 2900 |
| CON 5 | 2726 ± 15 | 0.17628 ± 7 | 0.05046 ± 4 | 1096 | 388 | 2208 | 2404 | 2570 |
| Z - 5L | 4400 ± 30 | 0.2498 ± 3 | 0.0994 ± 2 | 721 | 264 | 2290 | 2786 | 3170 |
| Z - 5H | 5780 ± 50 | 0.2435 ± 3 | 0.15061 ± 5 | 243 | 113 | 2791 | 2992 | 3130 |
| Z - 2 | 3827 ± 10 | 0.2573 ± 3 | 0.06799 ± 15 | 946 | 451 | 2844 | 3065 | 3210 |

*Constants used: $\lambda U\text{-}238 = 1.5513 \times 10^{-10} yr^{-1}$, $\lambda U\text{-}235 = 9.8485 \times 10^{-10} yr^{-1}$, $^{238}U/^{235}U = 137.88$.

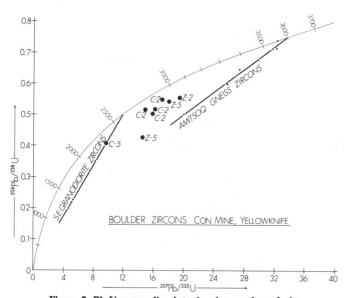

**Figure 5. Pb-U concordia plot, showing age boundaries.**

TABLE 4.     SULFUR-ISOTOPE ABUNDANCES IN YELLOWKNIFE
BOULDERS AND MATRIX

|  | Sample Description | $S^{34}$ ⁰/₀₀ | |
|---|---|---|---|
|  |  | Pyrite | Whole Rock |
| Z - 1 | Matrix | -1.7 | +5.2 |
| Z - 4 | Matrix | -1.1 | +2.1 |
| Z - 6 | Matrix | - | +4.5 |
| Z - 2 | Tonalite boulder | -6.7 | -2.7 |
| Z - 5 | Tonalite boulder | -4.1,-4.5 | -3.5 |
| Z - 6 | Tonalite boulder | -4.4 | -2.0 |
| Z - 7 | Tonalite boulder | -9.5 | -6.4 |
| Z - 8 | Tonalite boulder | - | -3.6 |
| Z - 3 | Vein material | -0.3 | +7.2* |
|  |  |  | +7.0* |
|  |  |  | +6.8* |

*mostly pyrite-free anhydrite portions.

feldspar porphyries (+2.9⁰/₀₀ to +6.8⁰/₀₀) and the silicified sericite schist ore zones (+2.9⁰/₀₀ to +4.3⁰/₀₀). Our whole-rock matrix results are in this latter range; this raises interesting possibilities about the relationship of diatreme to orebody—the matrix could be a gas-rich equivalent of the material that filled the gold-bearing shear zones.

The boulders are much more enriched in $S^{32}$ than any of the Yellowknife granitic rock studied by Wanless and others (1960) and do not compare with boulders from the Witwatersrand conglomerates (Hoefs and others, 1968) that yielded +1.0⁰/₀₀ to +4.0⁰/₀₀.

Three analyses of the carbonate vein material produced the highest $\delta S^{34}$ value, averaging +7.0⁰/₀₀. This is only slightly higher than the +3.4⁰/₀₀ value obtained from Precambrian bedded barites of the Barberton Mountain Land (see discussion in Sasaki, 1977, in reference to Perry and others, 1971) which may represent oceanic sulfate more enriched in $S^{32}$ than any other oceanic sulfate reported.

## DISCUSSION

Our data show that the tonalite boulders came from a basement much older than 3,210 m.y. and that the boulders have been variably altered and updated during the period of diatreme emplacement and granite series emplacement 2,570 m.y. ago. The boulders offer clear evidence that the volcanic rocks of the Slave craton at Yellowknife were underlain by a granitic basement correlative in age (Fig. 6) to the basement granites found in other parts of North America (Baadsgaard, 1973; Barton, 1975; Frith and Doig, 1975; Frith and others, 1974; Hurst and others, 1975; Krogh and Davis, 1971; Krogh and others, 1976; Krogh and others, 1974; Goldich and others, 1970; Cumming and Tsong, 1975), as well as in the Rhodesian and Kaapvaal cratons of Africa (Robertson, 1973; Condie and Hunter, 1976; Anhaeusser, 1971; Hunter, 1974).

Although a mechanism similar to that for the emplacement of diatremes can be invoked, a shear-zone system triggered by a

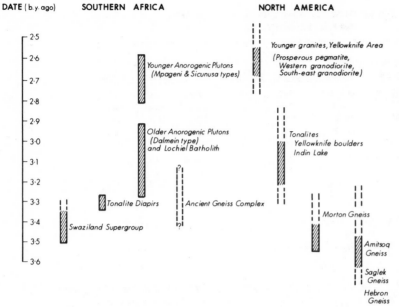

Figure 6. A transatlantic early Precambrian crustal comparison.

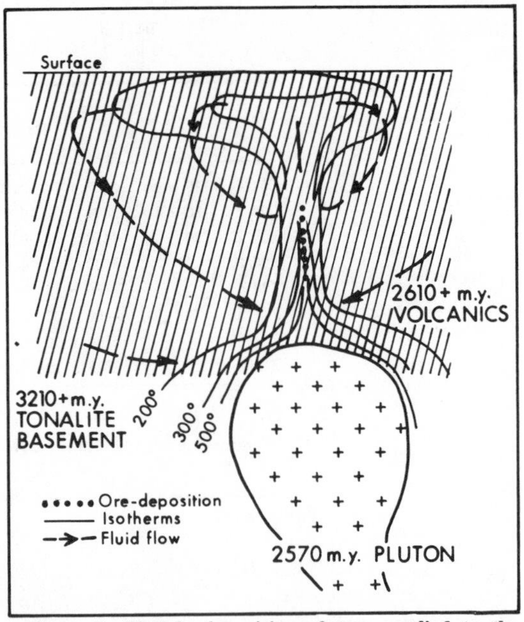

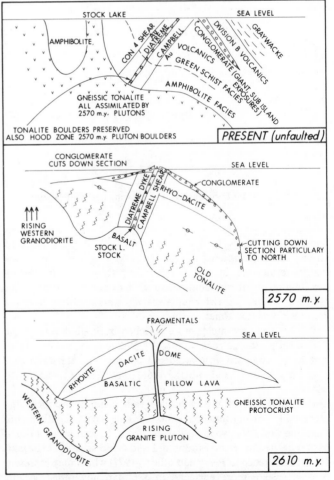

Figure 7. Model of a rising pluton, applied to the Yellowknife area (after Fyfe and Henley, 1973).

2,600-m.y.-old diapiric intrusion (Fyfe and Henley, 1973) appears an apt model (Fig. 7). The diapir would initially be intrusive into a tonalitic gneiss basement at least 3,210 m.y. old, and the shear zone would carry fluids derived from the overlying volcanics as well as fragments of the tonalite which would be rounded during transport in the same way that granite boulders in the Vredefort dome pseudo-tachylite are rounded.

The absence or scarcity of volcanic boulders in the diatreme suggests that what we see in the Con mine exposures are only boulders from depth and that a considerable part of the material once in the diatreme might have spewed out on the top of the volcanic pile to form the Yellowknife conglomerate, a rock unit which is limited to the area above the shear-zone system. If this interpretation is correct, it would have important and obvious implications for prospecting.

A sketch of the possible history of the area (Fig. 8) appears compatible with the existing surface geology of the Yellowknife area when reconstructed in its unfaulted position.

The model reasonably accounts for the ~3,200-m.y.-old zircons in the Con 2 boulder, zircons which are U and Pb rich, highly fractured, and expanded by radiation damage but not extensively recrystallized by transport in a low-temperature shear-zone diatreme matrix.

It accounts for the 2,900-m.y.-old Con 2 sink; zircons from the periphery of the boulder stewed enough in the diatreme matrix 2,600 m.y. ago to lose both Pb and U.

The 2,570-m.y.-old Con 5 zircons may represent complete resetting of 3,200-m.y.-old zircons, or they could be derived from fragments from the hood zone of the diapiric granite in the Fyfe model. At Yellowknife this diapir ultimately intruded the Yellowknife volcanics to give the Pud and Stock Lake stocks, and presumably during its earlier residence in a magma chamber, the incipient diapir was the source of the Yellowknife silicic volcanics, which were extruded 2,610 m.y. ago, a few million years before the diatreme developed.

Figure 8. Geologic history, Yellowknife Bay area.

## REFERENCES CITED

Anhaeusser, C. R., 1971, The geology of Jamestown Hills area of the Barberton Mountain Land, South Africa: Johannesburg, University of Witwatersrand, Economic Geology Research Unit, Information Circular no. 64.

Baadsgaard, H., 1973, U-Th-Pb dates on zircons form the Early Precambrian Amitsoq gneisses, Godthaab district, West Greenland: Earth and Planetary Science Letters, v. 19, p. 22–28.

Baragar, W.R.A., 1966, Geochemistry of the Yellowknife volcanic rocks: Canadian Journal of Earth Sciences, v. 3, p. 9–30.

Baragar, W.R.A., and McGlynn, J. C., 1976, Early Archean basement in the Canadian Shield—A review of the evidence: Geological Survey of Canada Paper 76-14.

Barton, J. M., Jr., 1975, Rb-Sr isotopic characteristics and chemistry of the 3.6 b.y. Hebron gneiss, Labrador: Earth and Planetary Science Letters, v. 27, p. 427–435.

Boyle, R. W., 1961, The geology, geochemistry and origin of the gold deposits of the Yellowknife district, N.W.T: Geological Survey of Canada Memoir 310.

Condie, K. C., and Hunter, D. R., 1976, Trace element geochemistry of Archean granitic rocks from the Barberton region, S. Africa: Earth and Planetary Science Letters, v. 29, p. 389–400.

Cumming, G. L., and Tsong, F., 1975, Variations in the isotopic composition of volatilized lead and the age of the western granodiorite, Yellowknife, N.W.T: Canadian Journal of Earth Sciences, v. 12, p. 558–573.

Folinsbee, R. E., and others, 1968, A very ancient island arc, in Knopoff, L., and others, eds. The crust and upper mantle of the Pacific area: American Geophysical Union Geophysical Monograph 12, p. 441–448.

Frith, R.A., and Doig, R., 1975, Pre-Kenoran tonalitic gneisses in the Grenville Province: Canadian Journal of Earth Sciences, v. 12, p. 844–849.

Frith, R. A., and others, 1974, Geology of the Indin Lake area (86B), District of Mackenzie: Geological Survey of Canada Paper 74-1, pt. A, p. 165–171.

Fyfe, W. S., and Henley, R. W., 1973, Some thoughts on chemical transport processes, with particular reference to gold: Minerals Science and Engineering, v. 5, no. 4, p. 302, Fig. 8B.

Goldich, S. S., Hedge, C. E., and Stern, T. W., 1970, Age of the Morton and Montevideo gneisses and related rocks, southwestern Minnesota: Geological Society of America Bulletin, v. 8, p. 3671–3696.

Green, D. C., and Baadsgaard, H., 1971, Temporal evolution and petrogenesis of an Archean crustal segment at Yellowknife, N.W.T.: Journal of Petrology, v. 12, p. 177–217.

Green, D. C., Baadsgaard, H., and Cumming, G. L., 1968, Geochronology of the Yellowknife area, N.W.T., Canada: Canadian Journal of Earth Sciences, v. 5, p. 725–735.

Henderson, J. B., 1972, Sedimentology of Archean turbidites at Yellowknife, N.W.T.: Canadian Journal of Earth Sciences, v. 9, p. 882–902.

Henderson, J. F., 1941, Gordon Lake South: Geological Survey of Canada Map 645A.

Henderson, J. F., and Brown, I. C., 1966, Geology and structure of the Yellowknife greenstone belt, District of Mackenzie: Geological Survey of Canada Bulletin 141.

Hoefs, J., Nielsen, H., and Schidlowski, M., 1968, Sulfur isotope abundances in pyrite from the Witwatersrand conglomerates: Economic Geology, v. 63, no. 8, p. 975–977.

Hunter, D. R., 1974, Crustal development in the Kaapvaal craton, part 1, The Archean: Johannesburg, University of Witwatersrand, Economic Geology Research Unit, Information Circular no. 83.

Hurst, R. W., and others, 1975, 3,600 m.y. Rb-Sr ages from very early Archean gneisses from Saglek Bay, Labrador: Earth and Planetary Science Letters, v. 27, p. 393–403.

Jolliffe, A. W., 1942, Yellowknife Bay, District of Mackenzie, N.W.T.: Geological Survey of Canada Map 709A.

——1946, Prosperous Lake, District of Mackenzie, N.W.T.: Geological Survey of Canada Map 868A.

Krogh, T. E., 1973, A low-contamination method for hydrothermal decomposition of zircon and extraction of U and Pb for isotopic age determinations: Geochimica et Cosmochimica Acta, v. 37, p. 485.

Krogh, T. E., and Davis, G. L., 1971, Zircon U-Pb ages of Archean metavolcanic rocks in the Canadian Shield: Carnegie Institute of Washington Year Book 70, p. 241–242.

Krogh, T. E., Ermanovics, I. F., and Davis, G. L., 1974, Two episodes of metamorphism and deformation in the Archean rocks of the Canadian Shield: Carnegie Institute of Washington Year Book 73, p. 573–575.

Krogh, T. E., Harris, N.B.W., and Davis, G. L., 1976, Archean rocks from the eastern Lac Seul region of the English River Gneiss belt, northwestern Ontario, part 2. Geochronology: Canadian Journal of Earth Sciences, v. 13, p. 1212–1215.

Krupicka, J., 1975, Early Precambrian rocks of granitic composition: Canadian Journal of Earth Sciences, v. 12, p. 1307–1315.

McGlynn, J. C., and Henderson, J. B., 1972, The Slave Province, in Price, R. A., and Douglas, R.J.W., eds., Variation in tectonic styles in Canada: Geological Association of Canada Special Paper 11, p. 505–526.

Perry, E. C., Jr., Monster, J., and Reimer, T., 1971, Sulfur isotopes in Swaziland system barites and the evolution of the Earth's atmosphere: Science, v. 171, p. 1015–1016.

Robertson, D. K., 1973, A model discussing the early history of the earth based on a study of lead isotope ratios from veins in some Archean cratons of Africa: Geochimica et Cosmochimica Acta, v. 37, p. 2099–2124.

Robertson, D. K., and Cumming, G. L., 1968, Lead- and sulfur-isotope ratios from the Great Slave Lake area, Canada: Canadian Journal of Earth Sciences, v. 5, p. 1267–1276.

Sasaki, A., 1977, Sulfur isotopic evolution in the Precambrian sea and stratabound sulfide deposits: Problems of ore deposition: International Association for the Genesis of Ore Deposits, 4th, Varna, 1974, Proceedings, v. 2, p. 102–109.

Thode, H. G., 1963, Sulfur isotope geochemistry, in Shaw, D. M., ed., Studies in analytical geochemistry: Royal Society of Canada Special Publication no. 6, p. 25–41.

Wanless, F. K., Boyle, R. W., and Lowdon, J. A., 1960, Sulfur isotope investigations of the gold-quartz deposits of the Yellowknife district: Economic Geology, v. 55, p. 1591–1621.

MANUSCRIPT RECEIVED BY THE SOCIETY APRIL 25, 1979
REVISED MANUSCRIPT RECEIVED NOVEMBER 26, 1979
MANUSCRIPT ACCEPTED JANUARY 11, 1980

Printed in U.S.A.